physica status solidi (a)

AKADEMIE-VERLAG · BERLIN

VOLUME 8 · NUMBER 2 · 1971

THEORIA CUM PRAXI

Classification Scheme

1. Structure of Solids
   1.1 Alloys. Metallurgy
   1.2 Solid-State Phase Transformations
   1.3 Surfaces
   1.4 Films
2. Non-Crystalline State
3. Crystallography
   3.1 Crystal Growth
   3.2 Interatomic Forces
4. Microstructure of Solids (Ferromagnetic Domains See 18.2 or 18.2.1; Ferroelectric Domains See 14.4.2)
5. Perfectly Periodic Structures
6. Lattice Mechanics. Phonons
   6.1 Mössbauer Investigations
7. Acoustic Properties of Solids
8. Thermal Properties of Solids
9. Diffusion in Solids
10. Defect Properties of Solids (Irradiation Defects See 11)
    10.1 Defect Properties of Metals
    10.2 Photochemical Reactions. Colour Centres
11. Irradiation Effects in Solids (X-Ray Diffraction Investigations See 1, 4, 5, 10, or 10.1)
12. Mechanical Properties of Solids (Plastic Deformations See 10)
    12.1 Mechanical Properties of Metals (Plastic Deformations See 10.1)
13. Electron States in Solids
    13.1 Band Structure. Fermi Surfaces
    13.2 Excitons
    13.3 Surface States
    13.4 Impurity and Defect States
14. Electrical Properties of Solids. Transport Phenomena
    14.1 Metals. Conductors
    14.2 Superconductivity. Superconducting Materials and Devices
    14.3 Semiconductors
        14.3.1 Semiconducting Films
        14.3.2 Semiconducting Devices. Junctions (Contact Problems See 14.4.1)
    14.4 Dielectrics
        14.4.1 High Field Phenomena, Space Charge Effects, Inhomogeneities, Injected Carriers (Electroluminescence See 20.3; Junctions See 14.3.2)
        14.4.2 Ferroelectric Materials and Phenomena
15. Thermoelectric and Thermomagnetic Properties of Solids
16. Photoconductivity. Photovoltaic Effects
17. Emission of Electrons and Ions from Solids
    17.1 Field Emission Microscope Investigations
18. Magnetic Properties of Solids
    18.1 Paramagnetic Properties
    18.2 Ferromagnetic Properties
        18.2.1 Ferromagnetic Films
    18.3 Ferrimagnetic Properties. Ferrites
    18.4 Antiferromagnetic Properties

(Continued on cover three)

physica status solidi (a)

applied research

Volume 8 · Number 2 · Pages 327 to 626, K71 to K140, and A9 to A16

December 16, 1971

AKADEMIE-VERLAG · BERLIN

Subscriptions and orders for single copies should be addressed to

AKADEMIE-VERLAG GmbH, 108 Berlin, Leipziger Straße 3 – 4 or to Buchhandlung KUNST UND WISSEN, Erich Bieber, 7 Stuttgart 1, Wilhelmstr. 4 – 6 or to Deutsche Buch-Export und -Import GmbH, 701 Leipzig, Postschließfach 160

Editorial Note:

"physica status solidi (a)" undertakes that an original paper *accepted for publication* before the 23rd of any month will be published within 50 days of this date unless the author requests a postponement. In special cases there may be some delay between *receipt* and *acceptance* of a paper due to the review and, if necessary, revision of the paper.

DAMW

Schriftleiter und verantwortlich für den Inhalt: Professor Dr. Dr. h. c. P. Görlich, 102 Berlin, Neue Schönhauser Str. 20 bzw. 69 Jena, Humboldtstr. 26. Chefredakteur: Dr. H.-J. Hänsch. Redaktionskollegium: Dr. W. Borchardt†, Prof. Dr. E. Gutsche, Dr. H.-J. Hänsch, Dr. H. Lange, Dr. S. Oberländer. Anschrift der Schriftleitung: 102 Berlin, Neue Schönhauser Str. 20, Fernruf: 42 33 80. Verlag: Akademie-Verlag GmbH, 108 Berlin, Leipziger Str. 3—4, Fernruf: 22 04 41, Telex-Nr. 112 020, Postscheckkonto: Berlin 350 21. — Die Zeitschrift „physica status solidi (a)" erscheint jeweils am 16. des Monats. Bezugspreis eines Bandes 120,— M. Bestellnummer dieses Bandes: 1085/8. Der Jahrgang 1971 erscheint in 5 Bänden. Jeder Band umfaßt drei bzw. zwei Hefte. Gesamtherstellung: VEB Druckerei „Thomas Müntzer" Bad Langensalza. — Veröffentlicht unter der Lizenznummer 1310a des Presseamtes beim Vorsitzenden des Ministerrates der Deutschen Demokratischen Republik.

Contents

Short Notes

Pre-printed Titles

Systematic List

The Author Index of Volume 8 Begins on Page 627

(It will be delivered together with Volume 9, Number 1.)

Review Article

phys. stat. sol. (a) 8, 335 (1971)
Subject classification: 14.4.1; 14.3.2; 20.3; 22.1.1; 22.1.2; 22.2.1; 22.2.2; 22.2.3

Departments of Physics and High Voltage Engineering,
Indian Institute of Science, Bangalore

Electrical Breakdown in Semiconductors

By
SUDHA MAHADEVAN, S. M. HARDAS, and G. SURYAN

Contents

1. Introduction

Ohm's law describing the linear dependence of electrical current through an electrical conductor on the applied voltage has its limitations. The deviations from Ohm's law can be small or large depending on the circumstances. We are here concerned with some situations in which the deviations are of very large magnitude even tending to catastrophic proportions. These effects occur when the conductor is subjected to high electric fields. In ordinary metallic conductors, it is not possible to maintain fields high enough because of the large power involved. Therefore the only materials in which large electric fields can be sustained are dielectrics [1] and semiconductors. In particular, in a reverse biased p–n junction, it is possible to attain large electric fields even for small applied voltages because of the small junction width. The breakdown can be of electrical or thermal origin, destructive or non-destructive.

In the following a brief review of the prominent features of the breakdown phenomenon as well as of the mechanisms invoked to explain them is given.

2. Breakdown in p–n Junctions

The physics of semiconductors is best understood on the basis of the band theory of solids which proposes that the energy levels of the electrons in a solid can be grouped into two categories called the valence and conduction bands, separated by a forbidden gap. Based on this model, electrical conduction can be brought about either by electrons in the conduction band (n-type) or defi-

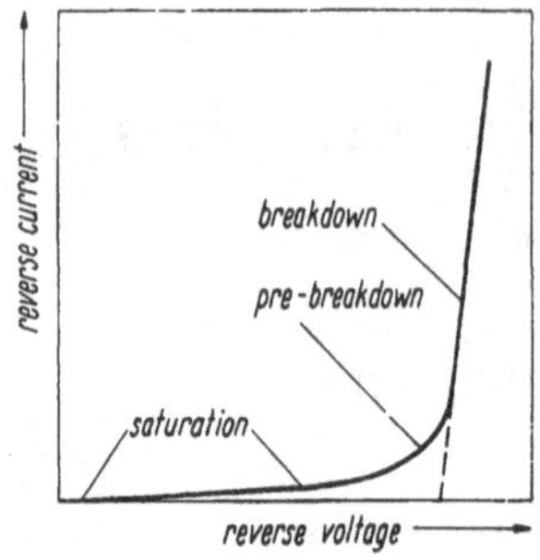

Fig. 1. Breakdown in a junction

ciency of electrons (holes) in the valence band (p-type). The junction between two parts of a semiconductor of which one is p-type and the other n-type offers low resistance to the passage of electrical current when the p-side is made positive with respect to the n-side (forward) and very high resistance if the n-side is made positive with respect to the p-side (reverse). As mentioned in the Introduction, it is possible to attain very large electrical fields (about 10^6 V/cm) in reverse biased p–n junctions, without excessive power dissipation because the depletion region is very narrow. But even in the reverse direction, p–n junctions do permit a sudden flow of electricity when the reverse voltage exceeds a certain limit called the breakdown voltage. The process may be reversible provided the current does not exceed a limit determined by the power handling capability of the device. Fig. 1 shows the current (I) versus voltage (U) characteristics of a reverse biased p–n junction which is typical of the effect and brings out the prominent features of the phenomenon under discussion. It is seen that the reverse characteristics can be roughly divided into three regions as indicated. In actual practice, the pre-breakdown region varies considerably and depending on its shape the breakdown is termed either as soft or as hard. The breakdown may be of electrical or/and thermal origin [2]. Here we are concerned with the electrical breakdown in semiconductor p–n junctions.

The above-mentioned deviations from Ohm's law are brought about by the field dependence of the number of carriers or of their mobility. Departure from Ohm's law which comes about because of the change in carrier mobility rather than by their number is referred to as the first class of high-field effects, and this aspect has been treated in detail by Gunn [3] and Conwell [4]. A second type of departure occurs when the field is high enough to cause ionisation of atoms leading to the break-up of lattice bonds or tunneling. The theories of hot electrons which are concerned with the field dependence of ionisation rate by electrons and holes have been reviewed in detail by Chynoweth [5]. As Chynoweth [5] and Kliene [2] have dealt with the atomistic details of the breakdown phenomenon, the treatment here is confined more or less to the phenomenological aspect of breakdown. Also, breakdown effects observed in heterojunctions have not been dealt with here.

3. Types of Breakdown

Two distinct mechanisms are used to explain the breakdown phenomenon observed in reverse biased p–n junctions. They are i) Zener effect or the internal field emission (IFE) and ii) the avalanche breakdown mechanism (AB).

Zener [6] investigated the problem of dielectric breakdown in insulators and concluded that at very high but experimentally realizable fields, particularly in thin junctions, in which the bands in the transition region are deeply tilted, it is possible for an electron to "long-jump" the forbidden gap because of the quantum mechanical tunnel effect and electrons could be torn from the valence band and sent to the conduction band at a rate sufficient to account for the large increase in current even with small terminal voltages. Zener's theory for the probability of excitation of electrons directly from the valence band to the conduction band in the presence of high electric fields was found applicable by McAfee et al. [7] to explain the breakdown in p–n junctions for which they obtained an expression for the current I per unit area for an applied voltage U as

$$I = U \exp\left(a - \frac{b}{E}\right),$$

where a and b are parameters depending on the semiconductor and E is the applied electric field. This form of the I–U curve was found to agree with this theory only for narrow junctions. However, it was found by McKay and McAfee [8] that breakdown in wide p–n junctions could be attributed to a process of carrier multiplication similar to the well known Townsend discharge phenomenon in gases which involves an avalanche mechanism.

In the presence of a strong electric field within the junction, consider an electron entering the junction from one side. The high electric field accelerates the electron. The electron interacts with the lattice vibrations of the crystal emitting and absorbing phonons and the net effect is a frictional force opposing the acceleration. Usually the electrons undergo collisions which do not allow them to build up sufficient energy within a mean free path to cause ionisation. But, if the field is very high, the electron gains energy faster than it loses, and within a mean free path it may acquire sufficient energy to enable it to knock a valence electron out of its bound state and raise it to the conduction band. This results in an electron in the conduction band and a hole in the valence band, both free to move. Each of these carriers may further take part in the pair production process if the field is high enough and the junction wide enough. Thus, the process initiated by the entry of a single free charge carrier into the high field region builds up itself into an avalanche with consequent high currents and this process is self-sustaining as long as the field exists. It should be borne in mind that the self-sustaining nature implies that both electrons and holes take part in the multiplication process thus leading to a positive feedback mechanism. Both IFE and AB mechanisms are diagrammatically depicted in Fig. 2 and 3 which are self-explanatory.

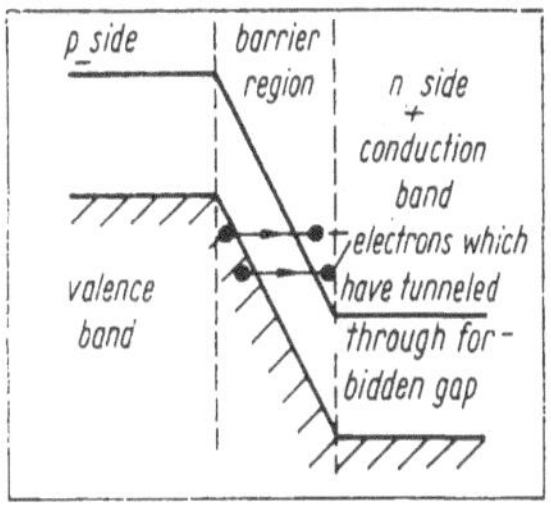

Fig. 2. Internal field emission. Thin junction resulting in a high field across the junction

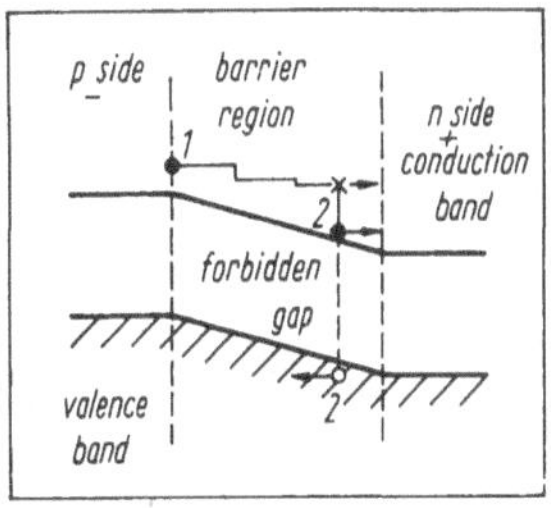

Fig. 3. Avalanche breakdown. × represents the place at which the initial electron 1 is capable of producing an electron–hole pair which undergoes further multiplication. Short vertical lines represent energy loss

Table 1
Criteria for the breakdown voltage

item	criterion	parameters involved and expressions used	limitations and/or applicability	references
i	where the current is increased by M times the saturation current	M as high as 18 at breakdown	uses the saturation current as the basis. In most cases except where the setting in of breakdown is very sudden, the saturation current is not easily experimentally determinable from the I–U characteristics	[8]
ii	by arbitrary choice when the reverse current reaches a certain magnitude	voltage at which the reverse current reaches 1) 1 μA 2) 10 μA 3) 100 μA 4) 10 mA 5) 100 mA	not universal. The magnitude of the current chosen depends on the particular set of diodes used by the various workers and also mainly on the dimensions of the diode. This is clearly brought out by the large range of magnitudes used by the various workers	[9 to 15]
iii	taken as the voltage at which the reverse current density reaches a certain magnitude	when the reverse current density reaches 1 A cm^{-2} in silicon in the examples given	though this eliminates the size of the diodes, this is not still universal. Moreover, the magnitude of the current density may not be suitable for all the materials	[12, 14 to 16]
iv	voltage intercept got by back extrapolation of the I–U characteristic in the breakdown region	—	applicable only to hard or well defined breakdown where the characteristics in the breakdown region can be approximated to a straight line	[17]

v	voltage at which the microplasma pulses begin to appear	—	requires noise pulse observations not mere I–U characteristics	[18]
vi	intercept on the voltage scale of the linear $\lg\left(1 - \frac{1}{M}\right)$ vs. $\lg U$ plot	uses the empirical formula of Miller that $M = \frac{1}{1 - (U/U_b)^n}$	not applicable to soft breakdown which does not yield a straight line for the $\lg\left(1 - \frac{1}{M}\right)$ vs. $\lg U$ plot	[19]
vii	for the IFE process, $U_b = 112.5\ R_n$ for $R_n < 0.5\ \Omega$cm. For the AB process $U_b = 83.4\ R_n^{0.6}$ for $R_n >$ $> 0.5\ \Omega$cm	based on a large number of earlier data, given for alloyed junctions of germanium. R_n refers to the resistivity of the base material	this empirical approach is questionable (see [21])	[20]
viii	for the IFE process breakdown voltage is defined as that point for which the ratio of the current density j to the voltage U has some nominal value $(j/U)_Z$ (Z for Zener)	chosen so that $(j/U)_Z$ is considerably above the saturation current density e.g. $(j/U)_Z = 3 \times 10^{-3}$ A/V cm^2 for germanium	the value of $(j/U)_Z$ chosen is not universal and the order of magnitude may not be suitable for all materials	[22]
ix	point of maximum curvature in the I–U characteristics	empirical	tedious though capable of analytical formulation and of universal applicability	—

Table 1 continued

item	criterion	parameters involved and expressions used	limitations and/or applicability	references
x	at breakdown $\alpha W_{\text{eff.}} = 1$ if electrons and holes have equal ionization coefficients and $\int\limits_0^W \alpha \exp\left[-\int\limits_0^x (\alpha - \beta)\,dx'\right] dx = 1$ if electrons and holes have different ionization coefficients	uses the relation $1 - \frac{1}{M} = \int\limits_0^W \alpha\,dx$ and letting $M \to \infty$ for breakdown. For unequal ionization coefficients of electrons and holes, the condition takes the form $\alpha' W_{\text{eff.}} = 1$ where $\alpha' = \frac{\alpha - \beta}{\ln \alpha/\beta}$ is the effective ionization coefficient	assumes a uniform field equal to the maximum field in the junction and an effective width, both being hypothetical. This is the most widely used criterion for calculations on AB	[23 to 25]
xi	for the IFE process the critical field is taken as the one for which the electron transition probability increases by a certain order of magnitude as the field changes over a certain narrow range	in the example quoted, the probability increases by a factor of 100 as the field changes from 1×10^6 to 1.1×10^6 V/cm	both the 100 fold increase in the probability as well as the range of field are arbitrary which may not be universally applicable	[6]
xii	the critical field is obtained by equating the rate of gain of energy from the field to the rate at which energy is lost to the lattice	for impact ionization of impurities, the condition gives $E_c = \frac{2C}{U}\left[\frac{\gamma I}{2kT} - 1\right]^{1/2}$	variation of mobility in the high field region is not explicitly taken into account	[27]

		where I is the ionization energy of neutral impurity centres, γ a quantity less than unity, C longitudinal velocity of sound, U drift mobility, T absolute temperature, and k Boltzmann's constant		
xiii	the critical field is equal to E_c for which $\overline{W}_i$, the impact ionization probability averaged over the distribution function equals $\overline{W}_r$, the probability of recombination averaged over the distribution function	expressions for the probabilities have been given by Keldysh [28]	not easily linked with experimental observations	[28]
xiv	the field at which the rate of ionization of an electron bound to an impurity centre equals the rate of capture of conduction electrons by empty donors	$\left(\frac{\partial f}{\partial t}\right) = \left(\frac{\partial f}{\partial t}\right)_F + \left(\frac{\partial f}{\partial t}\right)_{ac.} + \left(\frac{\partial f}{\partial t}\right)_i + \left(\frac{\partial f}{\partial t}\right)_n = 0$, where the suffixes mean F field, ac. acoustic scattering, i ionized impurity scattering, n neutral impurity scattering (explicit expressions for the four terms are given by Yamashita [29])	not easily linked with experimental observations	[29]

Whichever be the mechanism responsible for breakdown in a p–n junction, it is essential to know when the breakdown has set in, thus leading one to a discussion of the criteria for breakdown.

4. Experimental Criteria for the Breakdown Voltage

It is not easy to give a definite and clear-cut definition of the breakdown voltage or current, and this lack of clarity is reflected in the multiplicity of criteria found in the literature which are summarised in Table 1. We shall examine them critically. Some of the criteria (items ii, iii, viii, and x) are patently arbitrary in nature and others (items i and viii) depend on parameters like saturation current etc. which are not too well known or measured. Still others (items iv and vi) employ graphical extrapolation which depend on the choice of coordinate scale limiting the utility and are not at all capable of analytical definition for theoretical purposes. Moreover, in practice, these apply to extremely hard characteristics. Observations depending on the onset of microplasma (item v) are reasonably unambiguous but require auxiliary observations like noise pulses or visual observation of microplasma which may be precluded by the device structure in many cases.

Empirical rules and formulae (item vii) based on early measurements involving different base resistivities are useful in getting a rough estimate of the breakdown voltage but have no theoretical justification or general utility. The main difficulty is the formulation of a definition comprehensive enough to cover both the hard and soft characteristics and not dependent on the material.

So far we have been concerned with only a phenomenological definition shorn of the details of atomistic processes like ionisation etc., which are the ultimate cause of breakdown. In the latter part of Table 1 (from item x onwards) definitions or criteria based on the mechanism of breakdown are also included. In discussing the theoretical approaches, it is best to start with the criterion x which is actually midway between theory and experiment. Two experimentally measurable quantities, namely the ionisation coefficient, α (got from the multiplication factor M) and the barrier width, W, are involved in the expression. Both have good physical significance and can in turn be attacked from a more fundamental point of view as evidenced by the attempts reported in references [30 to 32] linking them with atomistic properties.

In discussing the IFE process (item xi) the link with experiment is not so well brought out. There is no independent means of experimentally determining the probability of electron transition. Besides, the breakdown is defined to take place at an arbitrary hundred-fold increase in the electron transition probability over a likewise arbitrary range of field, neither of which may apply for all materials.

Proceeding further down the table, the link with experimental parameters is lost altogether. Of all the criteria, the most physically reasonable one from the theoretical point of view is the one which defines the critical field as the field at which the impact ionisation probability averaged over the distribution function equals the probability of recombination also averaged over the distribution function. This definition is comprehensive to take into account hot electron effects. But, as with previous definitions, the link with experiment is rather tenuous because there is no straightforward method of determining the average ionisation probability and the average probability of recombination separately.

It would be most desirable to improvise experiments to determine these quantities, particularly under conditions of high fields.

From the point of view of an operational definition of a unique breakdown voltage, attention may be drawn to item ix proposed by the authors, whereby the breakdown voltage is taken as the voltage at which the I–U characteristics shows the maximum curvature. Of course it is possible to imagine I–U characteristics which are segments or arcs of circles in the breakdown region in which the curvature will be the same over a considerable range of voltage. Under such circumstances, a further proviso can be provided that the breakdown voltage is that corresponding to the middle point of the troublesome arc. The merit of this lies in its being capable of analytical expression which can be utilised for theoretical studies also.

In the final analysis, the difficulty in arriving at a unique criterion for the breakdown voltage is due to the lack of sharpness in the setting in of breakdown observed in some cases and this leads one to an inspection of fundamental differences between hard and soft breakdown, if any, as examined below.

5. Soft and Hard Breakdown

Considering the I–U characteristics alone, the distinction between soft and hard breakdown seems to be more of a quantitative than qualitative nature, for there is always a region where the I–U characteristic is curved. But there are indications from light emission patterns (hard—highly localised, soft—uniform glow) and temperature coefficient of the breakdown voltage (hard—positive sign for the coefficient, soft—negative sign), that the distinction may even be of a qualitative nature. Fig. 4 and 5 show typical soft and hard I–U characteristics, respectively. The main characteristics of soft breakdown may be summarised as follows. i) Soft breakdown is characterised by very high junction currents at voltages where the multiplication of photocurrent is negligible [33, 7]. ii) At the voltage where the soft breakdown becomes prominent, the sharpness of the rise is strongly dependent on the environment and the breakdown voltage has a negative temperature coefficient. Soft knee characteristics become less prominent at lower temperatures [16 to 18]. iii) With soft breakdown, whether caused by surface or bulk influences, a plot of lg $(1 - 1/M)$ vs. lg U does not yield a straight line [19]. It can be seen that the most distinctive features in soft breakdown are the absence of multiplication for photoinjected carriers and the negative temperature coefficient for the breakdown voltage.

Soft breakdown can also be caused by surface effects in many cases. The influence of surface effects has been demonstrated by careful experiments

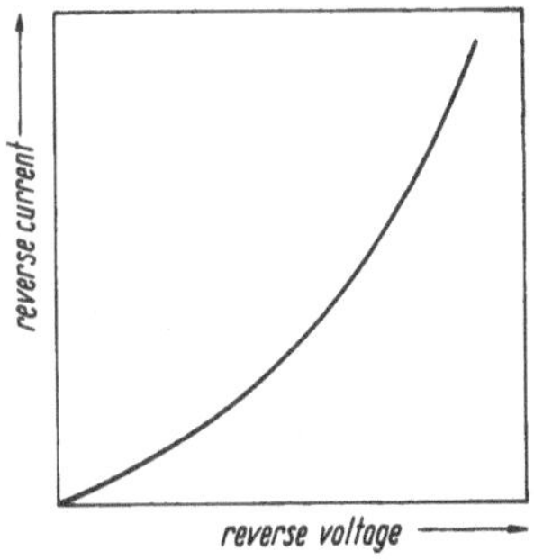

Fig. 4. Soft breakdown

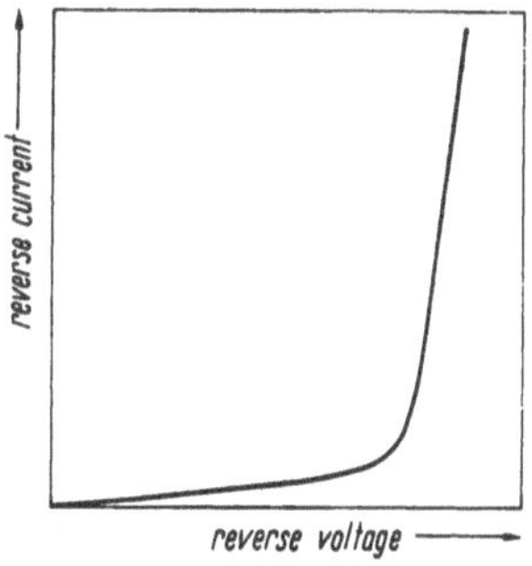

Fig. 5. Hard breakdown

involving surface treatments like etching [34], immersing in a fluid of high dielectric constant [35, 36], employing a mesa or bevelled structure for the diode geometry [37], incorporating guard electrodes [38] etc. Barring these, it has been shown that in very narrow junctions the built-in voltage itself is sufficient to cause fields high enough for IFE to take place even for small applied voltages resulting in soft characteristics [16]. Plummer [19] has observed the effect of heat treatment on breakdown characteristics. The heat treatment consists of repeated heating, quenching, and annealing and is found to cause substantial changes in the form of characteristics, leading to a softening of characteristics on sudden cooling and is found to be reversible on annealing. The effect is attributed to shallow traps. It has also been surmised that metal precipitates can lead to a softening of the I–U characteristics [39].

It is also known [40 to 42] that the effect of adsorbed water vapour can lead to considerable leakage currents because of ionic or channel conduction. A soft component in the reverse current can also come about because of surface damage like scratches [43]. On the basis that at low voltages in a p–n junction the diffusion current cannot be neglected in comparison with the total current, Rose [44] has shown that a soft breakdown must always precede a Townsend type breakdown in p–n junctions and its effect on characteristics depends on special conditions. Goetzberger and Shockley [45] have indicated that sometimes diffusion for a long time in a furnace leads to soft breakdown if the furnace tube is not extremely clean. But it is to be emphasised that in a given material, it is quite possible that a succession of localised hard breakdowns can together give the appearance of a soft breakdown [34, 46]. The succession of hard characteristics can in some cases be caused by local inhomogeneities also.

It is easy to see that the distinctions listed above are very similar to the distinctions between breakdowns in which avalanche and IFE mechanisms operate (discussed in detail later) and it is tempting to identify the two and say that IFE is associated with soft characteristics and AB with hard. In fact, in literature such an identification has often been made [16 to 18, 22]. But the most important distinction between IFE and AB seems to be their respective temperature dependence of the breakdown voltage. Also, it is known that there is a characteristic noise associated with AB mechanism [23].

6. Temperature Dependence of the Breakdown Voltage

By far the most important distinction between IFE and AB seems to be in the temperature dependence of the breakdown voltages. The temperature effect on AB results in a positive coefficient for the breakdown voltage, because, as the temperature increases, the mean free path of carriers decreases thereby requiring a higher electric field for acquiring sufficient energy for pair production. In IFE, the band gap decreases as the temperature increases because of thermal expansion and the consequent increase in lattice parameters. Because less energy is required for crossing the gap, the breakdown voltage is expected to decrease with increasing temperature. Incidentally, this mechanism of decreasing band gap is operative in the case of AB also. But the effect is overshadowed by the much larger inverse temperature dependence of the mean free path. It should be borne in mind that in the case of impurity-scattering, the temperature dependence of mobility is opposite to that of lattice scattering, and there is a temperature region where the mobility increases with increasing temperature. If p–n junctions were made with such material and studied in the appropriate temperature range, then it might be possible to find a negative temperature coefficient for the AB voltage. The above is, of course, a very simplified picture.

While there are many investigations reported on the temperature variation of the breakdown voltage in various substances, investigations with particular correlation with other phenomena like appearance of noise, microplasma are scarce. The most exhaustive report pertaining to the temperature variation of the breakdown voltage seems to be that of Chynoweth and McKay [16] and of Chynoweth et al. [47]. They have investigated the temperature variation of the breakdown voltage in silicon and germanium p–n junctions in the temperature range 20 to 400 °K. In the temperature range investigated, it appears that i) IFE normally has a small negative temperature coefficient of the breakdown voltage; occasionally even a small positive temperature coefficient has been reported for the breakdown field [16]. ii) The IFE current is not very sensitive to temperature changes. iii) The temperature variation of the IFE breakdown voltage does not obey a linear relationship over the entire temperature range investigated. Chynoweth et al. [47] have analysed the results on the basis of tunneling directly from band to band and indirectly involving a change in the momentum of electrons which goes under the name of phonon cooperation. There are two main factors responsible for the temperature behaviour of the breakdown voltage in IFE. As already mentioned, the first is the decrease of band gap with increasing temperature. The second is the increasing phonon density and the consequently increasing transition probability which in turn contributes to an increase of breakdown voltage with increasing temperature. The latter is overshadowed by the former. Phonons are operative only for the case of indirect transitions and the corresponding temperature variation is naturally steeper. This process takes place only for those substances whose conduction band minimum is away from the wave vector $\boldsymbol{k} = 0$, e.g. silicon.

In the case of AB the temperature coefficient of the breakdown voltage is substantial, positive, and of the order 10^{-4} to 10^{-3} deg^{-1}. In fact, this relatively large temperature coefficient has been used as a means of identifying AB, as already mentioned. Table 2 summarises the results of a wide number of papers on the temperature variation of breakdown voltage.

7. Distinctions between AB and IFE

While it is conceded that there are at least two mechanisms, namely AB and IFE operative in the electrical breakdown of p–n junctions, it is quite difficult in practice to isolate the dominant mechanism in individual cases. The different temperature dependence of the breakdown voltage, as mentioned above, is among other criteria the most useful means of distinguishing between the two. The next in importance is the absence of multiplication of photoinjected carriers in IFE in constrast to the pronounced multiplication in the AB process [18, 22]. The remaining distinguishing features which may be found distributed in the various tables in the article are of less value though they may supplement the above-mentioned observations. Further work is indicated for associating the observation of light emission, soft or hard knee, slope of the I–U characteristics, plots of the type $d(\lg I)/d(\lg U)$, $\lg (1 - 1/M)$ vs. $\lg U$, width of the junction, resistivity of the base material, dependence of the breakdown voltage on the net impurity concentration, magnitude of the field etc., with either one or the other of the two mechanisms. All the above-mentioned observations have been collected along with pertinent references in Table 3. Though it

Table 2

Temperature variation. β represents the temperature coefficient of breakdown voltage and U_b is the breakdown voltage

substance	type of breakdown	temperature range	main results and conclusions	references
GaAs	IFE	4.2 to 300 °K	breakdown voltage increases as the temperature decreases. Tunnelling current shows relatively little temperature dependence	[48]
GaAs	AB	77, 216, 300 °K	plot of breakdown voltage vs. temperature is available. β ranges from about 0.5×10^{-3} to about 1.6×10^{-3} deg^{-1} depending on V_b. Plot of β vs. V_b available	[10]
GaAs	AB	78, 196, 300 °K	$\beta = 5.3\times10^{-4}$ deg^{-1} for junctions which do not exhibit microplasmas. $\beta = 1.2\times \times10^{-3}$ deg^{-1} for junctions which exhibit microplasmas	[49]
GaAs	AB	77 to 540 °K	breakdown voltage has a positive temperature coefficient with $\beta = (1$ to $2)\times10^{-3}$ deg^{-1}. Temperature coefficient of critical field is also positive with a value of 10^{-3} deg^{-1}. Both show a linear temperature dependence in the range investigated	[50]
GaAs	AB	-20 to $+80$ °C	temperature coefficient of AB voltage is positive. β increases with increasing breakdown voltage and reaches a saturation at high voltages. Plot of β vs. U_b is available	[51]
GaP	AB	78, 196, and 300 °K	decrease of brakdown voltage as temperature is decreased	[52]
Ge	IFE	20 to 400 °K	plot of U_b vs. temperature not linear over the entire temperature range. The temperature studies indicate that direct transitions are dominant in Ge	[47]
Ge	IFE	77 to 300 °K	negative temperature coefficient for breakdown voltage	[11]
Ge	AB	-196 to -26 °C	positive temperature coefficient for the breakdown voltage with the value of $\beta = = 5\times10^{-4}$ deg^{-1}	[16]

Table 2 (continued)

substance	type of breakdown	temperature range	main results and conclusions	references
Ge	AB	−180 to 60 °C	positive temperature coefficient for the breakdown voltage with $\beta = 8 \times 10^{-4}$ deg^{-1}	[53]
Ge	AB	26 to −180 °C	voltage for ionization goes down to lower values as temperature is reduced	[54]
Ge	AB	20 to 300 °K	increase of U_b with increase of temperature	[18]
Ge	AB	77 to 300 °K	positive temperature coefficient for U_b	[11]
InSb	IFE	10 to 200 °K	decrease of U_b with increasing temperature. Plot of U_b vs. T available. The temperature studies give evidence of direct tunnelling transitions in InSb	[55]
Si	IFE	−80 to −160 °C	negative temperature coefficient of U_b. β is a function of both U_b and temperature	[12]
Si	IFE	−196 to +26 °C	i. Temperature dependence of the field required to produce a given emission current is small with either sign possible for the temperature coefficient. ii. Emission current is not sensitive to current changes. iii. Negative temperature coefficient of U_b	[16]
Si	AB	70 to 300 °K	U_b decreases linearly with decreasing temperature. Between 140 and 160 °K the slope changed indicating that above 140 °K less energy is required to cause ionization	[34]
Si	AB	0 to 204 °C	positive temperature coefficient of U_b; $\beta = 8.4 \times 10^{-4}$ deg^{-1}. (Reported as IFE before the AB mechanism was invoked)	[9]
Si	AB	−80 to +160 °C	highly temperature sensitive current. Positive temperature coefficient for U_b with $\beta > 7 \times 10^{-4}$ deg^{-1}	[12]
Si	AB	−100 to +100 °C	U_b increases linearly with increase of temperature. $\beta = (8 \text{ to } 9) \times 10^{-4}$ deg^{-1} for abrupt junctions and $\beta = (6.9 \text{ to } 7.5) \times$ $\times 10^{-4}$ deg^{-1} for graded junctions	[57]

Table 2 (continued)

substance	type of breakdown	temperature range	main results and conclusions	references
Si	IFE and AB	20 to 200 °C	negative temperature coefficient of U_b for IFE and positive for AB. The temperature coefficient of U_b varies with U_b, breakdown current, and ambient temperature	[56]
Si	IFE	20 to 400 °K	plot of U_b vs. temperature is not linear. Temperature studies indicate that in Si indirect transitions are dominant	[47]
Si	AB	−196 to 25 °C	positive temperature coefficient of U_b with $\beta = 8.8 \times 10^{-4}$ deg^{-1}. For the breakdown field, $\beta' = 4.4 \times 10^{-4}$ deg^{-1}	[23]
Si	AB	−25 to +100 °C	positive temperature coefficient of U_b with $\beta = 0.024$ V deg^{-1}	[38]
Si	AB	−196 to +130 °C	positive temperature cofficient of U_b	[58]
Si	AB and IFE	−60 to +150 °C	β positive for AB and negative for IFE. A theortical and experimental plot of β as a function of U_b available	[59]
Si	AB	4.2 to 300 °K	influence of temperature on breakdown voltage separated into three different regions. For $T > 100$ °K, breakdown voltage decreases as temperature decreases, being controlled by carrier mean free path in this temperature range; for $T < 30$ °K, breakdown voltage increases with decreasing temperature becaues of carrier freeze out. In the intermediate temperature range traps are observed	[61]
Ge and Si	AB	140 to 370 °K and 80 to 500 °K	a theoretical relationship has been derived which gives a linear rise of U_b with rise of temperature. Experiments with alloyed n-type Ge and Si have given confirmation of this theory in the temperature range mentioned	[60]

Table 3

Distinctions between AB and IFE types of breakdown

	IFE	AB	references
1. temperature dependence	negative temperature coefficient of the breakdown voltage	positive temperature coefficient of the breakdown voltage	see Table 2
2. behaviour towards photo-injected carriers	no multiplication of photo-injected carriers	there is multiplication of photo-injected carriers	[7, 18, 22]
3. class of breakdown	generally soft	generally hard	[16, 18, 19, 23, 39, 47, 49, 62, 71]
4. slope of the I–U characteristics in the breakdown region	$d(\lg I)/d(\lg U) =$ $= 1 + 2\,B/3\,E$ where B is a constant for a given material $d(\lg I)/dU = A/2\,E_m\,U$ where $A = (\pi^2/h\,e) \times$ $\times (m^*/2\,e_g^3)^{1/2}$ $(= 0.78 \times 10^7$ V/cm in Ge)	$d(\lg I)/d(\lg U) =$ $= 0.00070$ in Si and 0.0013 in Ge which are at least ten times smaller than those for IFE	[7, 8, 22]
5. typical plots	plot of $d(\ln I_m)/d(\ln U_a)$ vs. U_a is a straight line	plot of $\lg(1 - 1/M)$ vs. $\lg U$ is a straight line whereas for soft breakdown the above plot does not give a straight line	[19, 47]
6. junction width	observed in junctions of narrow width (less than 400 Å)	observed in broad junctions	[8, 24, 47, 63]
7. dependence on resistivity R of the base material	i. occurs for $R_n < 0.5$ Ωcm (Ge)	i. occurs for $R_n > 0.5$ Ωcm (Ge)	[20]
	ii. occurs for $R_n < 0.5$ Ωcm (Ge)	ii. occurs for $R_n > 2$ Ωcm (Ge)	[22]
	iii. occurs for $R_n < 0.14$ Ωcm (Ge)	iii. occurs for $R_n > 0.14$ Ωcm (Ge)	[11]
	iv. $U_b \sim R_n^{1.1}$ (Ge)	iv. $U_b \sim R_n^{0.38}$ (Si)	[64]
		v. $U_b \sim R_n^{0.67}$ (Ge)	[11]

Table 3 (continued)

	IFE	AB	references
8. dependence on net carrier concentration N_d	the breakdown voltage is proportional to $1/N_d$	for abrupt junctions, the breakdown voltage varies as $1/N_d^k$, where k is less than unity depending on the substance. For graded junctions U_b proportional to $1/a^{1/2}$ where a is the impurity concentration gradient	[9, 11, 22, 24, 59, 65, 67]
9. magnitude of field	magnitude of field required for IFE exceeds that for AB in a given material	magnitude of field is less than that for IFE	[24]
10. noise pulses	no noise pulses are observed	peculiar form of noise pulses observed at the onset of breakdown	[16, 23, 56]
11. magnitude of breakdown voltage	less than 3 V (Si) less than 4.5 V (Si) less than 2 V	higher than 14 V (Si) higher than 6.5 V (Si) higher than 19 V (Ge) higher than 7 V	[12] [55] [11] [68] *)
12. light emission pattern	faint, more or less uniform glow observed at current endsities greater than 2 A/cm² which on very close examinationis found to consist of a very large number of very tiny spots carrying currents of 10 μA or less	pattern consists of highly localised bright spots with each spot carrying a current of about 100 μA	[16] [69]
13. differential impedance r in the breakdown region	≈ 100 Ω (in Si alloyed diode)	10 Ω	[74]
14. dependence of U_b on energy gap E_g of the substance		$U_b = 60\left(\frac{E_g}{1.1}\right)^{3/2}\left(\frac{N_b}{10^{16}}\right)^{-3/4}$ for abrupt junctions, $U_b = 60\left(\frac{E_g}{1.1}\right)^{6.5} \times \times \left(\frac{a}{3} \times 10^{20}\right)^{-2/5}$ for graded junctions	[67]

*) As quoted in reference [7].

appears that IFE dominates as the breakdown mechanism in thin junctions and the avalanche process predominates in broad junctions, it is possible that both processes occur side by side in junctions of intermediate width as reported by Chynoweth and McKay [16], Aladinskii [59], and Tsu and Whitmore [70]. In his work on gallium arsenide, Williams [71] has indicated that depending on the choice of specimen thickness either tunneling or avalanche can be the predominant mechanism responsible for breakdown. Antula [72] has studied the contribution of impact ionisation and field emission processes to breakdown in step junctions and has given an expression for the breakdown current as a function of the breakdown voltage in terms of two special coefficients which characterise the two mechanisms. There have been experimental approaches towards separating the contributions from IFE and AB. Melnikov [73] has separated the individual contributions through the plots of calculated and measured values of the spectral density of noise in the reverse current for germanium junctions.

Weinerth [74], Aladinsky [59], and Tsu and Whitmore [70] on the other hand, measure the dU/dI vs. I and U characteristics and finally have arrived at a plot of resistivity vs. breakdown voltage in which three regions, namely, IFE, transition region, and AB are clearly distinguishable. The differential resistance dU/dI, which increases with increasing voltage both in the Zener and avalanche regions, shows a reduction with increasing voltage in the transition region [74, 14].

Calculations of avalanche breakdown voltages using available ionisation data and appropriate impurity distributions have been done for silicon by Kokasa and Davies [75], who have also given plots showing the dependence of the breakdown voltage i) of diffused junctions as a function of net impurity concentration using diffusion depths (10 to 100 μm) and surface concentrations (10^{17} to 10^{20} atoms/cm^3) as parameters and ii) of abrupt junctions as a function of net impurity concentration.

8. Theories for the Breakdown Phenomena

Theoretical approaches for the IFE process are concerned with the probability of tunneling of electrons from the valence band to the conduction band. It is well known [76 to 79] that both direct and indirect transitions occur in semiconductors, but for any material only one type is dominant, the direct ones if the semiconductor has its band gap minimum at $\boldsymbol{k} = 0$ and the indirect ones if the band gap minimum is not at $\boldsymbol{k} = 0$.

Chynoweth et al. [47] have analysed the reverse characteristics in the light of the tunneling mechanism and their conclusions are of value in determining whether the current in a junction is caused by tunneling and which type of transition (direct or indirect) is predominant in a given material. They have given an expression for the measured current as

$$I_m = A\, U_a^p\, E^q\, P\,,$$

where A is a constant for a given temperature, P is the appropriate (direct or indirect) tunneling probability, q is some power, E is the field, and U_a^p is a factor where p is some power ensuring that the measured current is zero when the applied voltage U_a is zero. They have also shown that to a good a approximation

$$\frac{d(\ln I_m)}{d(\ln U_a)} = p + \left(\frac{B}{2\,U_i}\right)^{3/2} V_a\,,$$

where

$$B = \frac{\alpha\, \varepsilon^{3/2}\, W_1}{2}.$$

ε is the forbidden energy gap appropriate to direct or indirect tunneling, α is either α_i or α_d where

$$\alpha_i = \frac{4\,(2\,m^*)^{1/2}}{3\,e\,\hbar} \quad \text{(for indirect transitions)}$$

and

$$\alpha_d = \frac{\pi (m^*)^{1/2}}{2\,e\,\hbar} \quad \text{(for direct transitions)}.$$

m^* is the reduced effective mass, $\hbar = h/2\,\pi$, h being Planck's constant and W_1 is the width constant (that is, the width at unit voltage) of the junction, and U_i is the built-in voltage. The above analysis shows that a plot of $d(\ln I_m)/d(\ln U_a)$ vs. U_a is a straight line if tunneling is the mechanism responsible for current in a junction.

The theories of hot electrons which are concerned with the field dependence of the ionisation coefficient have been dealt with in great detail by Chynoweth [5]. In a theoretical discussion of the AB mechanism the most important parameters linking theory with experiment is the ionisation coefficient, α, defined as the number of electron–hole pairs produced by one carrier (electron or hole) per cm travelling in the direction of field. In general, it is different for electrons and holes in given material and is strongly dependent on the electric field. In silicon for example, α for electrons is greater than for holes, whereas in germanium the reverse is true.

The ionisation coefficient is related to the multiplication factor M. M is defined as the number of carriers produced per one carrier entering the high field region where the multiplication takes place. The relation as derived by McKay [23] is

$$1 - \frac{1}{M} = \int_0^W \alpha\, dx\,. \tag{1}$$

W denotes the width of the junction.

If α is replaced by its value at maximum field in the junction and W by an effective width $W_{\text{eff.}}$, the relation simplifies to

$$\alpha(E_M) = \frac{1 - \frac{1}{M}}{W_{\text{eff.}}}.$$

Usually $W_{\text{eff.}}$ is set half the barrier width for graded junctions. The above approximation method to obtain α is referred to as the square field approximation. McKay [23] has given expressions connecting α and M by using the actual field distribution for abrupt and linearly graded junctions. For step junctions

$$\alpha(E_M) = \frac{2}{W_1^2}\frac{d}{dE_M}\left(1 - \frac{1}{M}\right) - \frac{4}{W_1^3}\left(1 - \frac{1}{M}\right)\frac{dW_1}{dE_M}$$

which is obtained by substituting the E vs. x relation for the junction geometry in the expression (1). If M is measured versus E_M (or total voltage as $E_M = 2\,U/W$)

on a single junction, W_1 is the width constant and therefore

$$\alpha(E_M) = \frac{2}{W_1^2} \frac{d}{dE_M}\left(1 - \frac{1}{M}\right)$$

or if the maximum field in a junction at breakdown, E_b, is measured for different junctions,

$$\alpha(E_b) = -\frac{4}{W_1^3} \frac{dW_1}{dE_M}$$

(using $M = \infty$ at breakdown). Similarly, for linearly graded junctions, using the appropriate relation in E and x, he obtains

$$1 - \frac{1}{M} = 2 \int_0^{W/2} \alpha \, dx = \frac{1}{2}\left(\frac{2}{3} W_1^3\right)^{1/2} \int_0^{E_M} \frac{\alpha(E) \, dE}{(E_M - E)^{1/2}}$$

and using Abel's integral transform

$$\alpha(E_M) = \frac{2}{\pi}\left(\frac{3}{2 W_1^3}\right)^{1/2} \frac{d}{dE_M} \int_0^{E_M} \frac{\left(1 - \frac{1}{M}\right)}{(E_M - E)^{1/2}} \, dE \, .$$

Depending on the availability of experimental data on M vs. E_M, the equation can be solved numerically. The above relations are valid under the assumption that electrons and holes have equal ionisation coefficients. In general, α differs for electrons and holes, though equality is often enough assumed. Miller [66] has distinguished the role of electrons and holes initiating multiplication in a junction and has given expressions connecting the ionisation coefficients of electrons and holes with the appropriate multiplication factors. To obtain the ionisation coefficient of electrons and holes separately, one requires complementary multiplication data from n^+–p and p^+–n junctions [66, 80 to 82]. Also, care should be taken to check the purity of electron or hole currents which initiate the multiplication in the junction [81]. Considering the possible differences in ionisation coefficients of electrons and holes, Lee et al. [81] have derived expressions connecting the multiplication factors of electrons and holes with α_n and α_p which are identical with those derived by Miller [66]. Values calculated from experimentally determined multiplication data for various substances are summarised in Table 4. In all cases the field distribution is assumed to be constant. However, the field in the depletion region must be modified by the free carriers resulting from the avalanche. These carriers make the ends of the depletion region more conducting, and thus the applied potential difference gets restricted to a narrower region leading to a further increase of the field. But the build-up of avalanche is lessened because of the narrower width. Thus the width of the high field region will stabilize to an equilibrium width which is somewhat smaller than the width without avalanche.

The experimental values of McKay and McAfee [8], McKay [23], and Miller [66] for the variation of α with electric field E have been replotted by Rose [13] who has given a power law of the form $\alpha = A\,E^n$ (A and n constants) for the dependence of α on E. Wul and Shotov [83] have given a similar form for the dependence of α on E with $n = 5.45$. By analogy with gas discharge theory, Chynoweth [84] has given an expression of the form $\alpha = a \exp(-b/E)$ (a and b

Table 4

Field Dependence of α

sub-stance	range of field (kV/cm)	range of α (cm^{-1})	method of calculation and conclusions	references
GaAs	80 to 150	1.2×10^4 to 35×10^4	α taken same for electrons and holes and computed from $1 - 1/M = 0.4\ W\ \alpha(E_M)$ where W junction width. Plot of lg α vs. $1/E_M^2$ is a straight line implying agreement with Wolff's theory	[88]
GaAs	200 to 400 from 77 to 373 °K	100 to 2×10^4	α calculated using exact field distribution for abrupt and linearly graded junctions and using that α is same for electrons and holes in GaAs. At a given field α decreases with increase of temperature	[87]
GaAs	330 to 500 from 20 to 80 °C	10^4 to 5×10^5	α obeys the law $\alpha = a \exp(-b/E)^2$. At a given field α decreases with increases of temperature. Temperature coefficients of a and b are estimated; $b = 10.4\times10^{-4}\,\mathrm{deg}^{-1}$; $a = 15.4\times10^{-4}\,\mathrm{deg}^{-1}$	[51]
GaP	450 to 1100	1000 to 4×10^4	from the relation $1 - 1/M = \alpha(E_M)\times W_{eff.}$ where $W_{eff.} = 0.6\ W_1 + 0.4\ W$, where W junction width and W_1 width constant of the junction	[52]
GaP	320 to 4500	250 to 1.3×10^6	from multiplication data using $1 - 1/M = m\ W\ \alpha(E_M)$. Agreement with Baraff's theory for $L_R = 40$ Å and $E_i = 1.5\ E_g$	[88]
Ge	100 to 250	0.1×10^4 to 0.8×10^4	calculated from $(1 - 1/M)/W_{eff.}$ taking uniform field as the maximum field in the junction and $W_{eff.}$ as half the junction width. A plot of α vs. E available	[8]
Ge	170 to 250 for both electrons and holes	5×10^3 to 6×10^4 for holes; 3.8×10^4 to 3.2×10^4 for electrons	α for electrons and holes obtained separately by breakdown and multiplication data on complementary junctions. α for electrons less than that of holes. With threshold energy (1.5 eV) and mean free path for electron and phonon collision of 130 Å there is agreement with the Wolff's theory	[66]

Table 4 (continued)

substance	range of field (kV/cm)	range of α (cm^{-1})	method of calculation and conclusions	references
Ge	100 to 600	40 to 1.5×10^4	from multiplication data on alloy junction transistors	[89]
Ge	140 to 260	2×10^3 to 6×10^4 for holes, 10^3 to 3×10^4 for electrons	from multiplication measurements on injected electrons and holes. For electrons $\alpha = 1 \times 10^{-25} E^{5.45}$, for holes $\alpha = 2 \times 10^{-25} E^{5.45}$	[83]
Si	150 to 1300	400 to 2×10^5	calculated using data on the variation of breakdown voltage with net impurity concentration. For E from 1.5×10^5 to 3×10^5 α is proportional to E^7. For E from 3×10^5 to 6×10^5 α varies with E as $\exp(-C/E)$ where C is a constant. For E greater than 6×10^5 α varies approximately linearly with E (V/cm)	[24]
Si	200 to 400	100 to 2000 for holes, 180 to 5×10^4 for electrons	using an improved technique for making diodes and extracting multiplication data of electrons and holes; agreement with Baraff's theory is obtained with E_g less than or equal to E_i, less than or equal to 1.5 E_g and L_R for holes between 50 and 70 Å. L_R for electrons between 30 and 45 Å	[81]
Si	290 to 470	0.3×10^4 to 5×10^4	$\alpha(E_M) = \frac{1 - 1/M}{W_{eff.}}$, E_M is the maximum field in the junction and $W_{eff.}$ half the junction width	[8]
Si	100 to 1000	10^2 to 10^5	calculated from multiplication and breakdown data on graded and step junctions using expressions derived for α. A semilogarithmic plot of α vs. E is given	[23]
Si	380 to 630	6×10^3 to 4×10^4 for holes, 1.5×10^4 to 2.5×10^5 for electrons	α for electrons greater than that for holes. Using 1.5 and 3.5 eV as the threshold energy for electrons and holes and 70 and 100 Å for mean free path of electron–phonon and hole–phonon collision, respectively, there is agreement with Wolff's theory	[90]

Table 4 (continued)

substance	range of field (kV/cm)	range of α (cm^{-1})	method of calculation and conclusions	references
Si	250 to 600 for holes 200 to 500 for electrons	200 to 8×10^4 for holes, 4×10^2 to 2×10^4 for electrons	α for electrons greater than that of holes. Dependence of α on E found to be of the form $\alpha = a \exp(-b/E)$ where a and b are constants different for electrons and holes. Plot of $\log \alpha$ vs. $1/E$ is available	[84]
Si	330 to 660 for electrons 320 to 800 for holes	10^4 to 1.25×10^5 for electrons 2×10^3 to 8×10^4 for holes	calculated from photomultiplication data on graded junctions. Values of E_i and L_R deduced from ionization data. α (holes) = 1/3 α(electrons). α dependence of the form $\alpha = a \exp(-b/E)$	[82]

constants depending on the semiconductor and the conduction type, i.e. electron or hole as the current carrier) for the variation of α with E. The constant b represents some characteristic field in the junction under study. It is usual to try the above type of plot for the variation of α with E, as many substances seem to obey this law for a considerable range of field. A more general form of the type $\alpha = a \exp(-b/E)^m$ has been given by Sze and Gibbons [67] for the variation of α with E, the values of a, b, and m are tabulated for Si, Ge, GaAs, and GaP. While $m = 1$ for Si and Ge, $m = 2$ for GaAs and GaP. For Ge and Si Aladinskii [85] has also reported a change from the dependence of the type $\alpha = a \exp(-b/E)$ to the type $\alpha = a \exp(-b'/E^2)$ at fairly high fields ($E > 300$ kV/cm for Ge and $E > 500$ kV/cm for Si).

A detailed understanding of AB requires the form of the dependence of α on the electric field E. For this, information on the threshold energy for pair production (E_i), the mean free path for pair production (L_i) as distinct from the mean free path for collision (L_R), (the latter serving to establish equilibrium with the lattice), and the distribution function of electrons in high electric fields are required. There are two types of phonon collisions, acoustical and optical. Acoustical phonon collisions are elastic and result in a change of momentum only, while the optical phonon collisions result in both energy and momentum changes.

There are three theories differing in the choice of the above parameters and the distribution function (by Wolff [91], Shockley [92], and Baraff [93]). The main points of comparison and contrast among these three approaches are listed in Table 5.

Comparing Wolff's theory with experiment, one finds that there is agreement in the high field region, where the average electron energy is comparable to the ionisation threshold. Wolff's method in essence consists in solving the Boltzmann equation for the motion of electrons (or holes) in a high field taking into account the effect of electron–phonon and pair-producing collisions on the distribution function. But Wolff's calculations are not readily amenable for

Table 5

Comparison of theories for α. L_R collision mean free path; E_i threshold energy; E applied field

	Wolff	Shockley	Baraff
probability of ionization	sharp. Zero below threshold and one above threshold	gradual increase from zero to one	same as Shockley
threshold value	somewhat greater than 1.5 times the band-gap	lower limit for the threshold is the band-gap itself	threshold treated as an adjustable parameter from 1 to 2 times E_g
validity	agreement in high field region. A straight line lg α vs. $1/E^2$ plot indicates agreement with this theory	agreement in low field region. A straight line lg (α/E) vs. $1/E$ plot indicates agreement with this theory	with proper choice of threshold energy and collision mean free path, plot of $\alpha \times L_R$ vs. $E_i/e\, E\, L_R$ gives fit throughout the field range
mean free path	assumes that pair production mean free path is less than the collision mean free path (15 and 200 Å in Si)	estimates that collision mean free path is much smaller than ionization free path (50 and 880 Å in Si)	collision mean free path treated as an adjustable parameter
velocity	Boltzmann type. Almost spherical symmetric	spiked in the direction of field	non-spherical being a superposition of spherical distribution with the spiked distribution
modification of basic theory	by Moll and Overstraeton [82] under less restrictive assumptions regarding the relative values of L_R and L_i	by Moll and Meyer [94]. By revising the probability of ionization taking multistage processes into account and giving correction term for Shockley's expression	by Crowell and Sze [95]. By taking into account phonon absorption also along with phonon generation

manipulation. Agreement with Wolff's theory is indicated if a plot of lg α vs. $1/E^2$ results in a straight line. His choice of the mean free path and the criterion that the electron–phonon free path is much larger than that for pair production has not however been borne out by later estimates [93, 82]. Shockley reverses the roles of the mean free paths in as much as that the free path for ionisation has been estimated to be sixteen times that for collision free path in silicon. In this connection, attention may be drawn to the work of Moll and Overstraeton [82] who have revised Wolff's theory by dropping the assumption that the mean free path for pair production is very small as compared to that for optical phonon collision. They have obtained an analytical expression for α in which, however, the overall α vs. E relationship is qualitatively unaltered, but the numerical details vary.

Shockley [92] calculated the ionisation rate as due only to the few electrons which escape phonon scattering, get up to the energy E_i and ionize in first try. This approach is likely to be dominant at low fields when the electron temperature is very small in comparison to the ionisation threshold. Moll and Meyer [94] attempted to improve Shockley's [92] theory by taking into account the alternative processes by which the carrier acquires the energy E_i necessary for pair production. Firstly suffering one or more phonon collisions and then reaching at E_i or secondly, having escaped phonon collision and obtaining energy greater than E_i and then coming back to E_i via phonon collisions. These processes were not considered important by Shockley because the carrier suffering phonon collision before obtaining E_i is scattered in a random direction and consequently has to travel longer to reach E_i than a carrier starting at rest. These multistage processes have been considered by Moll and Meyer [94] who have given correction terms to the expression

$$\alpha(E) = \frac{e\,E}{r\,E_r} \exp\left[-\frac{E_i}{e\,L_R\,E}\right]$$

of Shockley [92]. For the case of agreement with Shockley's theory, the plot of lg (α/E) vs. $1/E$ is a straight line. The same approach as Shockley's regarding the details of ionisation process is maintained in Baraff's theory except that the velocity distribution has been chosen to be a superposition of the types of distribution used by Wolff and Shockley individually. Baraff's theory assumes that i) phonon interaction results only in phonon generation and ii) the solution of Boltzmann's equation is feasible if the scattering is isotropic. The result of Baraff's theory relating α to E is a family of curves with two adjustable parameters, the threshold ionisation energy E_i and the mean free path L_R for scattering with an optical phonon of energy E_R being emitted. He has presented lg α vs. $1/E$ curves for a wide range of material parameters. His theory comprises both Wolff's and Shockley's theory by showing a dependence of α on E of the form exp $(-C/E^2)$ at high fields and of the form exp $(-b/E)$ at low fields. By proper choice of E_i and L_R agreement with Baraff's theory can be obtained over the entire field range. Crowell and Sze [95] take phonon absorption also into account and have been able to show that the mean free path for electron–phonon scattering is temperature dependent and reduces with increasing temperature. If L_{R0}, the value of L_R at 0 °K is known, then the normalising parameters of Baraff's theory are given by

$$\frac{\bar{E}_r}{E_r} = \frac{1}{2\,N+1} = \tanh\left(\frac{E_r}{2\,kT}\right) = \frac{L_R}{L_{R0}}\,,$$

where $\bar{E}_r$ is the average energy loss per collision taking both phonon generation and phonon absorption into account, whereas E_r is the average energy loss per collision taking phonon generation alone into account as was done by Baraff. To facilitate numerical analysis, Crowell and Sze [95] have given the following empirical relationship of α for comparison with Baraff's theory

$$\alpha L_R = \exp \{(11.5\, r^2 - 1.17\, r + 3.9 \times 10^{-4})\, x^2 + (46\, r^2 - 11.9\, r + 1.75 \times 10^{-2})\, x - 757\, r^2 + 75.5\, r - 1.92\},$$

where

$$r = \frac{\bar{E}_r}{E_i} \quad \text{and} \quad x = \frac{E_i}{e\, E\, L_R}\,.$$

Dumke [96] has treated the breakdown theory for low-band-gap polar substances like InSb and InAs.

It was found that measurements available prior to Baraff's theoretical investigations did not give satisfactory fit with theory with reasonable values for the parameters. Hence, Lee et al. [81] repeated the experiments with better methods also employing a new method of extracting ionisation coefficient from multiplication data. Their improvements upon the earlier ones consist in making junctions free of microplasma defects with considerable improvement in technology and control of carriers (electrons or holes individually) initiating the multiplication by using both penetrating and non-penetrating radiation for observing photoresponse. The results of Lee et al. were found to agree well with the theory of Baraff.

The values of mean free path and threshold energy deduced by a number of authors from theory and experiments on AB in various substances are listed in Table 6.

9. *I–U* Relation at Breakdown

There are not many calculations regarding the shape of the current–voltage relation in the breakdown region reported in literature, except the investigations by Armstrong [101, 102] and Rose [103] for impact ionisation and by Muss and Greene [22] for internal field emission.

Armstrong gives a relation of the form

$$J = \frac{J_0}{1 - (E/E_b)^{m + (b/E_b)}},$$

where J is the current density, J_0 is a quantity of the dimensions of current density varying very slowly with field, m is equal to zero for the intrinsic case, $m = 1/2$ for linear junctions, and $m = 1$ for abrupt junctions; b is the constant introduced in the ionisation coefficient–field relation by Chynoweth [84].

Armstrong [102] has further shown that

$$\frac{E}{E_b} = \begin{cases} \dfrac{U}{U_b} & \text{(intrinsic case)}\,, \\[2mm] \left(\dfrac{U}{U_b}\right)^{1/2} & \text{(abrupt junctions)}\,, \\[2mm] \left(\dfrac{U}{U_b}\right)^{2/3} & \text{(graded junctions)}\,. \end{cases}$$

Table 6

Experimental data of threshold energy and mean free paths

sub-stance	threshold energy E_i (eV)	mean free path for pair production L_i (Å)	mean free path for collision L_R (Å)	average energy loss per collision E_R (eV)	references
GaAs	1.7 ± 0.3		15 ± 2		[86]
				0.036	quoted in [95]
	$1.5\ E_g$		60 ± 6		[87]
			L_{R0} 58 $L_{R(300\,°K)}$ 35 $L_{R(uncorrected)}$ 45		[95]
GaP				0.05	quoted in [95]
			L_{R0} 42 $L_{R(300\,°K)}$ 32 $L_{R(uncorrected)}$ 38		[95]
			100		[52]
	$1.5\ E_g$			36 to 41	[88]
Ge	$0.7 < E_i < 2.9$				[8]
	1.5		130		[66]
				0.036	quoted in [95]
			L_{R0} 105 $L_{R(300\,°K)}$ 65 $L_{R(uncorrected)}$ 80		[95]
InSb	0.5				[55]

Table 6 (continued)

sub-stance	threshold energy E_i (eV)	mean free path for pair production L_i (Å)	mean free path for collision L_R (Å)	average energy loss per collision E_R (eV)	references
InAs	$1.2\ E_g < E_i <$ $< 1.6\ E_g$		70 to 105		[100]
Si	$1.1 < E_i < 3.6$				[8]
	$1.5 < E_i < 3.5$ probably 2.3	15	$160 < L_R < 260$ probably 200		[91]
	1.5 (electrons) 3.5 (holes)		70 (electrons) 100 (holes)		[90]
	2.25 (electrons)		140 to 230		[15]
	1.1	880	50		[92]
Si		190	60		[97]
	1.6		68 to 75		[81]
	1.8 ± 0.1 (electrons) 2.4 ± 0.1 (holes)	$(20 \pm 10)\ L_R$	70 to 80		[82]
				0.063	quoted in [95]
			L_{R0} 47 $L_{R\,(300\,°K)}$ 38 $L_{R\,(uncorrected)}$ 40		[95]
	1.8 (electrons) 2.5 (holes)				[99]

Hauser [98] has given a detailed treatment on threshold energy.

Therefore, in all cases

$$J = \frac{J_0}{1 - (U/U_b)^n}$$

which is similar to the empirical relation of Miller [66].

On the basis of his theory Armstrong [101] concludes that ln $(I - I_0)$ should be linear with respect to $|U|^{-1/2}$ in contrast to the relation proposed by Rose [103] which predicts ln $(I - I_0)$ linear with ln U. No definite conclusions regarding the validity of either one or the other could be made because of interfering effects of surface leakage etc. An attempt of comparing Armstrong's theory with experiment has been made by Rose [103] and there seems to be agreement provided surface effects are carefully avoided.

10. Microplasmas

Of all the phenomena associated with AB, the most exotic and tantalising is the existence of microplasmas which are small high current density discharges in the solid state. This was first observed by McKay [23] as peculiar noise pulses at the onset of breakdown. Subsequently, localised light emission associated with breakdown was observed by Newman et al. [104] and both phenomena were correlated by Chynoweth and McKay [105].

The possibility of studying a single microplasma was pointed out by Senitzky and Moll [63]. They proposed that this could be achieved by limiting the breakdown in the junction to a very small area of the order of magnitude of a single microplasma whose size has been estimated to lie anywhere between 500 to 3000 Å in diameter [16, 33, 34, 106, 107].

The important characteristics of a microplasma may be briefly summarised as follows. They are, as already mentioned small localised electrical discharges whose current densities can amount to 2×10^6 A cm^{-2} [106]. They appear to have at least two stable conditions, on or off, and randomly fluctuating between these states [106], the probability of being in one or the other of these states being determined by the voltage across the concerned microplasma region. It has been found that the current carrying capacity of a single microplasma, in silicon for example, is about 50 to 100 μA [23] which works out to be the formidable current density quoted above on the basis of the observed dimensions of a microplasma.

The on–off characteristics of microplasmas result in a considerable amount of electrical noise and, in fact, the observation of this noise led to their discovery [23]. There seems to be some correlation between the light emitting spots which are observed at breakdown and the microplasma spots. While complete correlation has been reported by Chynoweth and McKay [105] by simultaneous observation of noise pulses and light emission, Kikuchi and Tachikawa [108] have reported that a light emitting spot does not necessarily coicinde with a microplasma spot. The method of locating of a microplasma spot other than its probable light emission is by scanning by means of a very small light spot and observing the current multiplication due to photoinjected carriers which increase to large proportions, when the scanning spot is in the neighbourhood of a microplasma and is maximum when the scanning light spot is directly on the microplasma region. The ratio of light spot area to that of microplasma should not be too large.

The I–U characteristic of a single microplasma is found to be quite hard, the transition between the conducting and non-conducting states taking place in an voltage interval less than a fraction of a volt.

Depending on circuit parameters, it is possible for a negative resistance region to follow the onset of a single microplasma breakdown [58, 63, 106, 109, 110]. Recently, Misawa [111, 112] has indicated the possibility of negative resistance at AB in uniform junctions, and attributed it to the intrinsic nature of the avalanching electron–hole plasma. Microplasmas seem to prefer sites associated with crystal defects like etch-pits, dislocations, scratches, field inhomogeneities etc., the most important being the place where the junction intersects the surface [113 to 116]. Details of the role of imperfections like lattice defects, surface defects, metallic precipitates etc. in contributing to the origin of microplasma sites and elimination of microsplasma in order to yield uniform breakdown junctions have been dealt with by Kressel [117]. It is possible to produce microplasma free bulk AB if due care is taken to have structurally perfect junctions with suitable guard ring etc., to prevent surface effects [114, 92]. It is then observed that light emission is distributed uniformly over the entire area and the current also appears to be uniformly distributed [114]. Batdorf et al. [114] have reported the absence of noise in uniform junctions. Haitz [121] has given methods of producing artificial microplasma avalanche diodes with special geometry to avoid carrier diffusion from the bulk to the breakdown region and has indicated their use in studies of carrier generation mechanism in space charge layer.

In order to account for and describe the behaviour of microplasmas a number of models of varying complexity have been proposed by different authors. Earliest among them is that of Rose [106] who has proposed a model employing a p–i–n structure based on the analogy of the gas discharge phenomenon leading to strongly differing electric field distributions in either on or off condition. Consequently, the terminal voltages at the microplasma ends will differ rather markedly in the two conditions. Rose has attributed the turn-off phenomenon to statistical fluctuations of the number of carriers initiating the microplasma. The model does not provide for linear gradient or step junctions. Some experimental support to Rose's model is available in the work of Gibson and Morgan [118]. Kikuchi and Tachikawa [108] have proposed a model which is analogous to a p–n–p–n structure with its associated switching action. A local hole-emitting source is assumed without explicitly stating the origin or characteristics of the emitting source. According to Shockley [92] microplasmas are associated with structural imperfections and defects. He has proposed a model of structural imperfections capable of producing a locally enhanced electric field substantially in excess of the average, leading to breakdown and capable of immobilizing high densities of electric charge through traps. The actual structural imperfection considered for the specific case of breakdown in silicon is silicon dioxide in the silicon matrix which produces a local high electric field because of the large difference in the dielectric constants of the matrix and the precipitate. The effect of local metallic particles on the I–U charcateristics has also been discussed by Shockley [92].

Statistical models for the microplasma have been put forward by Burgess [110], Champlin [119], and McIntyre [120]. Electrical models for the microplasma have been proposed by Champlin [119], McIntyre [120], and Haitz [121]. All these are equivalent circuits with the addition of a switch with turn-on and turn-off probability which along with some series resistance and shunt capacitance provide a set of parameters to characterise the microplasma. Burgess's [110] model is based on a bistable switch with some turn-on and turn-off probability depending upon the voltage. He has analysed the problem by postulating both a power-law type and exponential-law type dependence of the transition probabilities on voltage which exist before each transition and also points out that the series impedance of the source has to play an important part and that the probabilities will neither be independent of each other or remain constant in time if the source impedance is large. Another aspect of interest is the possibility of a negative resistance region in the I–U characteristics if the source impedance is very large.

Champlin's [119] model is also similar and similar conclusions are drawn regarding the source series impedance and it is pointed out that the process is likely to be non-Markoffian if the source impedance is large. He concludes that the turn-on probability increases with increasing voltage but the turn-off probability also shows a continuous increase in constrast to that concluded on the basis of Rose's model [106]. Champlin has also reported that some microplasmas exhibit multiple states but this has later on been questioned by McIntyre [120] who has attributed this to possible parallel breakdown of more than one microplasma of nearly identical breakdown voltages. Ruge and Keil [122] have also indicated the possibility of mutual interaction between microplasmas leading to multistable levels. Champlin [119] concludes that turn-off always occurs in a narrow region around a definite voltage below which avalanche does not at all take place.

A quantitative statistical theory for the turn-on and turn-off probabilities has been worked out by McIntyre [120] who has also given experimental methods of getting the probability values. By analysing the effect of spreading resistance, space charge, and internal heating, which will limit the avalanche once it has commenced and which determine the shape of the I–U characteristic in the breakdown region, McIntyre has given an equivalent electric circuit of a avalanching diode using appropriate RC combinations to represent each one of the above effects. An experimental method of studying the turn-on of a microplasma has also been given by Kimura and Nishizawa [123].

11. Light Emission

Light emission that may be associated with electrical breakdown has been observed by numerous authors. Though the earlier works deal with silicon [104, 124 to 126] and germanium [127], much work in this aspect has been done on gallium arsenide [128] and gallium phosphide [129, 62] junctions.

Light emission may take place as a result of i) interband recombination [124, 125, 129], ii) free carrier transition from one level to another, i.e. intraband transition [124, 125, 129], or iii) bremsstrahlung of hot carriers in the coulomb field of charged centres [131].

The low energy threshold for the light emitted as a result of interband recombination of hot electrons and holes is the band gap of the material under consideration, whereas no such definite lower energy limit can be set for intraband transitions. In interband transitions which account for the radiation on the short wavelength side, the light output is expected to be proportional to the square of the current, whereas in the larger wavelength region due to intraband transitions the light output is linear in current [124, 125, 127]. While photons of highest energy most probably arise from interband recombination transitions, for the emission of photons with energies less than the band gap the additional mechanism of intraband transitions must be considered.

The spectrum is in general broad, extending to energies both considerably higher and less than the energy gap. In Si, Ge, SiC, and GaP, which have an indirect band structure, the emission spectrum is broad, whereas InP and GaAs, which have a direct band structure, exhibit a sharp line below the band gap [62]. The shape of the spectrum is not altered by current density variations but is altered by temperature changes.

Theories of the various mechanisms responsible for light emission have been dealt with by Wolff [130] and by Figielski and Torun [131]. Wolff [130] has based a theoretical model for interband and intraband transitions taking into account details of band structure and selection rules. While recombination and intraband transitions seem capable of accounting for the observed spectrum, Figielski and Torun [131] have put forward another hypothesis (for at least the high energy end of the spectrum) namely, bremsstrahlung of hot carriers in the coulomb field of charged centres.

12. Experimental Methods

Experimental methods in the study of breakdown in semiconductors are rather simple in principle and not too involved in practice. In fact, the preparation of samples is much more difficult than the actual measurements. Thus we may divide the experimental methods into two main groups, these pertaining to the preparation of samples and the junctions and those pertaining to the actual measurements. Only a very brief outline of the various methods and techniques employed are given here, with relevant references for further details.

12.1 Preparation of samples

While early measurements were performed using alloyed or diffused junction diodes with simple structures (circular or rectangular with ohmic contacts attached to the p and n sides), later on, complicated structures had to be used to avoid effects of surface leakage, microplasma, humidity, etc. Some of the methods currently employed are a) bevelling of the edges [37], b) mesa type of structure, c) single microplasma junction by isolating pyramidal etch pits and alloying only there [63], d) edge free junctions by planar diffusion methods involving a lightly doped surrounding layer which effectively increases the leakage path [114], e) concentric ring structure [132], f) truncated pyramid with an angle of 15° [25], g) guard-ring structure with deep annular diffusion along with SiO_2 mask for silicon [38]. The last mentioned, though complicated, seems to have yielded nearly ideal junctions.

12.2 Measurements

I–U characteristics. It is quite simple to obtain the dc *I–U* characteristics and need not be elaborated here. The only requirement is that the source of the e.m.f. has sufficient series impedance to limit the current. Measurements are carried out with the sample in the dark unless one is interested in photo-injection and photo-multiplication. Because of the very nature of the breakdown phenomenon, measurements have to be made under constant current conditions and it should also be remembered that a wide range of current capability (from μA to a few A) is required from the source. Excessive current in the region of breakdown may lead to device destruction or temperature rise; both can be avoided by a heat sink. When the effect is likely to be screened by a heat sink or other cooling methods, pulse methods are employed using low pulse repetition rates. The characteristics are usually displayed on a cathode ray oscilloscope [63, 107, 133 to 135].

Multiplication, photo-multiplication, and ionisation measurements. The ordinary I–U characteristics themselves can yield information on the multiplication factor M if the extrapolated saturation current is compared with the actual current at the required voltage [8, 19, 23]. Because considerable error will be involved in obtaining the exact value of the saturation current particularly when it is accompanied by additional currents due to surface leakage, it is the current practice to study the multiplication of carriers injected into the junction. Injection of carriers by light or by α-particles involves injection of both electrons and holes simultaneously. Forward biased emitter junctions were used by Miller [66] to inject pure electron or hole currents. Conditions for a more refined experimental measurement of multiplication factor have been compiled by Chynoweth [5] in the description of a model experiment. The same was practiced more or less completely by Lee et al. [81] whose results give a good fit to Baraff's theory [93]. In brief, the conditions require that measurements are made on junctions free from defects causing microplasma, by initiating multiplication with pure electron and pure hole injection. The spatial variation of the field in the junction should be known accurately.

Multiplication and capacitance data lead to the determination of the ionisation coefficient as already indicated earlier.

Microplasma, light-emission and noise observation: Visual observation of the microplasma is effected by having a thin base region through which light can pass. Electrical observation is effected by observing the associated noise pulses on a cathode ray oscilloscope [14]. Microplasma locations are discovered by scanning the junction with a minute light spot [63, 106]. The spectral distribution of the emitted light can be determined using suitable photo-multipliers and spectrometers [125, 135].

On–off probability. There are two main methods of estimating the on–off probabilities of microplasmas as indicated by McIntyre [120]. The number of pulses is counted electronically in conjugation with oscilloscopic measurement of the height of the current pulses $(I - I_0)$. The turn-off probability P_{10} and the turn-on probability P_{01} are given by

$$P_{01} = \frac{\nu}{\lambda}, \quad P_{10} = \frac{\nu}{1 - \lambda}, \quad \text{where} \quad \lambda = \frac{[I - I_0]_{\text{av.}}}{I - I_0},$$

ν is the pulse frequency. When the current pulses are extremely small, pulses are recorded photographically about 30 pulses at a time and their pulse length distribution is plotted on a semilogarithmic paper. The slope of the resulting line gives the probability. In addition, a noise spectrum method involving simultaneous measurement of the noise power spectrum, the pulse current, and the average pulse current is also given by McIntyre [120].

Capacitance measurements. The common bridge type measurements capable of measuring capacitances in the pF range are quite adequate for measuring the junction capacitance [136, 137]. But the junction capacitance measurement in the breakdown region is very complicated by the presence of a shunt conductance due to the increased conductance of the diode in the breakdown region. Bridge and heterodyne beat methods of obtaining the junction capacitance values in the presence of such conductance effects at, near, and through the breakdown region have been described by the authors elsewhere [138]. As against this point by point method, a more sophisticated method of automatically plotting the capacitance–voltage and conductance–voltage curves has been given in detail by Shewchun and Waxman [139].

Junction width. The width of the junction is an ever present quantity and determines the type of breakdown as already indicated. Measurement of the junction capacitance gives a direct determination of the width and the variation of the junction capacitance with reverse voltage gives an idea of the geometry (i.e. graded and step) of the junction [140], details of which need not be dealt with here.

13. Impact Ionisation of Impurities

In discussing the ionisation of electrons and holes, it may be noted that in general the electrons come from the valence band and the required energy for pair production is at least the gap energy. But it is pertinant to point out that electrons may be contributed from the impact ionisation of impurities also. There are many papers dealing with this topic few of which are indicated in the references [141 to 152]. The general current–field characteristic is interpreted in terms of Stratton's [153] theory of breakdown in polar crystals.

The effect is usually pronounced at low temperatures and manifests itself as a non-ohmic behaviour for quite low voltages and even for homogeneous samples. In the papers referred to, the investigations of Sclar and Burstein [133] deal with hydrogen-like impurities, those of Zylberstejn [151] and of Chapnin [152] with deep level impurities like zinc in germanium. Attempts of a theory for this type of low temperature breakdown in homogeneous samples have been made by Yamashita [154], Chuenkov [155], Zylberstejn [156], and Hrivnak [157]. In materials with compensated impurities, a complicated type of breakdown is observed which includes a negative resistance behaviour in the current–field characteristics. This was first reported by McWhorter and Rediker [158].

14. Time Required for the Formation of an Avalanche

There are only a few explicit references in the literature considering the time required for the formation of an avalanche. Even there, the time under discussion is not well defined. The earliest paper is that of McKay [159] who performed experiments with p–n junctions by initiating pulses with a narrow beam of α-particles. Each α-particle gives rise to one pulse and the charge collected is a measure of the number of pairs produced. (No multiplication is involved in this experiment, the object being to provide information on the threshold energy for pair-production. Incidentally, it may be remarked that this sets an upper limit to the threshold energy.) During this experiment, by comparing the rise time of the pulse with a calibrating pulse of known rise time, it was observed that the carriers are swept across the barrier (width 5×10^{-4} cm) in a time less than 2×10^{-8} s. By repeating the experiment with higher values of the reverse bias in the region of AB, McKay and McAfee [8] have estimated the total time required for multiplication to be less than 0.02 μs compared with the transit time of 10^{-10} s calculated on the basis of barrier width and velocity of carriers. A later investigation by Ascarelli [160] on the breakdown in silicon p–n junctions reports a time which is at least an order of magnitude higher than that of McKay and McAfee [8]. Ascarelli found the time for building up a specified carrier concentration after application of a sharp voltage pulse to the crystal by using a microwave probe and obtained times of the order of 1 to 10 μs depending inversely on the square of the applied electric field. Isaev [161] on the other hand found a delay in the formation of avalanche following the application of a square pulse with rise times not longer than 0.03 μs and with a flat top voltage higher than the static breakdown voltage. He reports delay times of the order of 0.6 to 0.03 μs, the delay time decreasing linearly with increasing excess voltage. Considering that there is a decrease of time with the excess-voltage, we may remark that the results of McKay and McAfee [8] may be a lower limit rather than an upper one, because the pairs are produced in high electric fields and the time is controlled by the energy of the pair after its generation. Isaev [161] has also presented data on the temperature variation of the time for formation of breakdown and found that the time increases with increasing temperature in the range 100 to 500 °K. It is to be expected that at very high fields the time approaches a constant value, but Isaev's [161] results do not show even a tendency towards a constant value in the range of his investigations.

Conwell [162], on the basis of a hydrogen-like model for impact ionisation in germanium, has concluded that even a value of 10^{-8} s could be possible.

15. Miscellaneous Effects

Effect of magnetic field. The effect of an applied magnetic field on breakdown has not been investigated in detail. Effects of a magnetic field have been studied for homogeneous samples by Darnell and Friedberg [163], Sclar and Burstein [133], and by Phelan and Love [164]. The results indicate that the magnetic field increases the critical electric field for breakdown. This is to be expected because in a magnetic field the carriers follow a spiral path thereby increasing the probability of collision and requiring a higher field for overcoming the collision loss. For compensated samples, Komatsubara and Kurono [165] find an angular dependence of the critical field and a linear dependence of the sustaining field on the magnetic field.

Effect of water vapour and ambients. There are pronounced effects of water vapour and other ambients in the formation and control of surface states, formation of channels, etc. which act as parallel conducting paths for current flow and because of an inversion layer formed lead to an increase in the reverse current and the general softening of the reverse characteristics. Details of this can be found in literature [166 to 177]. At the moment the above effect is spurious and does not contribute to an understanding of the basic process. But investigations in these lines are nevertheless important in order to identify and eliminate spurious effects.

Effects of strain. There have been many studies on the effect of hydrostatic and anisotropic stresses on the breakdown in p–n junctions and in bulk materials like silicon, first of all those of Rindner et al. [179 to 182]. Details of work in this direction are presented in [178 to 186]. The main feature observed is a reversible reduction in the reverse bias breakdown voltage, dU_b/U_b being

10^{-12} dyn^{-1} cm^{-2} in the pressure range and is related to the reduction in the band-gap of the material. Some effect associated with anisotropic stress has also been reported and is ascribed to the stress controlled generation–recombination of carriers. The temperature dependence of the anisotropic stress effect has also been studied which indicates an increase of the stress effect with increasing temperature.

16. Applications

As every physical phenomenon in semiconductors the breakdown phenomenon has also considerable practical applications, a few of which may be listed here: i) voltage regulators and voltage reference devices, ii) buck boost devices, iii) protective devices, iv) controlled rectifiers, v) solid state noise sources, vi) memory devices (cryosar), vii) high frequency oscillators, etc.

The first three are based on the enormous increase in current at the breakdown voltage in a p–n junction. With proper care, it is possible to have breakdown voltage defined within a few mV, thus serving as voltage reference device, aptly replacing their gaseous counterpart and establishing itself as an important circuit element.

Avalanche diodes are useful as noise sources. Both ordinary avalanche diodes as well as specially designed uniform avalanche diodes have been prepared as noise sources, particularly below 200 kHz. Haitz [187, 188] has discussed the design considerations. As noise sources they have the advantages of small size, extremely low temperature dependence of noise output, high spectral density, and wide operating conditions.

Cryosar is a semiconductor device intended for high speed computer switching memory applications, which utilises the low temperature AB produced by impact ionisation of impurities [158]. The discovery of a region of negative resistance between the high and low conducting states in the I–U characteristics of low-temperature breakdown of compensated germanium has opened a wide range of applications. This bistable cryosar can be used as a memory element, in multi-vibrator or flip-flop. Both the types are very fast, being limited by a turn-on time of 10^{-8} s or less, and have the advantage of extremely small size.

Silicon diodes can be used to obtain oscillations on a pulse basis. Read [189] has proposed the design of such an oscillator (for the frequency range 1 to 50 GHz) capable of oscillations in the microwave range when mounted on a suitable cavity. Electron–hole pairs are produced by AB in a reverse biased p–n junction. The electric field has a maximum at one edge of the space charge region. The build-up time of the avalanche and the transit time of the holes across the depletion region are used to delay the alternating current to approximately half a cycle relative to the ac voltage and power is delivered to the ac signal. If the diode is mounted on a suitable cavity, oscillations will build up. Johnston et al. [190] have given details of a silicon diode microwave oscillator.

Semiconductor p–n junction nuclear radiation detectors are preferred to other detectors using the ionisation phenomenon for the fact that the energy required to produce an electron–hole pair is very much smaller (for silicon (3.50 ± 0.07) eV, for germanium (2.94 ± 0.15) eV as compared to 30 eV for gases). Further the noise level is considerably reduced. For example [191] in silicon a 5 MeV particle produces 1.43×10^6 pairs, (in gases this may be smaller by an order of magnitude) with statistical fluctuations of $\pm 0.08\%$ (in gases this may be $\pm 0.23\%$ and in scintillation detectors $\pm 0.8\%$).

The radiation or charged particles are incident on the depletion region of the p–n (or p–i–n) junction which is highly asymmetric. The carriers produced by ionisation are swept by the reverse field applied and are collected by the electrode to produce an electrical signal (which may further be amplified). The number of carriers thus produced is a measure of the energy of the incident radiation. The wider the depletion region, the better is the stopping power for the incident particles. To achieve this, p–i–n structures with a central region preferably compensated by lithium (near to intrinsic resistivity) are used.

In semiconductor junctions, high reverse fields can conveniently be applied which result in a reduction of the collection time and the number of carriers lost in the recombination process. Fabrication techniques, electronic circuitry, and other details are available in literature [191 to 194].

Impact avalanche and transit time diodes (Impatt diodes) utilize the inherent time delay in avalanche process and the transit time through a drift region deliberately provided in the diode. The phase of the current in the avalanche process is 90° behind the field and a further 90° in drifting or transit time makes the overall phase difference of 180° of current and voltage in the device, which in turn makes it possible to pump power from dc biasing source and supply ac power at the desired frequency. The mathematical theory and details are available in literature [111, 112, 189].

Up to date achievements have been published in [195 and 196].

Trapped plasma avalanche triggered transit (Trapatt) diodes are associated with lower frequencies, but higher efficiencies as compared to the Impatt mode. This mode was revealed while seeking an explanation for the anomalously high efficiencies obtained by some authors [197, 198]. A computer analysis was presented by Bartelink and Scharfetter [199].

The Impatt mode operates for a few initial cycles and as soon as the RF voltage (Impatt mode) reaches a critical value (well in excess of breakdown voltage) a travelling avalanche zone sets in leaving behind a dense plasma. The travelling velocity of the avalanche zone being higher than the carrier velocity, the high field state (avalanching) is reduced to a low field one (plasma). With the charge removal the device recovers back to the high field state. Thus the operation involves the alternation of high field–low current and low field–high current states (obviously with less power dissipation in any state and subsequently higher efficiency). It may be noted that in contrast to the Impatt mode, this mode involves the drift velocity of carriers less than the saturation velocity and hence the lower frequency.

A comprehensive review of Impatt and Trappatt modes of avalanche diodes has been presented by De Loach [45].

References

[1] J. J. O'DWYAR, Theory of Dielectric Breakdown in Solids, Ed. LORD JACKSON of BURNLEY, H. FRÖHLICH, SIR N. MOTT, and E. C. BULLARD, Oxford University Press, 1964.

[2] N. KLIENE, Adv. Electronics and electron. Phys. **26**, 326 (1969).

[3] J. B. GUNN, Progr. Semicond. **2**, 213 (1957).

[4] E. M. CONWELL, Solid State Phys. **9**, Suppl. II (1967).

[5] A. G. CHYNOWETH, Semiconductors and Semimetals **4**, 263 (1968).

[6] C. ZENER, Proc. Roy. Soc. **A145**, 523 (1934).

[7] K. B. MCAFEE, E. J. RYDER, W. SHOCKLEY, and M. SPARKS, Phys. Rev. **83**, 650 (1951).

[8] K. G. McKay and K. B. McAfee, Phys. Rev. **91**, 1079 (1953).
[9] G. L. Pearson and B. Sawyer, Proc. IRE **40**, 1348 (1952).
[10] H. Kressel and A. Blicher, J. appl. Phys. **34**, 2495 (1963).
[11] T. Tokuyama, Solid State Electronics **5**, 161 (1962).
[12] S. M. Tyagi, Solid State Electronics **11**, 99 (1968).
[13] F. W. G. Rose, Proc. Internat. Coll. Semicond. and Phosphors, Garmisch-Partenkirchen (Germany) 1956.
[14] V. K. Aladinskii and S. I. Sergeeva, Soviet Phys. — Semicond. **1**, 650 (1967).
[15] A. G. Chynoweth and K. G. McKay, Phys. Rev. **108**, 29 (1957).
[16] A. G. Chynoweth and K. G. McKay, Phys. Rev. **106**, 418 (1957).
[17] J. W. Allen, J. Electronics and Control **7**, 254 (1959).
[18] M. Poleshuk and P. H. Dowling, J. appl. Phys. **34**, 3069 (1963).
[19] A. R. Plummer, J. Electronics and Control **5**, 405 (1958).
[20] R. D. Knott, I. D. Colson, and M. R. P. Young, Proc. Phys. Soc. **B68**, 182 (1955).
[21] A. G. Chynoweth, Progr. Semicond. **4**, 95 (1960).
[22] D. R. Muss and R. F. Greene, J. appl. Phys. **29**, 1534 (1958).
[23] K. G. McKay, Phys. Rev. **94**, 877 (1954).
[24] J. Shields, J. Electronics and Control **6**, 130 (1959).
[25] T. Ogawa, J. appl. Phys. **4**, 473 (1965).
[26] R. Hall, Internat. J. Electronics **22**, 521 (1967).
[27] N. Sclar and E. Burstein, J. Phys. Chem. Solids **2**, 1 (1956).
[28] L. V. Keldysh, Soviet Phys. — J. exper. theor. Phys. **37**, 509 (1960).
[29] J. Yamashita, J. Phys. Soc. Japan **16**, 720 (1961).
[30] J. Maserjian, J. appl. Phys. **30**, 1613 (1959).
[31] D. P. Kennedy and R. R. O'Brien, IRE Trans. Electron Devices **9**, 478 (1962).
[32] W. Fulop, Solid State Electronics **10**, 39 (1967).
[33] E. M. Pell, J. appl. Phys. **28**, 459 (1957).
[34] K. A. Maddex and M. R. P. Young, Electronics Letters **2**, 129 (1966).
[35] C. G. B. Garrett and W. R. Brattain, J. appl. Phys. **27**, 299 (1956).
[36] B. M. Vul and A. P. Shotov, Zh. tekh. Fiz. **27**, 2189 (1957).
[37] K. A. Maddex and M. R. P. Young, Electronics Letters **1**, 191 (1965).
[38] A. Goetzberger, B. McDonald, R. H. Haitz, and R. M. Scarlett, J. appl. Phys. **34**, 1591 (1963).
[39] A. Goetzberger and W. Shockley, J. appl. Phys. **31**, 1821 (1960).
[40] J. T. Law, Proc. IRE **42**, 1367 (1954).
[41] A. L. McWhorter and R. H. Kingston, Proc. IRE **42**, 1376 (1954).
[42] J. T. Law and P. S. Meigs, J. appl. Phys. **26**, 1265 (1955).
[43] M. Kikuchi, J. Phys. Soc. Japan **15**, 1822 (1960).
[44] F. W. G. Rose, J. Electronics and Control **3**, 396 (1957).
[45] B. C. De Loach, Jr., IEEE J. Solid State Circuits **4**, 376 (1969).
[46] R. H. Haitz, A. Goetzberger, R. M. Scarlett, and W. Shockley, J. appl. Phys. **34**, 1581 (1963).
[47] A. G. Chynoweth, W. L. Feldmann, C. A. Lee, R. A. Logan, G. L. Pearson, and P. Aigrain, Phys. Rev. **118**, 425 (1960).
[48] D. J. Dumin and G. L. Pearson, J. appl. Phys. **36**, 3418 (1965).
[49] R. A. Logan, A. G. Chynoweth, and B. G. Cohen, Phys. Rev. **128**, 2518 (1962).
[50] D. N. Nasledov and B. V. Tsarenkov, Soviet Phys. — Solid State **5**, 862 (1963).
[51] R. Hall and J. H. Leck, Internat. J. Electronics **25**, 539 (1968).
[52] R. A. Logan and A. G. Chynoweth, J. appl. Phys. **33**, 1649 (1962).
[53] J. Yamaguchi and Y. Hamakawa, Proc. IEE **B106**, Suppl. 15, 353 (1959).
[54] C. Yamanaka and T. Suita, Technol. Rep. Osaka Univ. **7**, 243 (1957).
[55] A. G. Chynoweth and R. A. Logan, Phys. Rev. **118**, 1470 (1960).
[56] A. E. Garside and P. Harvey, Proc. Inst. Electr. Eng. **B106**, Suppl., 982 (1959).
[57] K. Maeda and K. Suzuki, Japan. J. appl. Phys. **1**, 193 (1962).
[58] M. W. Nield, R. A. Reid, and J. H. Leck, J. Electronics and Control **15**, 537 (1963).

[59] V. K. ALADINSKII, Soviet Phys. — Radio Engng. and Electronics **10**, 81 (1965).
[60] ZH. I. ALFEROV, G. V. GORDEEV, and V. I. STAFEEV, Fiz. tverd. Tela **1**, Suppl. II, 104 (1959).
[61] K. I. NUTTALL and M. W. NIELD, Internat. J. Electronics **24**, 69 (1968).
[62] M. GERSHENZON and R. M. MIKULYAK, J. appl. Phys. **32**, 1338 (1961).
[63] B. SENITZKY and J. L. MOLL, Phys. Rev. **110**, 612 (1958).
[64] H. S. VELORIC, M. B. PRINCE, and M. J. EDER, J. appl. Phys. **27**, 895 (1956).
[65] R. YEE, J. MURPHY, A. D. KURTZ, and H. BERNSTEIN, J. appl. Phys. **30**, 596 (1959).
[66] S. L. MILLER, Phys. Rev. **99**, 1234 (1955).
[67] S. M. SZE and G. GIBBONS, Appl. Phys. Letters **8**, 111 (1966).
[68] J. L. MOLL, Physics of Semiconductors, McGraw-Hill Publ. Co., New York 1964.
[69] M. MIGITAKA, Solid State Electronics **8**, 295 (1965).
[70] R. TSU and F. C. WHITMORE, Phys. Letters (Netherlands) **24A**, 25 (1967).
[71] R. WILLIAMS, RCA Rev. **27**, 336 (1966).
[72] J. ANTULA, phys. stat. sol. **3**, 1829 (1963).
[73] A. I. MELNIKOV, Soviet Phys. — J. exper. theor. Phys., Letters **9**, 19 (1969).
[74] H. WEINERTH, Solid State Electronics **10**, 1053 (1967).
[75] R. A. KOKASA and R. L. DAVIES, IEEE Trans. Electron Devices **13**, 874 (1966).
[76] L. V. KELDYSH, Soviet Phys. — J. exper. theor. Phys. **6**, 763 (1958).
[77] L. V. KELDYSH, Soviet Phys. — J. exper. theor. Phys. **7**, 665 (1958).
[78] N. HORONYAK, Phys. Rev. Letters **3**, 167 (1959).
[79] L. ESAKI, Phys. Rev. **109**, 603 (1958).
[80] A. G. CHYNOWETH, J. appl. Phys. **31**, 1161 (1960).
[81] C. A. LEE, R. A. LOGAN, R. L. BATDORF, J. J. KLEIMACK, and W. WEIGMANN, Phys. Rev. **134**, A761 (1964).
[82] J. L. MOLL and R. VAN OVERSTRAETON, Solid State Electronics **6**, 147 (1963).
[83] B. M. VUL and A. P. SHOTOV, Internat. Congr. Phys. Solid State and Appl. Electronics and Telecomm., Brussels, Academic Press, **1**, 491 (1960).
[84] A. G. CHYNOWETH, Phys. Rev. **109**, 1537 (1958).
[85] V. K. ALADINSKII, Soviet Phys. — Semicond. **1**, 801 (1967).
[86] R. A. LOGAN, A. G. CHYNOWETH, and B. G. COHEN, Phys. Rev. **128**, 2518 (1962).
[87] Y. J. CHANG and S. M. SZE, J. appl. Phys. **40**, 5392 (1969).
[88] R. A. LOGAN and H. G. WHITE, J. appl. Phys. **36**, 3945 (1965).
[89] C. YAMANAKA and T. SUITA, Technol. Rep. Osaka Univ. **6**, 243 (1956).
[90] S. L. MILLER, Phys. Rev. **105**, 1246 (1957).
[91] P. A. WOLFF, Phys. Rev. **95**, 1415 (1954).
[92] W. SKOCKLEY, Solid State Electronics **2**, 35 (1961).
[93] G. A. BARAFF, Phys. Rev. **128**, 2507 (1962).
[94] J. L. MOLL and N. I. MEYER, Solid State Electronics **3**, 155 (1961).
[95] C. R. CROWELL and S. M. SZE, Appl. Phys. Letters **9**, 242 (1966).
[96] W. P. DUMKE, Phys. Rev. **167**, 783 (1968).
[97] D. J. BARTELIN, J. L. MOLL, and N. I. MEYER, Phys. Rev. **130**, 972 (1963).
[98] J. R. HAUSER, J. appl. Phys. **37**, 507 (1966).
[99] T. YA. PURITIS, Soviet Phys. — Semicond. **1**, 500 (1967).
[100] M. P. MIKHILOVA, D. N. NASLEDOV, and S. V. SLOBODCHIKOV, Soviet Phys. — Semicond. **1**, 94 (1967).
[101] H. L. ARMSTRONG, J. Electronics and Control **4**, 355 (1958).
[102] H. L. ARMSTRONG, J. Electronics and Control **5**, 97 (1958).
[103] F. W. G. ROSE, J. Electronics and Control **6**, 70 (1959).
[104] R. NEWMAN, W. C. DASH, R. N. HALL, and W. E. BURCH, Phys. Rev. **98**, 1536 (1955).
[105] A. G. CHYNOWETH and K. G. MCKAY, J. appl. Phys. **30**, 1811 (1959).
[106] D. J. ROSE, Phys. Rev. **105**, 413 (1957).
[107] B. SENITZKY and P. D. RADIN, J. appl. Phys. **30**, 1945 (1959).
[108] M. KIKUCHI and K. TACHIKAWA, J. Phys. Soc. Japan **15**, 835 (1960).
[109] A. GOETZBERGER, J. appl. Phys. **31**, 2260 (1960).

[110] R. E. Burgess, Canad. J. Phys. **37**, 730 (1959).
[111] T. Misawa, IEEE Trans. Electron Devices **13**, 137 (1966).
[112] T. Misawa, IEEE Trans. Electron Devices **13**, 142 (1966).
[113] A. G. Chynoweth and G. L. Pearson, J. appl. Phys. **29**, 1103 (1958).
[114] R. L. Batdorf, A. G. Chynoweth, G. C. Dacey, and P. W. Foy, J. appl. Phy. **31**, 1153 (1960).
[115] J. P. McCarthy, J. appl. Phys. **37**, 436 (1966).
[116] A. Goetzberger and C. Stephens, J. appl. Phys. **32**, 2646 (1961).
[117] H. Kressel, RCA Rev. **28**, 175 (1967).
[118] A. F. Gibson and J. R. Morgan, Solid State Electronics **1**, 54 (1960).
[119] K. S. Champlin, J. appl. Phys. **30**, 1039 (1959).
[120] R. J. McIntyre, J. appl. Phys. **32**, 983 (1961).
[121] R. H. Haitz, J. appl. Phys. **36**, 3123 (1965).
[122] I. Ruge and G. Keil, J. appl. Phys. **34**, 3306 (1963).
[123] C. Kimura and J. Nishizawa, Japan. J. appl. Phys. **7**, 1453 (1968).
[124] R. Newman, Phys. Rev. **100**, 700 (1955).
[125] A. G. Chynoweth and K. G. McKay, Phys. Rev. **102**, 369 (1956).
[126] M. Kikuchi and K. Tachikawa, J. Phys. Soc. Japan **14**, 1830 (1959).
[127] A. G. Chynoweth and H. K. Gummel, J. Phys. Chem. Solids **16**, 191 (1960).
[128] A. E. Michel, M. I. Nathan, and J. C. Marinace, J. appl. Phys. **35**, 3543 (1964).
[129] M. H. Pilkuhn, J. appl. Phys. **40**, 3162 (1969).
[130] P. A. Wolff, J. Phys. Chem. Solids **16**, 184 (1960).
[131] T. Figielski and A. Torun, Internat. Conf. Phys. Semicond., Exeter 1962 (p. 863).
[132] Y. C. Kao and E. D. Wolley, Proc. IEEE **55**, 1409 (1967).
[133] N. Sclar and E. Burstein, J. Phys. Chem. Solids **2**, 1 (1956).
[134] A. P. Shotov, Zh. tekh. Fiz. **26**, 1634 (1956).
[135] J. W. Allen and P. E. Gibbons, J. Electronics and Control **7**, 518 (1959).
[136] A. Hoffmann, Z. angew. Phys. **2**, 353 (1950).
[137] H. K. Henisch, Rectifying Semiconductor Contacts, Chap. VI, Clarendon Press, Oxford 1957 (p. 150).
[138] S. Mahadevan and G. Suryan, Indian J. pure appl. Phys. **5**, 201 (1967).
[139] J. Shewchun and A. Waxman, Rev. sci. Instrum. **37**, 1195 (1966).
[140] W. Shockley, Bell Syst. tech. J. **28**, 435 (1949).
[141] N. Sclar, E. Burstein ,W. J. Turner, and J. W. Davison, Phys. Rev. **91**, 215 (1953).
[142] N. Sclar, E. Burstein, and J. W. Davison, Phys. Rev. **92**, 858 (1953).
[143] M. Glicksman and M. C. Steels, Phys. Rev. **110**, 1204 (1958).
[144] E. J. Ryder, I. M. Ross, and D. A. Kleinman, Phys. Rev. **95**, 1342 (1954).
[145] A. C. Prior, J. Electronics and Control **4**, 165 (1958).
[146] S. H. Koling, Phys. Rev. **110**, 986 (1958).
[147] W. Kaiser and G. H. Weathley, Phys. Rev. Letters **3**, 334 (1959).
[148] M. C. Steele and M. Glicksman, J. Phys. Chem. Solids **8**, 242 (1959).
[149] M. C. Steele, J. Phys. Chem. Solids **9**, 93 (1959).
[150] J. C. Sohm, J. Phys. Chem. Solids **18**, 181 (1961).
[151] A. Zylberstejn, J. Phys. Chem. Solids **23**, 297 (1962).
[152] V. A. Chapnin, Soviet Phys. — Solid State **7**, 403 (1965).
[153] R. Stratton, Proc. Roy. Soc. **A246**, 406 (1958).
[154] J. Yamashita, Progr. theor. Phys. (Kyoto) **15**, 95 (1956).
[155] V. A. Chuenkov, Fiz. tverd. Tela **1**, Suppl. II, 200 (1959).
[156] A. Zylberstejn, Phys. Rev. **127**, 744 (1962).
[157] L. Hrivnak, Czech. J. Phys. **14**, 459 (1964).
[158] A. L. McWhorter and R. Rediker, Proc. IRE **47**, 1207 (1959).
[159] K. G. McKay, Phys. Rev. **84**, 829 (1951).
[160] G. Ascarelli, Nuovo Cimento **22**, 251 (1961).
[161] M. R. Isaev, Soviet Phys. — Solid State **7**, 2442 (1966).
[162] E. M. Conwell, Phys. Rev. **94**, 1068 (1954).

[163] F. J. DARNELL and S. A. FRIEDBERG, Phys. Rev. **98**, 1860 (1955).
[164] R. J. PHELAN, JR., and W. F. LOVE, Phys. Rev. **133**, 1134 (1964).
[165] K. KOMATSUBARA and H. KURONO, J. Phys. Soc. Japan **18**, 660 (1963).
[166] H. CHRESTENSEN, Proc. IRE **42**, 1371 (1954).
[167] A. L. MCWHORTER and R. H. KINGSTON, Proc. IRE **42**, 1376 (1954).
[168] J. T. LAW and P. S. MEIGS, J. appl. Phys. **26**, 1265 (1955).
[169] A. J. WAHL and J. J. KLEIMACK, Proc. IRE **44**, 495 (1956).
[170] V. G.MELINIK, L. I.KUZOVKINA, and G. A.YAGUNOVA, Soviet Phys.— Radio Engng. and Electronics **12**, 877 (1967).
[171] A. R. F. PLUMMER, Proc. Phys. Soc. **69B**, 539 (1956).
[172] W. T. ERIKSEN, H. STATZ, and D. A. DEMARS, J. appl. Phys. **28**, 133 (1957).
[173] R. O. LITVINOV and HAUI TUNG-LIANG, Soviet Phys. — Radio Engng. and Electronics **7**, 965 (1962).
[174] O. JANTSCH, Solid State Electronics **5**, 249 (1962).
[175] CHIH-TANG-SAH, IRE Trans. Electron Devices **9**, 94 (1962).
[176] B. S. MURAVSKII, Soviet Phys. — Radio Engng. and Electronics **8**, 143 (1963).
[177] C. P. COCKSHOTT and E. L. G. WILKINSON, British J. appl. Phys. **16**, 45 (1965).
[178] H. H. HALL, J. BARDEEN, and G. L. PEARSON, Phys. Rev. **84**, 129 (1951).
[179] W. RINDNER and R. NELSON, Proc. IRE **50**, 2106 (1962).
[180] W. RINDNER, J. appl. Phys. **33**, 2479 (1962).
[181] W. RINDNER and I. BRAUN, Internat. Conf. Phys. Semicond., Exeter 1962 (p. 167).
[182] W. RINDNER and I. BRAUN, J. appl. Phys. **34**, 1958 (1963).
[183] K. KOMATSUBARA, H. KURONO, and N. TAKASUGI, J. Phys. Soc. Japan **19**, 575 (1964).
[184] A. GOETZBERGER and R. H. FINCH, J. appl. Phys. **35**, 1851 (1964).
[185] W. RINDNER, J. appl. Phys. **36**, 2513 (1965).
[186] I. V. KUCHERNEKO, Soviet Phys. — Solid State **7**, 818 (1965).
[187] R. H. HAITZ, IEEE Trans. Electron Devices **12**, 198 (1965).
[188] R. H. HAITZ, IEEE Trans. Electron Devices **13**, 342 (1966).
[189] W. T. READ, Bell Syst. tech. J. **37**, 401 (1958).
[190] R. L. JOHNSTON, B. C. DE LOACH, JR., and B. G. COHEN, Bell Syst. tech. J. **44**, 369 (1965).
[191] W. J. PRICE, Nuclear Radiation Detection, Chap. VIII, 2nd edition, McGraw-Hill Publ. Co., 1964.
[192] G. DEARNALEY and D. C. NORTHROP, Semiconductor Counters for Nuclear Radiation, E. and F. N. Spon, Ltd., London 1963.
[193] G. L. MILLER, W. M. GIBSON, and P. F. DONOVAN, Ann. Rev. nuclear Sci. **12**, 189 (1962).
[194] A. J. TAVENDALE, Ann. Rev. nuclear Sci. **17**, 73 (1967).
[195] B. C. DE LOACH, JR., Advances in Microwaves, Vol. II, Ed. L. YOUNG, Academic Press, 1967 (p. 71).
[196] B. C. DE LOACH, JR., Microwave Semiconductor Devices and Their Circuit Applications, Ed. H. A. WATSON, Chap. 15, McGraw-Hill, 1969 (p. 489).
[197] D. J. I. GELSIAS and W. J. EVANS, Proc. IEEE **56**, 1610 (1968).
[198] R. L. JOHNSTON, D. J. BARTELINK, and D. L. SCHARFETTER, Proc. IEEE **56**, 1611 (1968).
[199] D. J. BARTELINK and D. L. SCHARFETTER, Appl. Phys. Letters **14**, 320 (1969).

(Received May 5, 1971)

Original Papers

phys. stat. sol. (a) 8, 375 (1971)

Subject classification: 9 and 11; 14.1; 21.1; 21.6

Institute of Metallurgy, Academy of Sciences of the USSR, Moscow

Non-Stationary Radiation-Enhanced Diffusion in Metals

Ordering in the Cold-Worked Silver-Zinc Alloys

By

L. N. Bystrov, L. I. Ivanov, and Yu. M. Platov

A model is proposed for the non-stationary radiation-enhanced diffusion in metals containing a significant concentration of unsaturable sinks for radiation point defects. By measuring the change of resistivity at −197 °C, ordering in the cold-worked samples of the Ag–8.75 at% Zn alloy was investigated under 2.3 MeV electron irradiation with flux of 9.35×10^{12} to 3.74×10^{13} cm^{-2} s^{-1} at temperatures from 10 to 35 °C. It is found that the radiation-enhanced diffusion in metals containing a considerable number of dislocations (unsaturable sinks) is a non-stationary process with a diffusion coefficient gradually decreasing with time in accordance with the equation derived for the non-stationary radiation diffusion. The validity of the model proposed for the non-stationary radiation-enhanced diffusion controlled by the motion of interstitials is confirmed also by the fact that the radiation ordering is temperature-independent and proportional to the square root of irradiation flux.

Es wird ein Modell vorgeschlagen für die nichtstationäre, strahlungsinduzierte Diffusion in Metallen, die eine beträchtliche Konzentration nichtsättigbarer Senken für Strahlungs-Punktdefekte enthalten. Durch die Messung der Widerstandsänderung bei −197 °C wurde die Fehlordnung in kaltbearbeiteten Proben von Ag–8,75 At% Zn-Legierungen bei einer 2,3 MeV Elektronenbestrahlung und Intensitäten von $9{,}35 \times 10^{12}$ bis $3{,}74 \times 10^{13}$ cm^{-2} s^{-1} bei Temperaturen zwischen 10 und 35 °C untersucht. Es wird gefunden, daß die strahleninduzierte Diffusion in Metallen mit beträchtlicher Zahl von Versetzungen (nichtsättigbare Senken) ein nichtstationärer Prozeß ist, mit einem Diffusionskoeffizienten, der allmählich abnimmt in Übereinstimmung mit der Gleichung, die für nichtstationäre Bestrahlungsdiffusion abgeleitet wurde. Die Gültigkeit des vorgeschlagenen Modells für nichtstationäre, strahleninduzierte Diffusion, die durch die Bewegung von Zwischengitterdefekten gesteuert wird, wird auch durch die Tatsache gestützt, daß die Strahlungsfehlordnung temperaturabhängig und proportional zur Quadratwurzel der Bestrahlungsintensität ist.

1. Introduction

In the analysis of the experimental results of radiation-enhanced diffusion in metals [1 to 8] usually the vacancy mechanism is preferred, which is quite natural for thermal diffusion. In this case the formation energy of interstitial atoms is rather high, and hence their concentration is negligibly small. During irradiation vacancies and interstitials are formed in approximately equal concentrations. Thus, the problem what type of defects is responsible for diffusion requires a special analysis.

It is generally accepted that a dumb-bell configuration is the most stable form of an interstitial. The motion of such an atom is consistent with consecutive exchanges between the interstitial and lattice atoms. The character of the atomic migration principally does not differ from that of vacancies: the migration of one interstitial by n lattice parameters is equivalent to that of n atoms per one lattice parameter in the crystal. Hence, the motion of interstitials in the crystal can lead to the microscopic diffusion-controlled effects, in particular to short-range ordering.

In our opinion the problem of the defect type responsible for the radiation-enhanced diffusion is closely connected with the so-called "stage III dilemma", since such diffusion is usually observed at temperatures of stage III. Nowadays there is no common viewpoint on the type of defects migrating at the stage III annealing. Most investigators suggest that either vacancies (Koehler et al. [9 to 11]) or interstitials in the dumb-bell configuration (Schumacher [12], Bauer and Sosin [13], Frank and Seeger [14], and Jackson and Hershbach [15]) migrate at stage III. Assuming that the motion of interstitials in the dumb-bell configuration results in the diffusion-rearrangement of the lattice, the absence of the radiation ordering effect at low temperatures (stage I) and its existence in the temperature range of stage III annealing can serve as evidence favouring the interstitial interpretation of stage III recovery.

In the present paper an attempt was made to discuss the possibility of participation of the interstitials in the radiation enhancement of diffusion in metals.

2. Theory

Diffusion enhancement by irradiation occurs due to an excessive concentration of point defects. The kinetic equations for the concentration of radiation defects can be written in the form:

$$\frac{dC_v}{dt} = K - K_v C_v - 4\pi r_v C_v C_i \nu_i , \tag{1}$$

$$\frac{dC_i}{dt} = K - K_i C_i - 4\pi r_v C_v C_i \nu_i , \tag{2}$$

where C_v and C_i are the concentrations of vacancies and interstitials, K is the introduction rate of point defects, r_v the radius of annihilation between interstitial and vacancy. ν_i is the jump frequency of the interstitial. The coefficients characterizing the annealing rate of point defects at dislocations are

$$K_v = \frac{2\pi\alpha}{\ln(r_1/r_0)} \lambda^2 \nu_v \approx \alpha \lambda^2 \nu_v , \tag{3}$$

$$K_i = \frac{2\pi\alpha}{\ln(r_1/r_0)} \lambda^2 \nu_i \approx \alpha \lambda^2 \nu_i , \tag{4}$$

where r_0 is the effective adsorption radius of a point defect by a dislocation, and $\pi r_1^2 = 1/\alpha$. It can be shown that if r_0 equals several interatomic distances the approximate values of K_v and K_i differ by no more than a factor of two from the actual values within a large range of dislocation densities ($10^5 \leqq \alpha \leqq$ $\leqq 10^{11}$ cm^{-2}).

Substituting (3) and (4) into (1) and (2) we obtain

$$\frac{dC_v}{dt} = K - \lambda^2 \alpha C_v \nu_v - 4\pi r_v C_v C_i \nu_i , \tag{5}$$

$$\frac{dC_i}{dt} = K - \lambda^2 \alpha C_i \nu_i - 4\pi r_v C_v C_i \nu_i . \tag{6}$$

The second term on the right-hand side of the equations characterizes the annealing of defects at dislocations. The third term defines the mutual annihilation between interstitials and vacancies.

At equilibrium conditions ($dC_v/dt = 0$, $dC_i/dt = 0$) the solution of equations (5) and (6) can be found, for example, in the paper of Dienes and Damask [3]. However, a steady-state solution is unacceptable for temperatures of stage III recovery if one assumes that at this stage the interstitials in the dumb-bell configuration are moving while the vacancies are immobile. Actually at equal introduction rates of vacancies and interstitials (for example under electron irradiation) a part of the interstitials will annihilate at dislocations. As a result vacancies which are practically immobile will accumulate. The concentration of sinks for interstitials (dislocations and vacancies) will increase while the active concentration of interstitials will decrease. The well-annealed metals with a very low dislocation density might be an exception. In this case the accumulation rate of uncompensated vacancies is very small, therefore the radiation-enhanced diffusion is practically steady and controlled by the mutual annihilation of interstitials and vacancies [16].

The system of equations (5) and (6) for the non-steady case has no analytical solution. An approximate solution can be found by neglecting the time of initial accumulation of interstitials and assuming that their concentration is defined only by the introduction rate of defects and the concentration of active sinks (quasi-stationary solution of equation (6)):

$$C_i = \frac{K}{\nu_i (\lambda^2 \alpha + 4 \pi r_v C_v)} . \tag{7}$$

In this case the vacancy concentration versus time can be obtained from a non-stationary solution of equation (5) by neglecting the second term on the right-hand side ($\nu_v = 0$):

$$\frac{dC_v}{dt} = K - \frac{4 \pi r_v C_v K}{\lambda^2 \alpha + 4 \pi r_v C_v} = \frac{K \lambda^2 \alpha}{\lambda^2 \alpha + 4 \pi r_v C_v} . \tag{8}$$

The solution of this equation has the form

$$(\lambda^2 \alpha + 4 \pi r_v C_v) = \lambda^2 \alpha \sqrt{\frac{8 \pi r_v K}{\lambda^2 \alpha} t + 1} . \tag{9}$$

The same approximate method for solving the system (5, 6) has been used by Dienes and Damask [17] in the analysis of kinetic changes of the concentration of radiation defects under irradiation at temperatures of stage I recovery.

The concentration of interstitials can be obtained by substituting (9) in (7):

$$C_i = \frac{K}{\nu_i \lambda^2 \alpha \sqrt{\frac{8 \pi r_v K}{\lambda^2 \alpha} t + 1}} . \tag{10}$$

The diffusion coefficient for the interstitial mechanism can be written as

$$D_i = C_i \nu_i \lambda^2 . \tag{11}$$

Substituting the relation for the interstitial concentration we obtain

$$D_i = \frac{K}{\alpha \sqrt{\frac{8\pi r_v K}{\lambda^2 \alpha} t + 1}}. \tag{12}$$

It can be seen that the radiation-enhanced diffusion coefficient is non-stationary and varies approximately as the reciprocal of the square root of irradiation time. Earlier we have derived a similar expression for D_i without considering the effective annihilation radius of vacancies and interstitials, r_v [18].

In order to elucidate the validity of the non-stationary radiation-enhanced diffusion in crystals with an increased dislocation density we investigated the radiation ordering in cold-worked Ag–8.75 at% Zn samples by measuring the resistivity.

We have already shown for annealed Ag + 8.75 at% Zn [16] that the resistivity at short-range ordering varies according to the second-order kinetics:

$$\frac{d\varrho}{dt} = -\frac{\Delta\varrho^2}{R\tau}, \tag{13}$$

where $\Delta\varrho$ is the deviation of resistivity from its equilibrium value, R the change of resistivity when passing from a completely disordered to an entirely ordered state, and τ the relaxation time of the process with a diffusion coefficient

$$D = \frac{\lambda^2}{Z}\frac{1}{\tau}, \tag{14}$$

where Z is the number of possible sites in the lattice providing a jump of the interstitial.

Substituting the relation for τ in (13) and accounting for the time dependence of the diffusion coefficient (12) after integrating we obtain

$$\frac{1}{\Delta\varrho} - \frac{1}{\Delta\varrho_0} = \frac{Z}{4\pi r_v R}\left(\sqrt{\frac{8\pi r_v K}{\lambda^2 \alpha} t + 1} - 1\right). \tag{15}$$

Since one measures the resistivity deviation from the starting value, $\Delta\varrho^*$ (instead from the equilibrium value, $\Delta\varrho$), it is convenient to substitute in equation (15)

$$\frac{\Delta\varrho^*}{\Delta\varrho_0 (\Delta\varrho_0 - \Delta\varrho^*)} = \frac{Z}{4\pi r_v R}\left(\sqrt{\frac{8\pi r_v K}{\lambda^2 \alpha} t + 1} - 1\right). \tag{16}$$

For the cold-worked Ag–8.75 at% Zn alloy the resistivity deviation from the equilibrium value ($\Delta\varrho_0/\varrho \approx 20\%$) usually exceeds that from the starting value ($\Delta\varrho^*/\varrho \approx 1\%$). In this regard one can simplify equation (16):

$$\Delta\varrho^* = \frac{Z\,\Delta\varrho_0^2}{4\pi r_v R}\left(\sqrt{\frac{8\pi r_v K}{\lambda^2 \alpha} t + 1} - 1\right). \tag{17}$$

From equation (17) it follows that at sufficiently long exposures ($t \gg (\lambda^2 \alpha / 8\pi r_v K)$) the $\Delta\varrho^*$ vs. $\sqrt{t}$ plot should be a straight line.

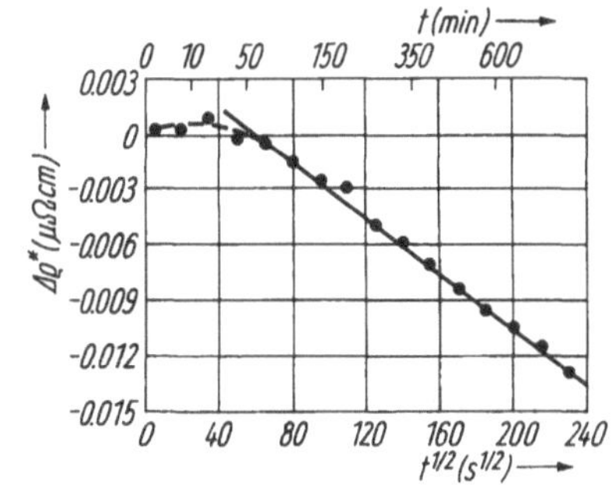

Fig. 1. Change of resistivity of the cold-worked alloy Ag–8.75 at% Zn under 2 MeV electron irradiation of 2.8×10^{13} electrons cm^{-2} s^{-1} flux at 10 °C

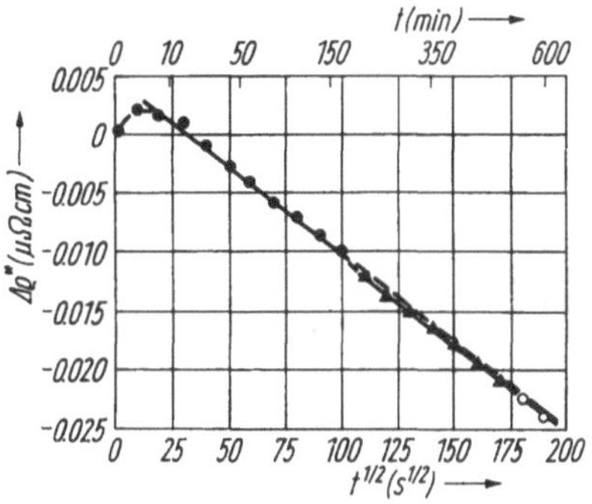

Fig. 2. Change of resistivity of the cold-worked alloy Ag–8.75 at% Zn under electron irradiation at different temperatures (● 10, ○ 25, ▲ 35 °C). Electron energy 2.3 MeV, flux intensity 3.74×10^{13} electrons cm^{-2} s^{-1}

3. Experimental Results and Discussion

The alloys were melted in vacuo and forged to plates 6 mm thick, then cold-rolled to plates 0.3 mm thin from which the samples were cut. The voltage and current connections were welded to them. In order to decrease the concentration of deformation vacancies the samples were annealed at 100 °C for 2 h prior to irradiation.

The irradiation was performed in a linear electron accelerator with a 2.3 MeV electron beam. The restistivity was measured by the potentiometer method at liquid nitrogen temperature.

Fig. 1 shows resistivity versus irradiation time at an intensity of $2.8 \times \times 10^{13}$ electrons cm^{-2} s^{-1} and a temperature of 10 °C. At exposures longer than 50 min the experimental points show a linear dependence $\Delta\varrho^*$ vs. $\sqrt{t}$ which favours a possible non-stationary radiation-enhanced diffusion. Some resistivity increase at the starting period of irradiation might be ascribed to the disordering action of dynamic crowdions.

It follows from equation (17) that the process should be temperature-independent and the slope of $\Delta\varrho^*$ vs. $\sqrt{t}$ should be proportional to the square root of the irradiation intensity.

Fig. 2 shows the kinetics of resistivity change upon irradiation at various temperatures. It seems that actually the slope does not depend on the irradiation temperature. A certain shift of the line observed in passing from one to another temperature is explained by an excess or deficiency of the real concentration of interstitials as compared with the quasi-equilibrium value at the given temperature.

Fig. 3 shows the results of irradiation at 25 °C and various intensities. Fig. 4 shows the respective slopes (tg φ) of the linear parts versus irradiation intensity (I) in logarithmic coordinates. The experimental data are well interpolated by the relation tg $\varphi \sim I^{1/2}$. Thus, the dependence of the diffusion process upon the irradiation intensity is also in agreement with the equation derived for the non-stationary radiation-enhanced diffusion.

In order to elucidate the validity of the mechanism suggested for the radiation-enhanced diffusion it was rather interesting to estimate numerically the coeffi-

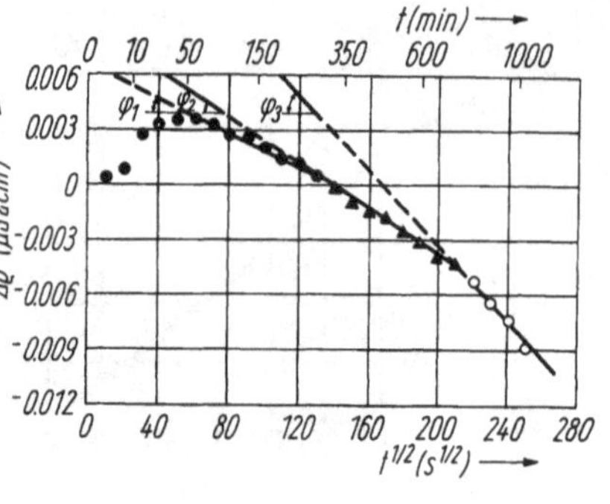

Fig. 3. Change of resistivity of the cold-worked alloy Ag–8.75 at% Zn at different electron irradiation fluxes (● 9.35×10^{12}; ▲ 1.87×10^{13}; ○ 3.74×10^{13} electrons/cm² s). Electron energy 2.3 MeV, irradiation temperature 25 °C

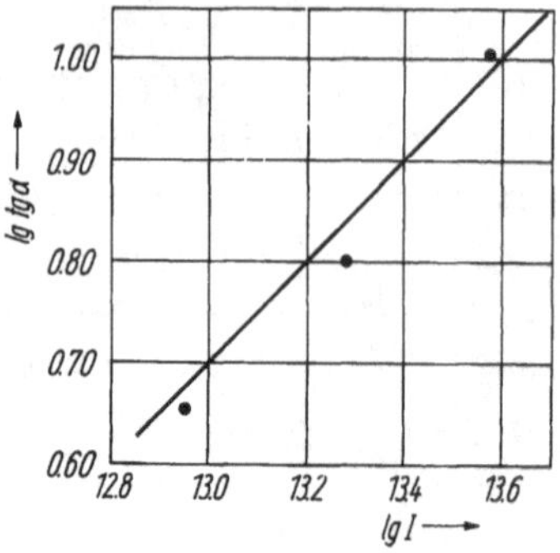

Fig. 4. The "process rate" ($d\varrho/d\sqrt{t} = \mathrm{tg}\,\varphi$) versus irradiation flux. The solid line represents $\mathrm{tg}\,\varphi \approx I^{1/2}$

cients in equation (17). Such an estimation is desirable for r_v, R, and α. Two values can be determined from the experimental data using the equation of the straight line, while the third value must be found independently. Most definitely the radius of annihilation between interstitial and vacancy, r_v, can be determined independently. The radius of spontaneous (athermic) annihilation is usually supposed to be 2 to 3 lattice parameters (the volume of spontaneous annihilation comprises 100 to 400 atoms). The range of correlated annihilation extends to 10 lattice parameters. Therefore, a mean effective radius of annihilation between interstitial and vacancy can be taken as $r_v = 5$.

The parameter R (the resistivity change of the alloy in passing from the completely disordered to the entirely ordered state) can be found by extrapolating the linear part of the plot (Fig. 1) to $t = 0$. Accounting for an increase of resistivity due to the initial disordering the extrapolated value is equal to $\Delta\varrho_e^* = 4\times10^{-9}\ \Omega$ cm. From equation (17) it follows that

$$\Delta\varrho_e^* = \frac{Z\,\Delta\varrho_0^2}{4\,\pi\,r_v\,R}. \tag{18}$$

Upon annealing the resistivity of disordered cold-worked samples changes by $6.4\times10^{-7}\ \Omega$cm. Assuming that this change is on the whole associated with the effect of ordering ($\Delta\varrho_0 = 6.4\times10^{-7}\ \Omega$cm) and taking $Z = 8$, $r_v = 5$ we obtain

$$R = \frac{Z\,\Delta\varrho_0^2}{4\pi r_v \Delta\varrho_e^*} = \frac{8\times41\times10^{-14}}{4\,\pi\times5\times4\times10^{-9}} = 1.3\times10^{-5}\ \Omega\text{cm}\,.$$

It is natural that the resistivity change caused by the unit concentration of Frenkel defects in the f.c.c. lattice ($\approx 10^{-4}\ \Omega$cm) can be taken as an upper limit for R since the lattice distortions due to disordering are probably less than those due to Frenkel defects. A possible lower limit for R is the maximum resistivity change observed in annealing the cold-worked alloy ($\Delta\varrho_0 = 6.4\times\times10^{-7}\ \Omega$cm). Thus, the value obtained is quite applicable as it fits the range of possible R-values.

The dislocation density in the irradiated samples can be estimated from the slope of the linear part of Fig. 1 ($\mathrm{tg}\,\varphi = 7.5\times10^{-11}\ \Omega\text{cm/s}^{1/2}$). Assuming that an electron beam of one electron/cm² generates one seperate Frenkel pair per cm³, $K \approx 5 \times 10^{-10}\ \mathrm{s}^{-1}$ and

$$\alpha = \frac{(\Delta\varrho_e^*)^2\, 8\,\pi\, r_v\, K}{\lambda^2\, \mathrm{tg}^2\, \varphi} = \frac{16 \times 10^{-18} \times 8\,\pi \times 5 \times 5 \times 10^{-10}}{10^{-15} \times 7.5^2 \times 10^{-22}} \approx 2 \times 10^{11}\ \mathrm{cm}^{-2}\,.$$

This value is quite reasonable for the cold-worked samples.

Thus, for the cold-worked Ag– 8.75 at% Zn alloy the experimental dependence of diffusion ordering upon the external irradiation parameters (temperature and intensity) along with the estimated internal parameters of the irradiated sample (R, α) confirms the mechanism suggested for the non-stationary radiation-enhanced diffusion in metals with increased dislocation density.

References

[1] H. P. Bonzel, Acta metall. **13**, 1084 (1965).
[2] D. Heitkamp, W. Biermann, and T. S. Lundy, Acta metall. **14**, 1201 (1966).
[3] G. J. Dienes and A. C. Damask, J. appl. Phys. **29**, 1713 (1958).
[4] A. C. Damask, Effects of Neutron Irradiation on Non-Fissionable Alloys, in: Radiation Damage in Solids, Vol. II, IAEA, Vienna 1962 (p. 3).
[5] M. C. Wechsler and R. H. Kernohan, The Effect of Radiation on Diffusion-Controlled Reactions in Copper-Base Alloys, in: Radiation Damage in Solids, Vol. II, IAEA, Vienna 1962 (p. 81).
[6] I. Ya. Dekhtyar, V. V. Pilipenko, and A. M. Shalaev, Izmenenie podvizhnosti atomov v metallakh pri komnatnoi temperature pod deistviem ioniziruyshchikh izluchenii, in: Izuchenie defektov kristallicheskogo stroeniya metallov, Naukova Dumka, Kiev 1966 (p. 58).
[7] R. E. Larsen and A. C. Damask, Acta metall. **12**, 1131 (1964).
[8] D. Frenzel, Z. phys. Chem. N.F. **51**, 67 (1966).
[9] C. Lee and J. S. Koehler, Phys. Rev. **176**, 813 (1968).
[10] R. K. Sharma, C. Lee, and J. S. Koehler, Phys. Rev. Letters **19**, 1379 (1967).
[11] F. Dworschak, K. Herschbach, and J. S. Koehler, Phys. Rev. **133**, A293 (1964).
[12] D. Schumacher, The Recovery of Electron and Neutron Irradiated F.C.C. Metals, X. Colloque de Métallurgie (Fragilité et effets del'irradiation), Saclay 1966, Gif-sur-Yvette/Paris 1967 (p. 49).
[13] W. Bauer and A. Sosin, Phys. Rev. **136**, A474 (1964).
[14] W. Frank and A. Seeger, Radiat. Eff. **1**, 117 (1969).
[15] J. J. Jackson and K. Herschbach, Phys. Rev. **164**, 951 (1967).
[16] L. N. Bystrov, L. I. Ivanov, and Yu. M. Platov, phys. stat. sol. (a) **7**, 617 (1971).
[17] G. J. Dienes and A. C. Damask, Phys. Rev. **128**, 2542 (1962).
[18] L. N. Bystrov, L. I. Ivanov, and Yu. M. Platov, Dokl. Akad. Nauk SSSR **185**, 309 (1969).

(Received July 8, 1971)

phys. stat. sol. (a) 8, 383 (1971)

Subject classification: 20.1; 2; 22.8

A. F. Ioffe Physico-Technical Institute, Academy of Sciences of the USSR, Leningrad

Infrared Absorption Spectra in Glasses on the Basis of Tetrahedral Phases

By

N. A. GORYUNOVA†, G. S. KUZMENKO, B. KH. MAMEDOV, and E. O. OSMANOV

Results are presented of a study of absorption spectra of binary and ternary compounds and their intermediate alloys in vitreous state. The resemblence is shown of spectra of glasses ($CdAs_2$ to $CdGeAs_2$; $CdGeP_2$ to CdP_2) with the same "anion" — the element of V-th group. The values of characteristic absorption band frequencies in the spectra of glasses $CdGeP_2$ (1000, 930, and 725 cm^{-1}), CdP_2 (1000, 930, and 730 cm^{-1}) and values of absorption in them are more similar than in spectra of the glasses $CdGeAs_2$ and $CdAs_2$. The presence of analogous absorption bands in glass spectra is considered as a favoured manifestation of certain chemical bonds, namely A^2–C^5.

Es werden die Ergebnisse einer Untersuchung der Absorptionsspektren von binären und ternären Verbindungen und ihrer Legierungen für den flüssigen Zustand mitgeteilt. Es werden Ähnlichkeiten der Spektren der Gläser ($CdAs_2$ bis $CdGeAs_2$; $CdGeP_2$ bis CdP_2) mit demselben „Anion" — dem Element der V. Gruppe — aufgezeigt. Die Werte der Frequenzen der charakteristischen Absorptionsbande in den Gläsern $CdGeP_2$ (1000; 930; 725 cm^{-1}), CdP_2 (1000; 930; 730 cm^{-1}) und deren Absorptionswerte sind ähnlicher, als in den Spektren der Gläser $CdGeAs_2$ und $CdAs_2$. Die Erscheinung von analogen Absortionsbanden in Glasspektren wird als bevorzugtes Auftreten gewisser chemischer Bindungen, nämlich A^2–C^5, angesehen.

The $A^2B^4C^5_2$ compounds are of tetrahedral structure. Two A^2 atoms and two B^4 atoms are connected with each C^5 atom. The $A^2C^5_2$ compounds ($CdAs_2$ and CdP_2) crystallize in a similar structure with the difference that they have C^5–C^5 bonds [1]. $CdGeAs_2$, $CdGeP_2$, and the above compounds have been produced earlier [2 to 4]. The presence of identical atoms in ternary and binary compounds may assist the acquisition of information about chemical bonds primarily in glasses, where the peculiarities of the crystalline structure of the initial compound show up to a less degree. In this connection it seems reasonable to study the IR spectra of glasses on the basis of the above compounds and to compare them with spectra of crystals of these compounds.

In this paper a comparison is made between absorption spectra of ternary and binary compound glasses, their intermediate alloys ($CdGeAs_2$–$CdAs_2$; $CdGeP_2$–CdP_2), and complex compositions obtained by gradual substitution of "cations" (groups II, IV elements) for "anions" (group V elements). Besides, the absorption spectra of $CdGeAs_2$ and $CdGeP_2$ crystals are adduced.

The initial elements for synthesis were of high purity. The spectra were recorded with the UR-10 device in the 400 to 5000 cm^{-1} range at room temperature. The sample thickness is ≈ 1 mm, their surfaces being polished mechanically. Absorption bands are in the 400 to 1200 cm^{-1} range; in the 1200 to 5000 cm^{-1} range impurities introduced give rise to absorption "tails" connected probably with light scattering by glass heterogeneities.

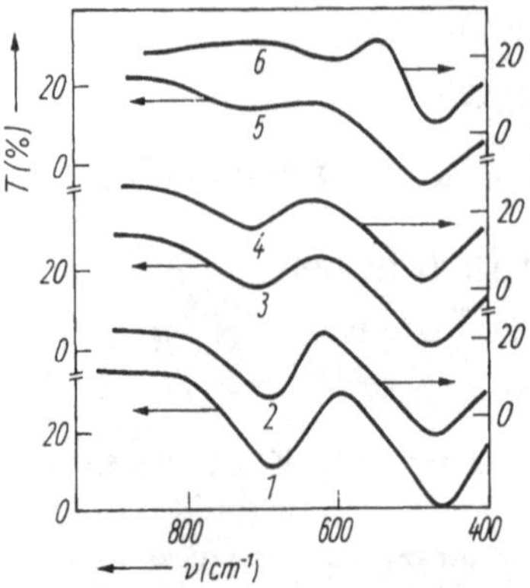

Fig. 1. Transmission spectra of CdGe$_x$As$_2$ glasses. (1) $x = 1$; (2) $x = 0.75$; (3) $x = 0.5$; (4) $x = 0.25$; (5) $x = 0.1$; (6) $x = 0$

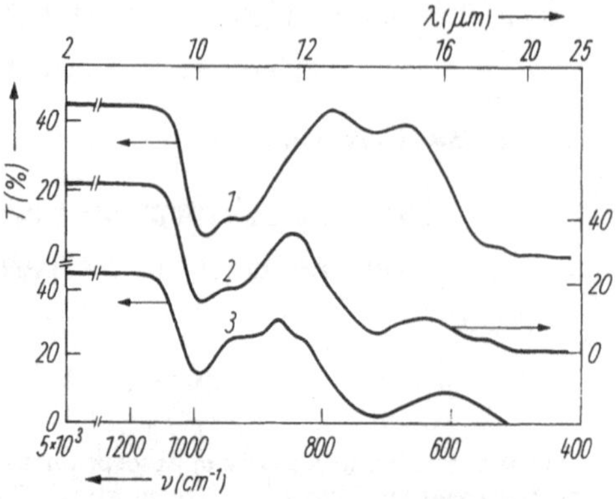

Fig. 2. Transmission spectra of CdGe$_x$P$_2$ glasses. (1) $x = 0$; (2) $x = 0.5$; (3) $x = 1$

An investigation of the effects of impurities on the absorption spectrum of glasses suggests that the absorption band in the interval 400 to 1200 cm^{-1} is connected with the eigenoscillations of the glass network.

The spectra of binary and ternary glasses shown in Fig. 1 (curves 1 and 6) and Fig. 2 (curves 1 and 3) are characterized by the following absorption bands: $CdGeAs_2$: 675 and 450 cm^{-1}; $CdAs_2$: 625 and 475 cm^{-1}; $CdGeP_2$: 1000, 930, and 725 cm^{-1}; CdP_2: 1000, 930, and 730 cm^{-1}.

The absorption coefficient for these bands is ≈ 10 cm^{-1}. There is a marked resemblence between the absorption spectra of binary and ternary glasses which have the identical "anion". Taking into account the fact that the glasses have identical atoms and retain their tetrahedral coordination, the absorption bands observed can be considered as a display of definite chemical bonds. The dependence of band intensity and band position in the spectra on the chemical bond content can be followed, for instance, in the system $CdGe(As_xP_{1-x})_2$ (Fig. 3). The progressive substitution of phosphorus for arsenic atoms leads to a reduction of absorption in the 675 cm^{-1} band. We failed to observe the 450 cm^{-1} band since it is overlapped with a more intensive absorption band in $CdGeP_2$. A rise in the phosphorus content leads to an increase of absorption in the bands characteristic of $CdGeP_2$.

On the other hand, partial substitution of silicon or tin (compositions $CdGe_{0.8}Sn_{0.2}As_2$; $CdGe_{0.8}Si_{0.2}As_2$) for germanium atoms does not lead to an essential change of glass spectra at constant ratio of Cd to As atoms. It should be only mentioned that in the spectrum of $CdGe_{0.7}Si_{0.3}As_2$ glass, bands and weak-absorption peaks appear attesting to the onset of crystallization of CdSiAs (the frequency values of this peak coincide with the position of the intensive bands of the absorption spectrum of $CdSiAs_2$ crystals [5]).

Thus, for the glasses investigated the band position and the absorption level are likely to be determined basically by the concentration of A^2–C^5 bonds. This indicates that in the spectral range under study the A^2–C^5 bonds are optically more active than the B^4–C^5 bonds in ternary glasses and the C^5–C^5 bonds in binary glasses. The optical activity is liable to depend on the degree of bond ionicity. According to Suchet the A^2–C^5 bonds are considerably more ionic than the B^4–C^5 bonds [6].

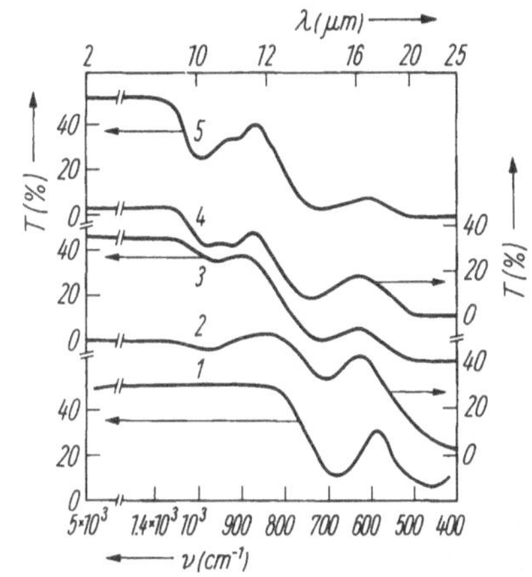

Fig. 3. Transmission spectra of $CdGe(As_xP_{1-x})_2$. (1) $x = 1$; (2) $x = 0.8$; (3) $x = 0.4$; (4) $x = 0.2$; (5) $x = 0$

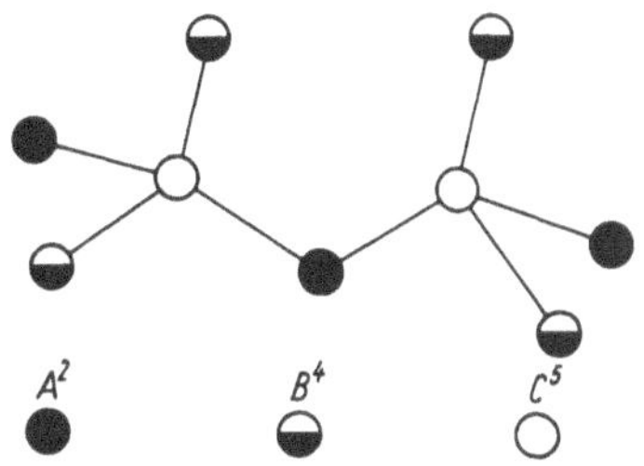

Fig. 4. Structural unit of $A^2B^4C_2^5$ glasses

Approaching the absorption spectra of binary and ternary glasses from the viewpoint of elucidating atomic short-range structure it is anticipated that there is a likeness in the structural unit of the glasses referred to. The structural unit of $A^2B^4C_2^5$ glasses is presented, by the way of example, in the form A 1/2 A 2/4 B 4/4 C_2 (Fig. 4).

A minor difference between $CdGeAs_2$ and $CdAs_2$ spectra is apt to be caused by a fair discrepancy between the structural units of these glasses. The structural discrepancy between the atomic arrangement in the vitreous system often assists microliquation. Division of glass into two vitreous phases can be traced in the IR spectra if the network oscillations of these phases fall into different spectral regions. A microstructural analysis of the $CdGe_xAs_2$ glasses subjected to annealing at softening temperatures evidences the presence of spherical formations in samples. Such heterogeneities may be thought of as an effect of metastable liquation interrupted by the crystallization of $CdGeAs_2$. Formations of such kind have not been revealed in $CdGe_xP_2$ glasses. The structural units of $CdGeP_2$ and CdP_2 glasses may be more alike. The positions of the main absorption bands and their intensities in $CdGeP_2$ and CdP_2 do not differ much from one another most likely due to the iso-structural nature of these vitreous phosphides.

Fig. 1 shows the shift of absorption bands and the dependence of their intensity on composition for $CdGe_xAs_2$ glasses. Major changes in the glass spectra take place when incorporating a small amount of germanium. It seems that when germanium is supplemented, the As–As bond in $CdAs_2$ glass switches over to germanium atoms. These latter go into the glass structural network and affect the vibrational spectra of glass.

Fig. 5 displays the transmission spectra of $CdGeAs_2$ and $CdGeP_2$ crystals and glasses. The absorption bands of the corresponding crystal spectrum are also in the wave number range 400 to 1200 cm^{-1}. The spectral position of bands, their temperature dependence, and the corresponding absorption coefficients (≈ 50 cm^{-1}) are characteristic of two-phonon processes appreciable in crystals with ion-covalent bonds.

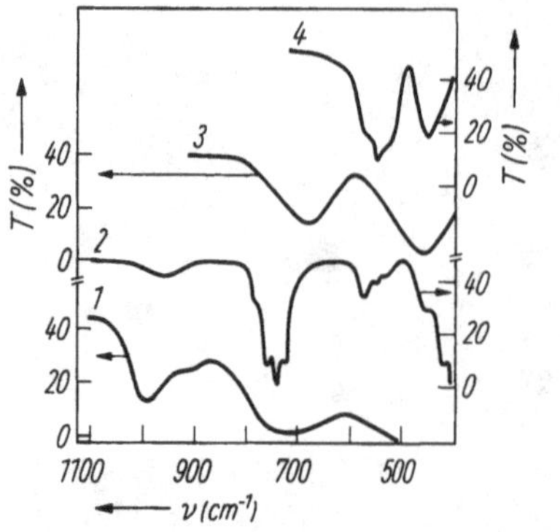

Fig. 5. Transmission spectra of glasses and crystals. $CdGeP_2$ (1) glass; (2) crystal, $CdGeAs_2$ (3) glass; (4) crystal

The comparison of the phonon spectrum of $A^2B^4C_2^5$ crystals with chalcopyrite type structure with that of A^3B^5 with sphalerite type structure seems to be expedient since these compounds are similar in the type of chemical bond. The presence of two different atoms in the cation sublattice vastly complicates the spectrum of $A^2B^4C_2^5$ crystals. Nevertheless, the values of peak frequencies tabulated in Table 1 are close to one another.

Thus, the absorption bands observed in glass and crystal spectra are formed by like characteristic phonons.

Table 1

Absorption peaks (cm^{-1})

$CdGeP_2$	GaP [7]	$CdGeAs_2$	GaAs [7]
950	—	545	524
752	756	535	517
733	739	520	510
518	722	450	501
575	575		
555	558		
535	—		
450	444		
430	427		

References

[1] E. PARTHE, Crystal Chemistry of Tetrahedral Structures, Gordon & Breach, New York/London 1964.

[2] A. A. VAIPOLIN, N. A. GORYUNOVA, E. O. OSMANOV, and YU. V. RUD, Dokl. Akad. Nauk SSSR **160**, 633 (1965).

[3] YA. A. UGAY, T. A. ZYUBINA, and K. B. ALEINIKOVA, Izv. Akad. Nauk SSSR, Ser. neorg. Mater. **4**, 17 (1968).

[4] V. S. KOVAL, V. K. MITYUREV, I. I. TYCHINA, and A. B. FEDOTOVSKII, V. Nauchn.-Tekh. Conf. Kishinevsk. Politekh. Inst., Kishinev 1969 (p. 163).

[5] G. K. AVERKIEVA, R. K. KHARYMSHAKOV, V. D. PROCHUKHAN, and M. SERGINOV, Fiz. Tekh. Poluprov. **4**, 591 (1970).

[6] J. P. SUCHET, Chemical Physics of Semiconductors, Van-Nostrand Co., 1965.

[7] R. K. WILLARDSON and A. C. BEER (Ed.), Semiconductors and Semimetals, Vol. 3, New York/London 1967.

(Received February 8, 1971)

phys. stat. sol. (a) 8, 387 (1971)

Subject classification: 20.3; 13.4; 22.8

A. F. Ioffe Physico-Technical Institute, Academy of Sciences of the USSR, Leningrad

Luminescence of $CdSnP_2$ Doped with Different Impurities

By

E. I. LEONOV, V. M. ORLOV, V. I. SOKOLOVA, and YU. G. SHRETER

An investigation was made of the influence of a fast electron beam on the luminescence at 80 °K of $CdSnP_2$ single crystals doped with various impurities and photoluminescence spectra of crystals doped with copper were measured at 4.2 and 80 °K.

Исследованы спектры люминесценции под действием пучка быстрых электронов монокристаллов $CdSnP_2$ легированных различными примесями при 80 °K. Измерены спектры фотолюминесценции кристаллов легированных медью при 4,2 и 80 °K.

1. Introduction

The ternary semiconducting compounds of the $A^2B^4C^5_2$ type including $CdSnP_2$ have band energy spectra very similar to that of materials of A^3B^5 type whose analogues they are. The direct transitions determining the width of the forbidden gap in these compounds are maintained at least up to 1.6 eV [1].

$CdSnP_2$ is the isoelectronic direct-band analogue of InP and is characterized by a series of interesting properties: high mobility of carriers [2], multi-valley structure of the conduction band giving reasons for observing the Gunn effect [3], photoelectric properties of particular interest [4]. In a previous paper the authors reported on the laser radiation under the influence of a beam of fast electrons observed in undoped monocrystals of $CdSnP_2$ at $T = 80$ °K [5].

The threshold current density and the generation wavelength are 0.3 A/cm² and 10112 Å, respectively. The radiation linewidth was ≈ 2 Å. In [6] also the stimulated radiation produced in $CdSnP_2$ under the action of an electron beam and of an He–Ne laser has been reported.

The studies of photoluminescence of undoped $CdSnP_2$ crystals performed in this work display the availability of lines of bound excitons and impurity bands adjacent to the main exciton band in the radiation spectra.

The present work studies the luminescence spectra under action of a beam of fast electrons in $CdSnP_2$ doped with different impurities, in order to ascertain the distinctive features of the spectrum of impurity levels and the influence of different impurities on the edge radiation.

2. Experimental Procedure

The single crystals of $CdSnP_2$ were grown according to the method described in [2]. The dopings were introduced directly into the solution melt. The luminescence spectra were investigated at $T = 80$ °K. The experimental equipment comprised an electron gun, the spectrometer ИКС-12 with prism Ф-1, and a synchronous detection system. The luminescence spectra were taken at the beam current of 1 A/cm² on samples of 3 to 4 mm diameter, fastened on a copper block cooled by liquid nitrogen.

3. Experimental Results and Discussion

Fig. 1 (curve 1) presents the cathodoluminescence spectrum of specially undoped samples of $CdSnP_2$. The spectra of spontaneous radiation in these samples are characterized by the presence of a long-wave "wing" (in some samples — a radiation band), with a distance from the radiation maximum of the order of 0.02 to 0.03 eV. This "wing" was observed practically in all undoped samples under study, which indicates the presence of shallow centres in $CdSnP_2$ determining the electronic character of the $CdSnP_2$ conductivity.

If $CdSnP_2$ is doped with germanium the spontaneous radiation maximum is shifted by 0.002 eV towards the short-wavelength direction, and if doped with arsenic the maximum is shifted towards the long-wavelength region by 0.06 eV (Fig. 1, curves 2 and 3, respectively). It was previously demonstrated that in the $CdSnP_2$–$CdSnAs_2$ system there exist regions of solid solutions, both on the side of $CdSnAs_2$ and on the side of $CdSnP_2$ [7]. Therefore, the observed shift of the spontaneous radiation maximum in $CdSnP_2$ doped with arsenic in the long-wavelength direction is caused by a decrease of the forbidden gap, as a solid solution $CdSn(P_xAs_{1-x})_2$ is formed. The displacement of the edge radiation maximum in $CdSnP_2$ doped with germanium towards the short-wavelength direction is due to the formation of a solid solution $CdGe_xSn_{1-x}P_2$ near $CdSnP_2$, as there exists a continuous series of solid solutions in this system [8].

The $CdSnP_2$ crystals doped with germanium are also characterized by a wide impurity band with the radiation maximum at 1.08 eV (Fig. 1, curve 2). The studies of the spectral distribution of photoconductivity of $CdSnP_2$(Ge) reveal the same band.

Fig. 2 shows the cathodoluminescence spectra of $CdSnP_2$ crystals doped with Ag, Al, and Au (curves 1, 2, and 3, respectively). As it is apparent from Fig. 2, doping with Al and Au results in the appearance of an intense impurity band in the luminescence spectrum the intensity of which exceeds the edge radiation. In $CdSnP_2$ samples doped with B and Ga only one radiation band is observed shifted to the long-wavelength region of the spectrum relative to the band edge of the undoped crystals (Fig. 1, curves 4 and 5). The energy displacement of this band amounts to hundreds of eV, which apparently indicates the formation of small hydrogen-like centres in $CdSnP_2$ due to B and Ga.

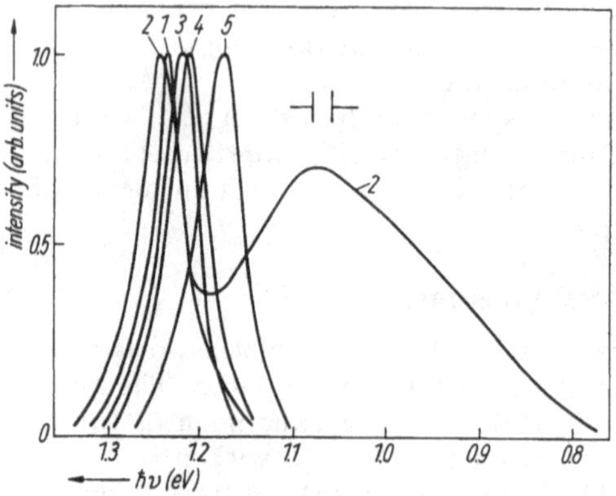

Fig. 1. Cathodoluminescence spectra of $CdSnP_2$ at 80 °K. (1) Undoped samples; (2) to (5) doped crystals ((2) Ge, (3) As, (4) B, (5) Ga)

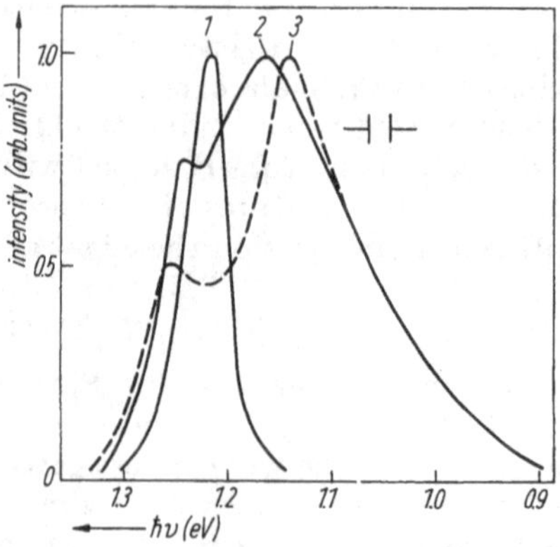

Fig. 2. Cathodoluminescence spectra of $CdSnP_2$ samples doped with Ag (1), Al (2), and Au (3)

Table 1

Energetic position and width of extrinsic radiation bands of $CdSnP_2$ monocrystals doped with different impurities

impurity	energy $E_g - E_{imp.}$ (eV)	half-width of extrinsic radiation band (eV)	remarks
B	0.02	0.04	hydrogen-like
Ga	0.03	0.05	impurities
Ag	0.02	0.04	
Ge	−0.01	0.04	solid solution
As	0.06	0.06	solid solution
Cu	0.16	0.15	acceptor level
Al	0.07	0.15	
Au	0.09	0.10	

Table 1 summarizes the energetic positions of extrinsic bands of $CdSnP_2$ doped with different impurities and the widths of these bands determined at half-intensity of the luminescence peak. It should be noted that at the present time it is difficult to give an unambiguous interpretation of the character of these levels, as the doping occurs against the "background" of donor centres (which moreover are compensated) steadily introduced in the process of growth. Consequently, it is not possible to establish distinct regularities in the changes of Hall constant and conductivity, when different impurities are introduced.

As previously mentioned, the specially undoped $CdSnP_2$ displays an n-type conductivity with low resistivity (0.1 to 0.05 Ωcm). When copper is added, the resistivity of the $CdSnP_2$ sample increases and attains (10^3 to 10^4) Ωcm ($T = 300$ °K). Fig. 3 presents the luminescence spectra of $CdSnP_2$ doped with different quantities of copper. As it is shown in Fig. 3, with increasing copper concentration the intensity of the extrinsic radiation ($h\nu_{max.} = 1.07$ eV) increases and the intensity of the edge radiation simultaneously decreases. In the photoconductivity spectra of $CdSnP_2$(Cu) among other levels a transition was revealed with a quantum energy of 1.07 eV from a level located in the lower half of the forbidden gap of $CdSnP_2$ [4]. The presence of a wide band at $h\,\nu_{max.} = 1.07$ eV in the luminescence spectrum of $CdSnP_2$(Cu) apparently indicates the formation of a complex centre in $CdSnP_2$ doped with copper. This centre has acceptor properties and can also include a lattice defect (complex copper–lattice defect).

The detection of a negative photoconductivity [9] at quantum energies of 0.5 to 0.9 eV and of optical quenching with a threshold energy of 0.9 eV in $CdSnP_2$ crystals compensated with different quantities of copper indicates the multi-charge character of the centre introduced into $CdSnP_2$ by doping with copper. The level ($E_c - 1.07$) eV corresponds apparently to the ground state of this centre.

This work was also concerned with the investigations of photoluminescence in $CdSnP_2$(Cu) at 80 and 4.2 °K. As excitation source the pulse tube ИСШ-100 was used; the detecting system was identical with that described above.

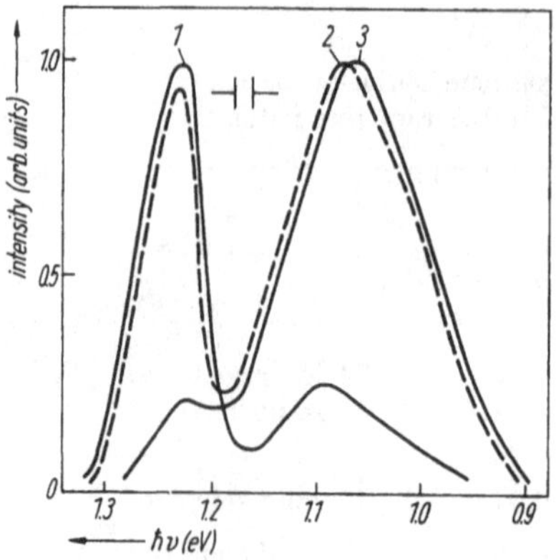

Fig. 3. Cathodoluminescence spectra of $CdSnP_2$ samples doped with different quantities of copper. The concentration of copper introduced into the solution melt increases from curve 1 to curve 3

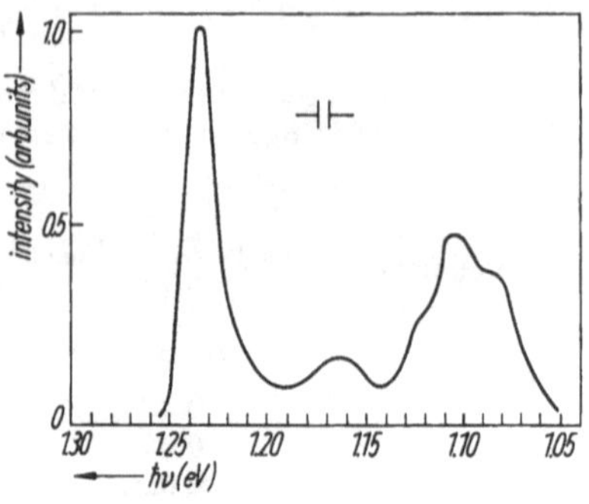

Fig. 4. Photoluminescence spectrum of $CdSnP_2$ doped with copper at 4.2 °K

Fig. 4 shows the photoluminescence spectrum of $CdSnP_2$(Cu) at 4.2 °K. It is apparent from Fig. 4 that an impurity band with $h\nu_{max.} = 1.11$ eV appears in the photoluminescence spectrum which corresponds to the extrinsic band $h\nu_{max.} = 1.07$ eV at 80 °K, taking into account the increase of the forbidden gap of $CdSnP_2$. A variety of maxima, whose nature remains still unexplained, has been revealed in the photoluminescence spectra of $CdSnP_2$(Cu) at 4.2 °K.

4. Conclusion

In the luminescence spectra of undoped $CdSnP_2$ crystals there exists beside the radiation edge band an impurity band with $h\nu_{max.} = 1.19$ eV. The appearance of this band is due to centres introduced during crystal growth. The luminescence spectra of $CdSnP_2$ doped with As and Ge suggest that in the systems $CdSnP_2$–$CdSnAs_2$ and $CdSnP_2$–$CdGeP_2$ solid solutions exist, which is confirmed by other investigations.

Such impurities as B, Ga, and Ag create shallow hydrogen-like centres in $CdSnP_2$. However, it does not seem possible to determine their character because of the presence of a great quantity of "inherent" donor centres in all samples.

In $CdSnP_2$ doped with copper a radiation band with $h\nu_{max.} = 1.07$ eV appears in the luminescence spectrum. The data of electrical and photoelectrical studies of $CdSnP_2$(Cu) attest that the introduction of copper causes a multi-charge acceptor centre resulting in an increase of resistance and photosensitivity. The impurity band with $h\nu_{max.} = 1.07$ eV appears also in the photoluminescence spectrum of $CdSnP_2$ crystals doped with copper.

References

[1] N. A. GORYUNOVA, Proc. IX. Conf. Phys. Semicond., Moscow 1968, Vol. 2, Izd. Nauka, Leningrad 1969 (p. 1267).

[2] V. I. SOKOLOVA, V. M. ORLOV, G. P. SHPENKOV, E. V. TSVETKOVA, I. K. POLUSHINA, V. A. RADUL, and N. A. GORYUNOVA, Rep. XXVI. Conf. LISI (Phys.), Izd. LISI, Leningrad 1968 (p. 14).

[3] N. A. GORYUNOVA, E. I. LEONOV, V. M. ORLOV, L. F. RODIONOV, V. I. SOKOLOVA, and V. P. SONDAEVSKII, Phys. Letters (Netherlands) **31A**, 393 (1970).

[4] I. P. Akimchenko, N. A. Goryunova, E. I. Leonov, and V. M. Orlov, phys. stat. sol. (a) **3**, K51 (1970).
[5] F. M. Berkovskii, S. M. Rivkin, N. A. Goryunova, V. I. Sokolova, V. M. Orlov, G. P. Shpenkov, and E. V. Tsvetkova, Fiz. Tekh. Poluprov. **2**, 1218 (1968).
[6] I. L. Shay, R. L. Leheny, E. Buchler, and I. Wernick, Appl. Phys. Letters **16**, 357 (1970).
[7] V. M. Orlov, E. P. Trifonova, V. A. Radul, A. A. Vaipolin, E. V. Tsvetkova, and N. A. Goryunova, Rep. XXVII. Conf. LISI (Phys.), Izd. LISI, Leningrad 1969 (p. 18).
[8] A. A. Vaipolin, V. M. Orlov, and V. I. Sokolova, Rep. VI. Scient. Tech. Conf. Kishinev Polytech. Inst., Kishinev 1970 (p. 204).
[9] I. P. Akimchenko, N. A. Goryunova, E. I. Leonov, and V. M. Orlov, phys. stat. sol. (a) **3**, K149 (1970).

(Received July 22, 1971)

phys. stat. sol. (a) 8, 393 (1971)

Subject classification: 10.1; 21; 21.1

*Institute of Solid State Physics,
Academy of Sciences of the USSR, Chernogolovka (a), and
A. F. Ioffe Physico-Technical Institute, Academy of Sciences of the USSR, Leningrad (b)*

Change in Creep Mechanism of B.C.C. Metals at Transition from Low to High Temperatures

By

M. M. Myshlyaev (a), W. A. Stepanov (b), and V. V. Shpeizman (b)

The creep behaviour of Cu, Al, and Pb has been studied in a wide range of temperatures and stresses. A change both in dislocation structure and creep characteristics has been found which indicates the change in creep mechanism as the temperature rises and the applied stress decreases.

Es wurde das Kriechverhalten der Metalle Cu, Al und Pb in einem breiten Intervall von Temperatur und Zugspannung untersucht. Dabei wurden Änderungen der Versetzungsstruktur des Metalls und der Charakteristiken des Kriechens festgestellt, die auf eine Änderung des Kriechmechanismus bei Temperaturerhöhung und Abnahme der Zugspannung zurückzuführen sind.

1. Introduction

In a medium temperature range the steady creep rate of many metals was found to be an exponential function of applied stress σ and absolute temperature T:

$$\dot{\varepsilon} = \dot{\varepsilon}_0 \exp\left(-\frac{U_0 - \gamma\,\sigma}{kT}\right), \tag{1}$$

where k is the Boltzmann constant, $\dot{\varepsilon}_0$, U_0, γ are parameters of the material which were discussed at length by Zhurkov and Sanfirova in [1, 2]. However, at high temperatures equation (1) is not valid giving both overstated (e.g. in the case of Al, references [1]) and understated (Cu, Pb in [3]) values of the strain rate as compared to experimental ones. Sometimes under different ambient conditions even for the same metal the above mentioned deviations from equation (1) arise (see e.g. Pb in [3]).

The nature of these deviations is not clear so far. According to the electron microscopic data [4] one can only say that equation (1) describes the creep of subgrain structures which is accompanied by gradual disintegration of misoriented subgrain boundaries due to the motion of dislocation segments. At high temperatures subgrain structures were not revealed.

The study of high temperature deviations from (1) is also of interest in connection with Indenbom's hypothesis in [5 to 7], predicting the substitution of the vacancy mechanism of creep by an interstitial one as the test temperature decreases and the applied stress rises.

The present paper deals with the deviations from equation (1) alongside with an electron microscopic study of the structure of the deformed specimens.

2. Material and Procedure

The experiments were done on copper of commercial purity (99.95% Cu), pure aluminium (99.99% Al), aluminium of commercial purity (99.3% Al), and lead (99.992% Pb). Tensile tests in the creep rate range 10 to 10^{-6} s^{-1} were performed on machines described in [8, 9]. The samples used had the shape of cylinders (2 to 3 mm diameter, gauge length 5 or 10 mm) and of flat plates ($22 \times 6 \times \times 0.15$ mm³).

Before the test Al, Cu, and Pb specimens were annealed for one hour at 500, 680, and 50 °C, respectively. Copper and aluminium were annealed in vacuum while lead in air. Grain diameters of copper, aluminium, and lead were 0.1, 0.2, and 0.2 μm, respectively. Aluminium single crystals (99.99% Al) were investigated too. Copper specimens were tested in helium purified from oxygen by blowing through copper shavings, the remaining ones in air.

Before electron microscopic test the specimens were cooled under load to avoid changes in dislocation structure. The structure obtained was examined by means of transmission electron microscopy on JEM-150. The foil thinning technique was described in [10].

3. Experimental Results

Experimental data for copper are presented in Fig. 1a. Deviations from equation (1) were found at temperatures appreciably higher than $T = 200$ °C. At first they manifest themselves as breaks on the curves, the creep rate thus

a b c d

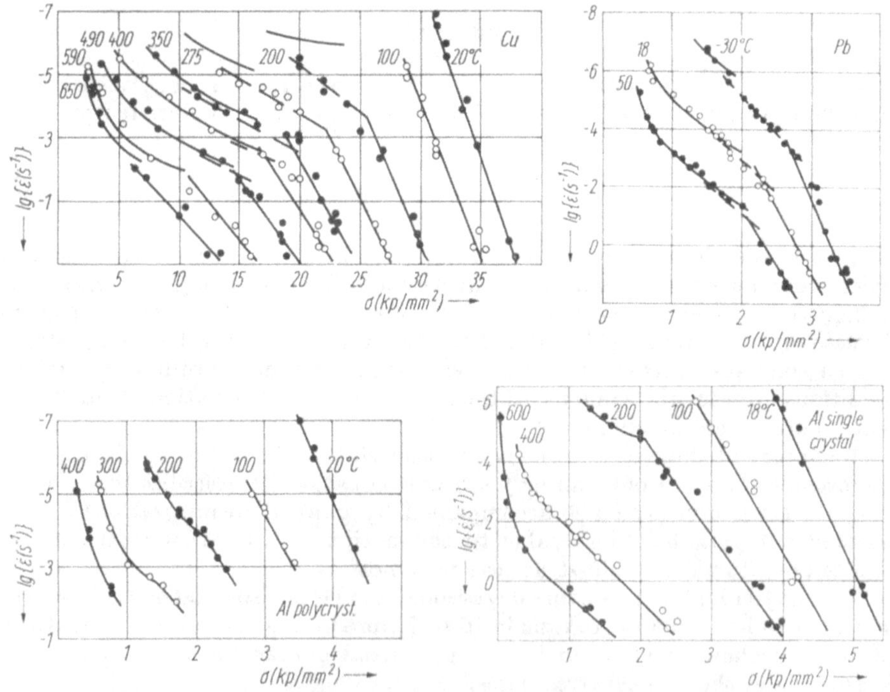

Fig. 1. Stress dependence of the steady creep rate at various temperatures: solid straight lines — according to (1), curves — according to (2), dashed line — intermediate range; a) Cu; b) Pb; c) pure Al, polycrystalline; d) Al, single crystal

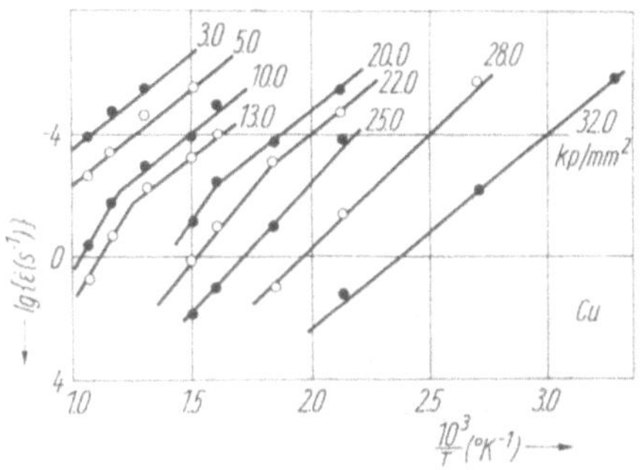

Fig. 2. Temperature dependence of steady creep rate for Cu

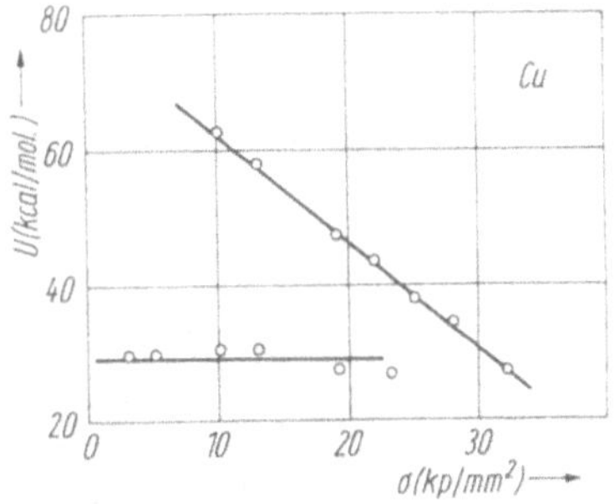

Fig. 3. Stress dependence of activation energy of the process controlling steady creep deformation of Cu

exceeding the one calculated from equation (1), whereas at high temperatures the dependences $\dot{\varepsilon}(\sigma)$ deviates from (1) towards smaller creep rates.

Fig. 2 shows the temperature dependence of the steady creep rate. Conspicuous are two ranges of temperature and stress. In one range the straight lines in lg $\dot{\varepsilon}$ vs. $1/T$ coordinates can be approximately considered parallel, while in the other range the slope of the straight lines varies with stress.

From the slope of these lines (Fig. 2) the activation energy for the process controlling the steady creep of Cu was determined (Fig. 3). Within the first range the energy can be considered as stress independent. It is equal to ≈ 30 kcal/mol (1.3 eV). In the second range the energy is a linear function of stress ($U = U_0 - \gamma\,\sigma$), U_0 being equal to 77 kcal/mol (3.3 eV). A similar dependence of the activation energy on stress is reported in [11].

The data available show equation (1) to be fair in the case of large stresses and low temperatures. High-temperature creep is well described by Weertman's equation for creep controlled by dislocation climb [12 to 14]

$$\dot{\varepsilon} = A\,\sigma^n \exp\left(-\frac{Q}{kT}\right). \tag{2}$$

A being a constant and the exponent n is determined by the slope of the curve in Fig. 4 and equals 4.9.[1])

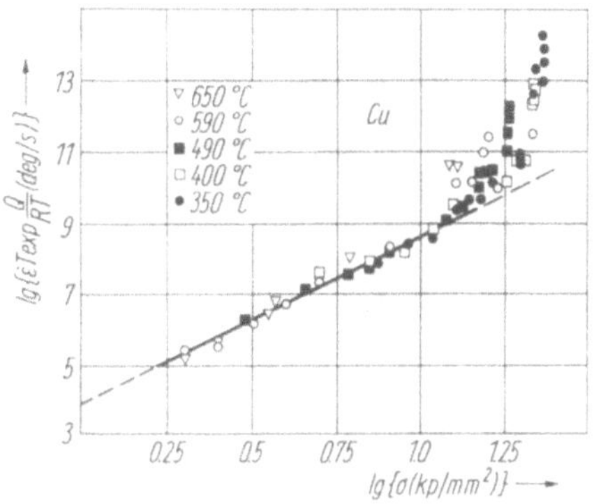

Fig. 4. Stress dependence of the value $\dot{\varepsilon}\,T \exp(Q/kT)$ within the range of high temperatures and low stresses (Cu). On the ordinate axis deg has to be replaced by °K

[1]) According to Weertman, $A = B/T$, B being constant. The inclusion of the temperature dependence of the pre-exponential factor results in an insignificant ($\approx 5\%$) increase in energy.

Fig. 5. Dislocation structure of deformed Al specimens within the range of moderate temperatures and high stresses: a) pure Al, $T = 200$ °C, $\sigma = 2.2$ kp/mm²; b) Al of commercial purity, $T = 200$ °C, $\sigma = 4.2$ kp/mm²

In the middle range (Fig. 1a, 200 and 275 °C) the creep rate can be described by equation (1) with a coefficient $\dot{\varepsilon}_0$ exponentially depending on stress, $U_0 = 30$ kcal/mol and γ near to zero.

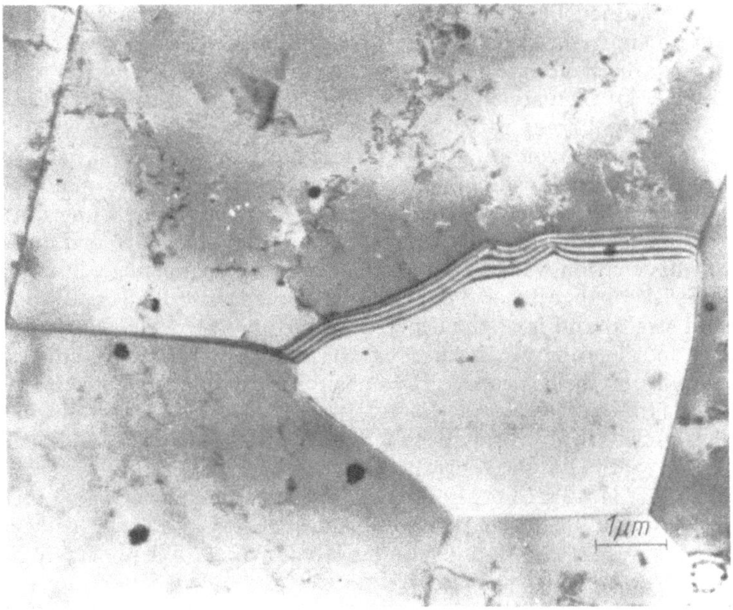

a

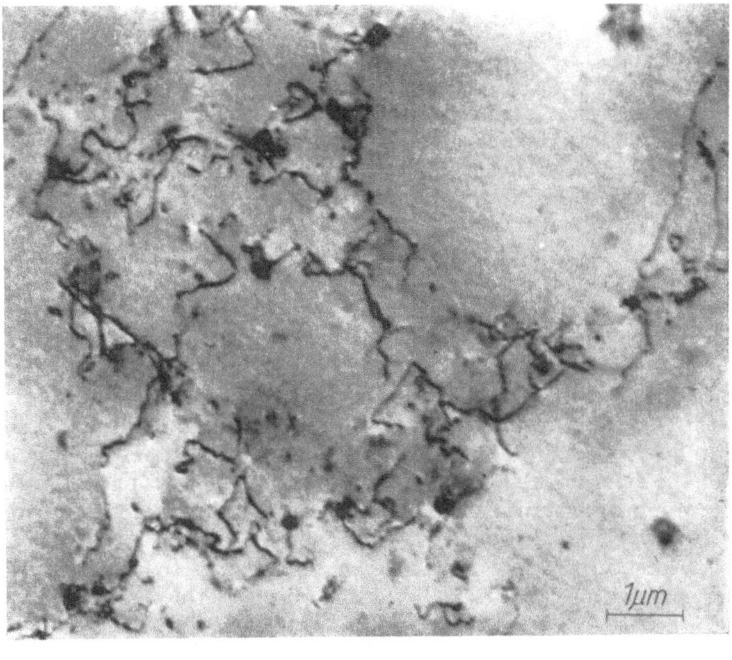

b

Fig. 6. The same as shown in Fig. 5 for the range of high temperatures and low stresses: a) Al of commercial purity, $T = 300$ °C, $\sigma = 0.75$ kp/mm²; b) pure Al, $T = 200$ °C, $\sigma = 1.3$ kp/mm²

The result obtained on Pb (Fig. 1b), polycrystalline aluminium (Fig. 1c)[2]), and aluminium single crystals (Fig. 1d) are similar to those on Cu. The values of U_0, Q, and n are given in the table. In the case of Al the deviation from equation (1) towards an increase in creep rate are less obvious than for Cu and Pb and are seen at small creep rates only (Fig. 1c, d).

The results of an electron microscopic study have shown that the steady creep in the range of moderate temperatures is characterized by a pronounced polygonized structure formed by misorientated subgrains (Fig. 5a). Subgrain boundaries consist of simple hexagonal dislocation net-works (Fig. 5b) and more rarely of single row dislocation walls. There often occur breaks in subgrain boundaries. The characteristic subgrain size depends on the specific test conditions (temperature and stress) and is of the order of some microns. The average dislocation density in the subgrain interior is 10^7 to 10^8 cm^{-2}, i.e. the subgrains are virtually free from dislocations. The mean dislocation spacing in subgrain boundaries is (100 to 200) Å.

Fig. 6 shows the structure of specimens deformed at high temperatures and low stresses. Practically there are no subgrains. If subgrains were observed, their size was comparable to that of grains, the sub-boundaries being formed by complicated non-equilibrium nets with insignificant dislocation density as compared with that in subgrain boundaries mentioned above. A relatively homogeneous distribution of both arbitrarily oriented detached dislocations and dislocation tangles (similar to the net-works shown in Fig. 6b) was peculiar to the grains.

In the specimens tested in the transition range elements of the dislocation structure typical for the moderate temperature range and high stresses as well as for the high-temperature range and low stresses were present.

4. Discussion and Conclusions

The data obtained convincingly show that the dislocation structure peculiar to steady creep as well as the stress and temperature dependences of creep rate are quite different in the range of moderate temperatures and high stresses from that of high temperatures and low stresses. It is believed that the rate-controlling mechanisms are different too.

As seen from Table 1 there is good agreement between the values of activation energy for creep calculated from our experimental results and the sublimation energy E or the energy of self-diffusion for interstitials W in the former range and the energy of diffusion of vacancies (the energy of self-diffusion Q_1 for Pb and Al and that of vacancy migration Q_2 for Cu) in the latter one. The form of the stress and temperature dependence of the creep rate in the high-temperature range and the fact that the activation energy for creep is equal to that of self-diffusion are to be expected if climb of dislocations controls the creep rate. The low value of the activation energy obtained on Cu may be connected with considerable vacancy over-saturation. Such values of activation energy were observed on Ni [25], Cu [26], and Zn [27].

The low-temperature creep with activation energy close to that of sublimation may be compared with Orlov's model [28], according to which creep is controlled

[2]) The character of the creep rate dependence on stress and temperature was identical for commercial and pure aluminium and was not affected by the shape of specimens (cylinders or plates).

Table 1

Comparison of the values of activation energies for creep (U_0 and Q) with those of energies of sublimation (E) and point defect diffusion (W, Q_1, Q_2)

	activation energy (eV)						n
	U_0	E	W	Q	Q_1	Q_2	
Al polycryst.	2.2	2.3 [15] 3.4 [16]	2.6 [21]	1.4	1.4 [17] 1.5 [18]	0.7 [18, 24]	4.9
Al single crystal	2.3			1.4			4.3
Cu	3.3	3.5 [15, 16]	3.4 [22]	1.3	2.0 to 2.4 [17] 2.1 [18]	0.8 [18] 1.1 [19] 1.3 [20]	4.9
Pb	1.7	2.0 [16] 1.8 [15]	—	1.1	1.1 to 1.2 [17, 18]	0.5 to 0.6*)	5.5

*) The value Q_2 was obtained by subtracting the energy of vacancy formation in Pb (0.58 ± 0.13) eV [23] from Q_1.

by crack propagation, as well as with Indenbom's hypothesis [5 to 7] assuming the creep over the whole range under study to be determined by nucleation and migration of point defects: vacancies at high and interstitials at low temperatures. Indenbom has predicted that as temperature decreases and applied stress rises there occurs a change both in creep mechanism and in macroscopic creep behaviour due to the fact that sinks of vacancies are surpressed and then convert to sources of interstitials. According to Indenbom, let us consider a sink of vacancies (and consequently a source of interstitials). Then the equation for the total flux of atoms to the sink has the form

$$V = \frac{D}{r_0}\left[\eta - \exp\left(-\frac{\gamma\,\sigma}{kT}\right)\right] + \frac{D'}{r_0'}\left[\exp\left(\frac{\gamma\,\sigma}{kT}\right) - 1\right]. \tag{3}$$

Here $D = D_0 \exp(-Q_1/kT)$ and $D' = D_0' \exp(-W/kT)$ are self-diffusion coefficients, the former one for vacancies, the latter for interstitials; Q_1 and W are activation energies for these processes; r_0 and r_0' are radii of the areas absorbing vacancies and emitting interstitials, respectively, η is the vacancy over-saturation far from the sink.

Assuming $\gamma\,\sigma \gg kT$, we obtain

$$V = \frac{D}{r_0}\,\eta + \frac{D'}{r_0'}\exp\left(\frac{\gamma\,\sigma}{kT}\right). \tag{4}$$

As long as the stresses are small, the first term prevails in equation (4) (as $D' \ll D$). The activation energy for this process does not depend on stress and should be equal to that of self-diffusion, Q_1. For stresses larger than σ^* the second term prevails, the activation energy being equal to W and decreases with stress. As is seen from Table 1 and Fig. 3 both the magnitude of activation energy for creep and its stress sensitivity as predicted by the interstitial hypothesis are consistent with those obtained by the present authors. Equating the terms in equation (4) we can find the conditions for a transition from the interstitial diffusion mechanism to the vacancy one. Then we obtain

$$W - \gamma\,\sigma - Q_1 = A\,kT\,, \tag{5}$$

where $A = \ln\,(D_0'\,r_0/D_0\,r_0'\,\eta)$ which has been estimated macroscopically in [7] to

$$A = \ln \frac{R^2\,\dot{\varepsilon}_0\,\omega}{6\,\gamma\,D_0}\,, \tag{6}$$

where R is the grain size and ω the atomic volume.

Calculations of A according to (6) give the values 11.5 ($R = 0.2$ mm, $\dot{\varepsilon}_0 = 10^{12}\,\text{s}^{-1}$, $D_0 = 1.7\,\text{cm}^2/\text{s}$, $\gamma = 280\,\omega$) for Al and 15 ($R = 0.1$ mm, $\dot{\varepsilon}_0 = 10^{13}\,\text{s}^{-1}$, $D_0 = 0.6\,\text{cm}^2/\text{s}$, $\gamma = 90\,\omega$) for Cu. This value can be compared with that obtained experimentally when treating the data by equation (5). The energy difference $W - \gamma\,\sigma - Q_1$ should be taken then from Fig. 3 (for Cu) while the stresses correspond to the break on the lg $\dot{\varepsilon} = f(\sigma)$ curve at the temperature T. At different temperatures A lies between 14 and 17 which is in good agreement with the estimation by equation (6).

If within the range $\sigma \lesssim \sigma^*$ (the intermediate range) the creep rate dependence on stress is rather weak, then, as further decrease in stress and increase in temperature occur, i.e. when approaches kT, the curves lg $\dot{\varepsilon} = f(\sigma)$ deviate to the opposite side (upwards in Fig. 1). The intermediate range includes the stresses (1 to 2) $kT/\gamma < \sigma < \sigma^*$ and according to (5) shrinks as the temperature rises. For $kT > (A + 1)^{-1}\,(W - Q_1)$ this range disappears, Fig. 1, and only well-known upward bend is observed on the curves lg $\dot{\varepsilon} = f(\sigma)$ [1, 2].

Thus, the present paper confirms the conclusion drawn by many authors (e.g. [29, 30]) that creep deformation at high temperatures and low stresses is controlled by diffusion. The steady-state creep rate is in this case a parabolic function of stress. As temperature decreases and applied stress rises the parabolic dependence is changed to an exponential law. Simultaneously the magnitude of the activation energy changes. The complicated creep rate dependence on stress at high temperatures as well as deviations from (1), (see Fig. 1) can be explained by the existence of two creep mechanisms controlling the low and high-temperature creep deformation. At high temperatures and low stresses the stationary creep is described by a well-known model of dislocation climb [12 to 14]. In the range of high stresses and moderate temperatures the creep deformation is connected either with thermally activated crack propagation [1, 28], or with a mechanism the rate of which is controlled by interstitial diffusion. This mechanism is suggested in what follows. As has been mentioned above, steady-creep strain develops due to successive disintegration of subgrain boundaries. As dislocations escape from the hexagonal net-works (Fig. 5b) consisting of screw dislocations [4], jogs are being formed along the dislocation line and move non-conservatively, the strain rate being thus controlled by this process. According to Cottrell's theorem [31], the jogs in question are expected to be mainly of interstitial type. The jog spacing corresponds to the dislocation density in sub-boundaries which leads to a correlation between the parameter γ in equation (1) and the dislocation spacing in sub-boundaries [4]. Indenbom's hypothesis [5 to 7] thus predicts the general form of the stress dependence of the steady creep rate with two types of deviations from equation (1), giving also the correct dependence of activation energy on temperature and stress, as well as the temperature sensitivity of the critical stress value. However, to finally elucidate the low temperature creep mechanism it seems necessary to carry out experiments on materials whose sublimation energy strongly differs from that of interstitial diffusion, which, unhappily, is difficult at present because the data available concerning diffusion parameters of interstitials are not sufficient.

Acknowledgement

The authors express their appreciation to Prof. V. L. Indenbom for fruitful discussions and valuable remarks.

References

[1] S. N. Zhurkov and T. P. Sanfirova, Zh. tekh. fiz. **28**, 1719 (1958).
[2] T. P. Sanfirova, Thesis, Leningrad Politech. Inst., 1961.
[3] W. A. Stepanov and V. V. Shpeizman, Fiz. Metallov i Metallovedenie **29**, 375 (1970).
[4] M. M. Myshlyaev, Fiz. tverd. Tela **7**, 591 (1965); Thesis, Crystallography Inst., USSR Acad. Sci., Moscow 1968.
[5] V. L. Indenbom, in: Phys. Alkali Halide Crystals, Latv. State University, Riga 1962 (p. 449); Internat. Conf. Crystal Lattice Defects, Abstr. 1—11, Kyoto 1962.
V. L. Indenbom and A. N. Orlov, Uspekhi fiz. Nauk **76**, 557 (1962).
[6] V. L. Indenbom, Zh. eksper. teor. Fiz., Pisma **12**, 526 (1970).
[7] V. L. Indenbom and A. N. Orlov, Probl. Prochnosty **12**, 3 (1970).
[8] S. N. Zhurkov and E. E. Tomashevskii, Zh. tekh. fiz. **25**, 66 (1955).
[9] V. V. Shpeizman, in: Machines and Instruments for Testing Materials, Metallurgy, Moscow 1968.
[10] M. M. Myshlyaev, Fiz. tverd. Tela **9**, 1669 (1967).
[11] V. A. Pavlov, M. G. Gaidukov, N. I. Noskova, and V. V. Melnikova, Fiz. Metallov i Metallovedenie **12**, 97 (1961).
V. A. Pavlov, M. G. Gaidukov, V. V. Melnikova, ibid. **12**, 748 (1961).
[12] J. Weertman, J. appl. Phys. **26**, 1213 (1955).
[13] J. Weertman, J. appl. Phys. **28**, 1185 (1957).
[14] J. Weertman and J. R. Weertman, in: Phys. Metallurgy, Chap. 15, North-Holland Publ. Co., Amsterdam 1965.
[15] E. V. Britske, A. F. Kapustinskii, B. K. Veselovskii, L. M. Shamovskii, L. G. Tchentsova, and B. I. Anvajer, in: Thermal Constants of the Inorganic Substances, USSR Acad. Sci., Moscow 1949.
[16] R. Hultgren, R. L. Orr, P. D. Anderson, and K. K. Kelley, in: Selected Values of Thermodynamic Properties of Metals and Alloys, Wiley, New York/London 1963.
U. D. Veryatin, V. P. Mashirev, N. G. Ryabtsev, V. I. Tarasov, B. D. Rogozkin, and I. V. Korobov, in: Thermodynamic Properties of Inorganic Substances, Atomic Energy, Moscow 1965.
[17] J. E. Dorn, in: Creep and Recovery, Seminar of Amer. Soc. Metals, Cleveland 1957.
[18] J. Friedel, in: Dislocations, Pergamon Press, 1964.
[19] A. Seeger, in: Radiation Damage in Solids, Vol. 1, IAEA, Vienna 1962 (p. 101).
[20] G. Airoldi, G. L. Bacchella, and E. Germagnoli, Phys. Rev. Letters **2**, 145 (1959).
[21] K. Isebeck, F. Rau, W. Schilling, K. Sonnenberg, P. Tischer, and H. Wenzl, phys. stat. sol. **17**, 259 (1966).
[22] J. Diehl, in: Moderne Probleme der Metallphysik, Vol. 1, Ed. A. Seeger, Springer-Verlag, Berlin/Heidelberg/New York 1965 (p. 225).
[23] A. J. Leadbetter, D. M. Newsham, and N. H. Picton, Phil. Mag. **13**, 371 (1966).
[24] J. Bass, Phil. Mag. **15**, 717 (1967).
[25] A. F. Orlov and K. A. Osipov, in: Diffusion Processes in Metals, Naukova Dumka, Kiev 1966.
[26] P. Feltham and J. D. Meakin, Acta metall. **7**, 614 (1959).

[27] R. Kamel and T. A. Bessa, J. Phys. D (Brit. J. appl. Phys.) **14**, 459 (1963).
[28] A. N. Orlov, Fiz. tverd. Tela **3**, 500 (1961).
[29] G. M. Bartenev, L. I. Ivanov, and E. A. Surova, Izv. Akad. Nauk, Met. **3**, 134 (1966).
[30] V. M. Rozenberg, in: Creep in Metals, Metallurgy, Moscow 1967.
[31] A. H. Cottrell, in: Dislocations and Mechanical Properties of Crystals, Wiley, New York 1958.

(Received August 12, 1971)

phys. stat. sol. (a) 8, 403 (1971)

Subject classification: 13.4 and 14.4.1; 13.1; 22.2.3

Department of Applied Mathematics and Mathematical Physics, University College, Cardiff

Electric Field Dependence of a Two-Electron Transition Process

By

G. JONES[1]) and A. R. BEATTIE

The effect of an electric field on a transition process where a fully occupied two-level centre simultaneously loses both electrons, one to the conduction band and the other to the valence band, is considered. The field dependencies of the related capture constants are calculated using drifted Maxwellian distribution functions for the carrier distributions. Spherical energy surfaces are assumed although a non-parabolic conduction energy band and an energy dependent electron effective mass are allowed for. The resulting field dependencies of the capture constants are discussed with particular reference to their effect on the formation of bulk electrical instabilities.

On traite de l'effet d'un champ électrique sur un processus de transition dans lequel un centre complètement occupé perd simultanément ses deux électrons, l'un passant à la bande de conduction et l'autre à la bande de valence. La dépendance des coefficients de recombinaison combinés est calculée en utilisant une fonction de distribution Maxwellienne déplacée pour les porteurs de charge. On suppose des surfaces d'énergie sphérique bien qu'on adopte une bande de conduction non-parabolique et donc une masse effective qui dépend de l'énergie. Ces coefficients de recombinaison sont alors discutes en rapport avec leur effet sur la formation d'instabilites électriques dans un cristal.

1. Introduction

In a previous paper [1], hereafter called I, we considered an electron collision process involving a two-level centre in semiconductors, and the likely effect of this process on the lifetimes of charge carriers in InSb. This process is different from those usually considered in that a fully occupied (empty) centre loses (gains) two electrons simultaneously one to (from) the conduction band and the other to (from) the valence band. The present work is concerned with the field dependence of this 2–0 transition.

As in I the transition is considered to take place by Auger collision only. The theory is based on one-electron band state wave functions, perturbation theory, non-interacting impurities and non-degeneracy of the carrier distributions. While the valence band is taken to be parabolic the conduction band is not and its E–k relation is that given by Kane [2]. The major effect of the electric field on the capture constants is assumed to enter through a drifted Maxwellian distribution function, where for the conduction band the field dependencies of the two parameters calculated in the paper [3] (hereafter called II) are used. Such a distribution function is not strictly a solution of the transport equation and its use here is discussed in Section 3.

---

[1]) Present address: Département de Physique, Université de Montreal, Montreal, Canada.

Using the notation of I the net transition rate for the process 2–0 may be written as:

$$R^F_{2-0} = G\,p\,n_{t_2} - G'\,n\,n_{t_0}\,, \tag{1.1}$$

where n and p are the concentration of electrons and holes respectively and n_{t_i} is the concentration of two-level centres in state t_i where t_i is the centre state when occupied by i electrons. G is the capture constant related to a loss of both electrons from a t_2 state centre and G' the capture constant related to a gain of two electrons by a t_0 state centre.

Expressions for the capture constants are derived in Section 2 and the numerically evaluated results for the case of InSb are discussed in Section 3. The concluding section deals with the possible effect of the transition mechanism on the propagation of stable negative resistance domains.

2. Electric Field Behaviour of *G* and *G'*

In calculating the field dependencies of the capture constants G and G' the quantum mechanical perturbation treatment of I is used and the assumptions of spherical energy surfaces, quasi-Fermi levels and non-degenerate semiconductors are made. While parabolicity is assumed for the valence energy band, the conduction band is taken to be nonparabolic and of the form given by Kane [2] (equation (13)), which is applicable to III–V compounds.

In I the expression for the net transition rate includes only one capture constant, but the distribution functions of the carriers in the case of an electric field prohibit such a reduction and the two capture constants of equation (1.1) are expressed as

$$G = \frac{1}{t\,p} \int \theta(\mathrm{c})\,\theta(\mathrm{v})\,T_{\mathrm{i.f.}}\,\mathrm{d}S\,, \tag{2.1}$$

$$G' = \frac{1}{t\,n} \int \varrho(\mathrm{c})\,\varrho(\mathrm{v})\,T_{\mathrm{i.f.}}\,\mathrm{d}S\,, \tag{2.2}$$

where $T_{\mathrm{i.f.}}$ is the transition probability from initial to final state of the system (as in I) and the integration is over all band states. $\theta(j)$ $(j = \mathrm{c}, \mathrm{v})$ is the appropriate probability of electron occupation or vacancy of state j, while $\varrho(j) = 1 - \theta(j)$ denotes the probability for the reverse process.

$$\varrho(\mathrm{c}) = \exp\left[F_{\mathrm{c}} - \eta_{\mathrm{c}} - \frac{E\,[(\boldsymbol{k} - \boldsymbol{k}_{\mathrm{d_c}})^2]}{k_{\mathrm{B}}\,T_{\mathrm{c}}}\right], \tag{2.3}$$

$$\theta(\mathrm{v}) = \exp\,[\eta_{\mathrm{v}} - F_{\mathrm{v}} - \sigma_{\mathrm{v}}\,(\boldsymbol{k} - \boldsymbol{k}_{\mathrm{d_v}})^2]\,, \tag{2.4}$$

$$\theta(\mathrm{c}) = \varrho(\mathrm{v}) = 1\,, \tag{2.5}$$

where

$$\sigma_{\mathrm{v}} = \frac{\hbar^2}{2\,m_{\mathrm{v}}\,k_{\mathrm{B}}\,T_{\mathrm{v}}}\,, \qquad k_{\mathrm{B}}\,T_i\,\eta_i = E_i\,.$$

$k_{\mathrm{B}}\,T_i\,F_i$ are the quasi-Fermi levels for the conduction and valence bands $(i = \mathrm{c}, \mathrm{v})$, m_{v} is the apparent density of states hole effective mass [4], k_{d_i} and T_i are the electron $(i = \mathrm{c})$ and hole $(i = \mathrm{v})$ drifted Maxwellian parameters, E_{c} and E_{v} are the conduction and valence band edge energies respectively and k_{B} is Boltzmann's constant. $U_{\mathrm{i.f.}}$ involves the initial and final wave functions

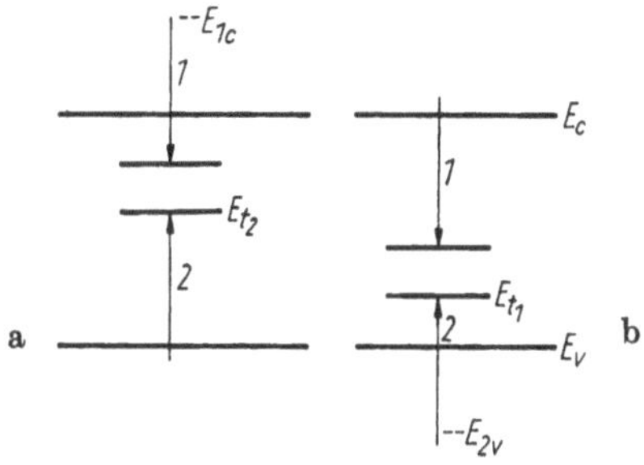

Fig. 1. Diagrammatic representation of transitions related to the capture constant G' for a V-series case (a) and a C-series one (b). E_{1c} and E_{2v} are the energies of electrons taking part in the transitions depicted for a V- and a C-series position of the centre energy levels respectively

which should, in the electric field case, be field dependent [5]. It is assumed here, however, that the major effect of the electric field can be taken into account by the distribution functions, so the expression for $U_{i.f.}$ calculated in I is used in the present work.

In the evaluation of the capture constants G and G', the parabolic E–k relation for the valence band and the non-parabolic one for the conduction band result in the change of energy in the transition being expressed as:

$$w = \frac{\hbar t}{2 m_c} \left[\frac{m_c}{\hbar^2} [(E_G^2 + a k_1^2)^{1/2} - E_G] - \mu k_2^2 \mp \xi \right], \tag{2.6}$$

where

$$\mu = \frac{m_c}{m_v}, \qquad \xi = |E_G - 2(E_c - E_{t_2}) - \Delta|, \qquad a = \frac{2 \hbar^2 E_G}{m_c},$$

$$\Delta = E_{t_2} - E_{t_1}.$$

E_{t_i} is the energy level of the centre occupied by i electrons, E_G is the forbidden band gap energy, m_c is the electron effective mass for the conduction band, k_1 is the wave vector related to conduction band states and k_2 the wave vector of valence band states.

The negative sign in equation (2.6) applies when

$$(E_c - E_{t_2}) + (E_c - E_{t_1}) < E_G \tag{2.7}$$

and the positive sign when

$$(E_c - E_{t_2}) + (E_c - E_{t_1}) > E_G. \tag{2.8}$$

As energy is conserved in the transition ($w = 0$) it follows, see Fig. 1, that when the negative sign applies

$$E_{2v}^{\text{max.}} = E_v, \qquad E_{1c} > E_c \tag{2.9}$$

and for the positive sign

$$E_{1c}^{\text{min.}} = E_c, \qquad E_{2v} < E_v. \tag{2.10}$$

Due to this behaviour, which necessitates a different order of integration for equations (2.1) and (2.2), the evaluation of the capture constants G and G' are divided into two regions; the V-series when (2.9) applies; the C-series when (2.10) applies. The subscript v or c is attached to the capture constants to denote to which region they apply, expressions for them being given in the Appendix.

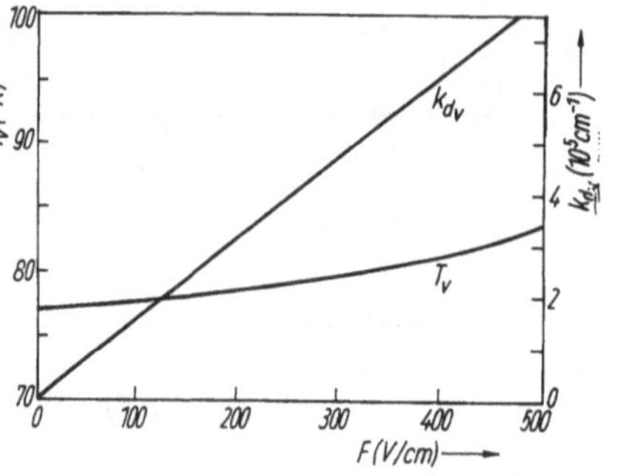

Fig. 2. The field dependent behaviour of the drifted Maxwellian parameters calculated for the valence band using InSb at a lattice temperature of 77 °K as the semiconductor model. A constant hole mobility $\mu_p = 10^4$ cm²/Vs is used in the calculation [6]

3. Numerical Results for G and G'

The expressions for the capture constants, given in the Appendix, require a knowledge of the field dependent behaviour of the parameters k_{d_i} and T_i before the dimensionless integrals may be computed. In the case of conduction band electrons the results calculated in II for InSb at 77 °K lattice temperature are used. In the determination of the behaviour of the parameters for valence band holes it is assumed that polar scattering is the dominant energy loss mechanism [4], and the valence band is parabolic with a constant apparent density of states effective mass. The results for InSb, using a constant hole mobility [6], are shown in Fig. 2 for a lattice temperature $T_0 = 77$ °K. The behaviour of the parameters demonstrates that the applied field has little effect on the hole distribution function.

Numerical results for the capture constants are presented in the form of graphs, where the energy level splitting is $\Delta = 0.015$ eV. The main features of the results can best be explained by considering Fig. 1, where it is important to notice that the condition $w = 0$ determines that the energy paths (1) and (2) are equal. Because of this the threshold energy level in the conduction band for the V-series, $E_{1\mathrm{c}}^{\mathrm{thresh.}}$, is given by

$$E_{1\mathrm{c}}^{\mathrm{thresh.}} - E_{\mathrm{t}_2} = E_{\mathrm{t}_1} - E_{\mathrm{v}} \tag{3.1}$$

and for the C-series the threshold energy level in the valence band, $E_{2\mathrm{v}}^{\mathrm{thresh.}}$, is given by

$$E_{\mathrm{c}} - E_{\mathrm{t}_2} = E_{\mathrm{t}_1} - E_{2\mathrm{v}}^{\mathrm{thresh.}} . \tag{3.2}$$

For the capture constant G' the valence band non-degenerate electron probability function is unity and it is therefore the behaviour of the conduction band electron probability function which is important. In the V-series case (Fig. 1a) the electron distribution function spreads out, as the field increases from zero, with the maximum moving away from the close proximity of E_{c} towards $E_{1\mathrm{c}}^{\mathrm{thresh.}}$. This increase in the density of high energy electrons gives rise to an increase of the capture constant G'. As the maximum of the distribution function passes $E_{1\mathrm{c}}^{\mathrm{thresh.}}$ the capture constant traverses a maximum and then starts to decrease (e.g. Fig. 3) due to the minimum energy change weighting of the matrix element $U_{\mathrm{i.f.}}$. Centre energy levels much closer to the conduction band edge than the valence band, as in Fig. 4, lead to a particularly rapid increase of G' with field.

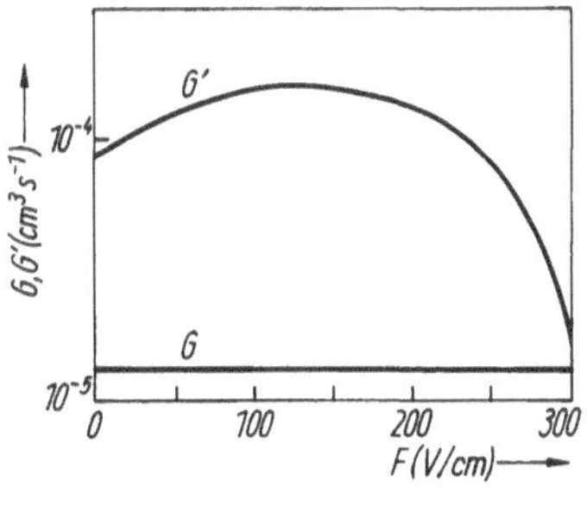

Fig. 3. Calculated curves of capture constant versus field for InSb at a lattice temperature of 77 °K, with $E_c - E_{t_1} = 0.1$ eV and $\Delta = 0.015$ eV

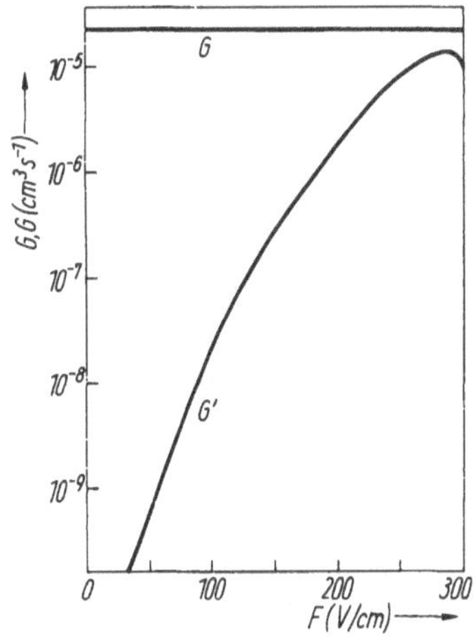

Fig. 4. Calculated field dependencies of the capture constants G and G' for InSb at a lattice temperature of 77 °K, with $E_c - E_{t_2} = 0.05$ eV and $\Delta = 0.015$ eV

In the C-series case the capture constant G' does not exhibit a maximum since the electron distribution peak always moves away, with increasing field, from the threshold transition region. The resulting decrease of capture constant with increase in field is again brought about by the weighting of the matrix element $U_{\text{i.f.}}$.

The behaviour of G (Fig. 1 with reversed arrows) is governed by the hole distribution function, the probability of an electron vacancy in the conduction band being taken as unity. The hole distribution function, however, alters little with respect to electric field and, in both V- and C-series positions, the variation of G is very small (Fig. 3 and 4).

Although the drifted Maxwellian distribution function is a good approximation only at high electron densities, it continues to be used at intermediate densities to give reasonable agreement with experiment (e.g. [7]). It has also been pointed out [8] that two-phonon processes are comparable with first-order (inelastic) processes and that the elastic part of the two-phonon processes tends to drive the electrons to a drifted Maxwellian distribution. A recent Monte Carlo calculation [9] determines that the drifted Maxwellian approach underestimates the number of high energy electrons, but there is much closer agreement on values for the mean energy and drift velocity. The difference between the calculated mobility values at low field strengths is likely to be less important in the present problem since the parameters of the drifted Maxwellian were determined from experimental values of the mobility (II).

Despite this argument as to how reasonable the drifted Maxwellian distribution is in any given problem, we still expect the same qualitative behaviour of the capture constants for centre energy levels closer to the conduction band edge than the valence band. The reasons for this conclusion have been given earlier in this section and are based on the fact that the capture constant G' involves only electrons in the conduction band above $E_{1c}^{\text{thresh.}}$. Provided the number of electrons at and above this energy increases with field, the captive constant G' also increases.

Thus a similar behaviour of G' with electric field is to be expected in semiconductors other than InSb provided the two-level centre energy levels are suitably positioned.

4. Discussion

One of the effects of the two-electron transition mechanism is in the propagation of electrical domains. As pointed out by Ridley and Wisbey [10], in their theory of domain propagation in the presence of trapping by a single impurity level, it is necessary that the capture rate is field dependent but that the generation rate is solely due to field independent processes. There is also the tacit assumption made that the material possesses a differential negative resistance over a certain field range. If one considers a very simple model where only the 2–0 transition process is operable it can be shown [11] that such a behaviour of the current field characteristics can be obtained.

A more realistic model should include other transition mechanisms where electrons are captured or generated only one at a time. Nevertheless, despite these complications, the 2–0 transition process can be expected to contribute to domain propagation by virtue of the increase of G' with electric field.

Consider, as an example, the case of gold-doped Ge for which it has been shown that thc practical result of a large increase of capture rate with field for the Au^{2-} level at 0.04 eV is a specific differential negative resistance [12]. From this follows the formation and propagation of domains as a necessary consequence of electrical stability [13]. Results were subsequently reported on specimens containing enough antimony to partially fill the Au^- level. Thus both the Au^- and Au^{2-} levels at 0.2 and 0.04 eV are active traps [14] and can be effectively considered as two-level centres. In obtaining the field dependencies of the capture constants from the measured time constants, Pratt and Ridley considered only one-electron transition processes. Bearing in mind the discussion of Section 3 we expect the strong field dependence of the capture constant for the Au^{2-} level (α_1 of [14]) to contain a component due to the two-electron transition 2–0 as well as the usual tunneling component due to one-electron transitions. The size of this two-electron contribution will depend strongly on the number of unfilled Au^- centres present in the specimen.

In summary, the two-electron transition discussed in this paper is expected to contribute to the creation of bulk electrical instabilities in semiconductors which have two-level centres with energy levels closer to the edge of the conduction band than that of the valence band.

Appendix

(i) The expressions for the capture constant G in the two regions can be written

$$G_i = \frac{1024\,\pi^2\,e^4\,|F_3|^2}{m_v^{1/2}\,\varepsilon_0^2\,M_v^*\,z_{d_v}\,[2\,\pi\,k_B\,T_v]^{3/2}}\,I(i) \qquad (i = c, v)\,, \tag{A.1}$$

where

$$M_v^* = \frac{\hbar^2\,\alpha^2}{2\,m_v\,k_B\,T_v}\,, \qquad z_{d_v} = \frac{k_{d_v}}{\alpha}\,, \qquad \alpha = \frac{8\,\varepsilon_0}{5\,e^2}\,\Delta\,.$$

ε_0 is the static dielectric constant and the derivation of the expression for α appears in paper I. Under the assumption that the overlap functions vary more slowly with wave vector than the probability functions, they are taken outside the integral, evaluated for the most probable transition, and denoted by F_3.

In the V-series case

$$I(\mathrm{v}) = \int\limits_{p_\mathrm{v}}^{\infty} \left[4\,\Phi_0^2 + \frac{1}{3\,x^4\,z_2^4}\,\Phi_1^2\right] \exp\left[-M_\mathrm{v}^*\,(z_2^2 + z_{\mathrm{d_v}}^2)\right] \sinh\,(2\,M_\mathrm{v}^*\,z_2\,z_{\mathrm{d_v}})\,x^2\,\mathrm{d}x\,, \tag{A.2}$$

where Φ_0 and Φ, are expressions in x and z_i ($i = 1, 2$) first introduced in I.

The lower limit of integration in the above equation arises from the condition (2.9) which results in

$$p_\mathrm{v} = \left[\beta\left(1 + \frac{A\,\beta}{4}\right)\right]^{1/2},$$

where

$$A = \frac{a\,\alpha^2}{E_\mathrm{G}^2}\,, \qquad \beta = \frac{\xi}{\alpha^2}\,. \tag{A.3}$$

Furthermore, from the condition $w = 0$, z_2 may be expressed as

$$z_2 = \left[\frac{\frac{2}{A}\left[(1 + A\,x^2)^{1/2} - 1\right] - \beta}{\mu}\right]^{1/2}. \tag{A.4}$$

In the C-series case

$$I(\mathrm{c}) = \mu \int\limits_{p_\mathrm{c}}^{\infty} \left[4\,\Phi_0^2 + \frac{1}{3\,x^4\,z_1^4}\,\Phi_1^2\right] \exp\left[-M_\mathrm{v}^*\,(x^2 + z_{\mathrm{d_v}}^2)\right] \times$$

$$\times \sinh\,(2\,M_\mathrm{v}^*\,x\,z_{\mathrm{d_v}})\,x\,z_1\,(1 + A\,z_1^2)^{1/2}\,\mathrm{d}x\,, \tag{A.5}$$

where

$$z_1 = \left[(\mu\,x^2 - \beta)\left[\frac{A}{4}\,(\mu\,x^2 - \beta) + 1\right]\right]^{1/2}$$

and

$$p_\mathrm{c} = \left(\frac{\beta}{\mu}\right)^{1/2}.$$

(ii) The expressions for the capture constant G' are more involved, due to the form of the electron distribution function. They can be written

$$G_i' = \frac{256\sqrt{2}\,e^4\,\pi\,|F_3|^2}{m_\mathrm{c}^{1/2}\,\varepsilon_0^2\,A\,z_{\mathrm{d_c}}\,n^*\,E_\mathrm{G}^{3/2}}\,I'(i) \qquad (i = \mathrm{c}, \mathrm{v})\,, \tag{A.6}$$

where

$$z_{\mathrm{d_c}} = \frac{k_{\mathrm{d_c}}}{\alpha}\,.$$

The expression

$$n^* = \left[\mathrm{K}_1(C) + \frac{c}{2}\,\mathrm{K}_0(C)\right], \tag{A.7}$$

where

$$C = \frac{E_G}{2\,k_B\,T_c}$$

arises from the work reported in II. $K_j(C)$ is a Bessel function of the third kind of order j.

In the V-series case

$$I'(\mathrm{v}) = \frac{1}{\mu} \int\limits_{p_v}^{\infty} \left[4\,\Phi_0^2 + \frac{1}{3\,x^4\,z_2^4}\,\Phi_1^2\right] \times$$

$$\times\;[(1 + C_1)\exp(-\,C_1) - (1 + C_2)\exp(-\,C_2)]\;x\,z_2\,\mathrm{d}z\;, \qquad \text{(A.8)}$$

where

$$C_{1,2} = C\,[1 + A\,(x \mp z_{\mathrm{d_c}})^2]^{1/2}\,.$$

In equation (A.8) p_v and z_2 are given by (A.3) and (A.4), since the form of w is the same.

In the C-series case

$$I'(\mathrm{c}) = \int\limits_{p_c}^{\infty} \left[4\,\Phi_0^2 + \frac{1}{3\,x^4\,z_1^4}\,\Phi_1^2\right][(1 + C_3)\exp(-\,C_3) - (1 + C_4)\exp(-\,C_4)] \times$$

$$\times\;x^2\,(1 + A\,z_1^2)^{1/2}\,\mathrm{d}x\;, \qquad \text{(A.9)}$$

where

$$C_{3,4} = C\,[1 + A\,(z_1 \mp z_{\mathrm{d_c}})^2]^{1/2}$$

with z_1 and p_c defined by (A.5).

Acknowledgements

One of us (G. J.) is grateful to the Ministry of Defence, U.K. for support and to the National Research Council of Canada for assistance while carrying out this work.

References

[1] G. Jones and A. R. Beattie, phys. stat. sol. (a) **4**, 193 (1971).
[2] E. O. Kane, J. Phys. Chem. Solids **1**, 249 (1957).
[3] G. Jones, G. Smith, and A. R. Beattie, phys. stat. sol. **20**, K135 (1967).
[4] O. Madelung, Physics of III–V Compounds, J. Wiley & Sons, Inc., New York 1964.
[5] J. Callaway, Energy Band Theory, Academic Press, New York 1964.
[6] M. Glicksman and W. A. Hicinbothem Jr., Phys. Rev. **129**, 1572 (1963).
[7] R. C. Curby and D. K. Ferry, Phys. Rev. B **3**, 3379 (1971).
[8] A. Sher and K. K. Thornber, Appl. Phys. Letters **11**, 3 (1967).
[9] W. Fawcett, A. D. Boardman, and S. Swain, J. Phys. Chem. Solids **31**, 1963 (1970).
[10] B. K. Ridley and P. H. Wisbey, Brit. J. appl. Phys. **8**, 761 (1967).
[11] G. Jones, Ph.D. Thesis, unpublished 1967.
[12] B. K. Ridley and R. G. Pratt, J. Phys. Chem. Solids **26**, 21 (1965).
[13] B. K. Ridley, Proc. Phys. Soc. **82**, 954 (1963).
[14] R. G. Pratt and B. K. Ridley, Proc. Phys. Soc. **85**, 293 (1965).

(Received August 18, 1971)

phys. stat. sol. (a) 8, 411 (1971)

Subject classification: 13 and 14.4.1; 13.4; 22

Research Institute for Technical Physics of the Hungarian Academy of Sciences, Budapest

Recombination of Hot Electrons on Dislocations and the Possibility of Negative Differential Conductivity[1])

By

G. PATAKI

In the first part of the paper, the transient and stationary lifetimes of recombination at arbitrary concentration of dislocations are given, using the model proposed by Glaenzer and Jordan [14]. In the second part, the hot electron recombination is considered if the distribution function is determined by: (i) acoustic phonon scattering, (ii) combined acoustic and impurity scattering, and (iii) combined optical and acoustic phonon scattering. In cases (i) and (ii) tunnelling of electrons through the potential barrier is assumed. The mobility increase due to hot electron scattering on charged dislocations is calculated. The analysis shows that the lifetime of hot electrons generally decreases but NDC is possible only for case (iii) in a special range of dislocation density and temperature, where the tunnelling is not a rate-limiting process. Numerical estimates are given for Ge.

В первой части работы рассчитаны постоянные времени переходного и стационарного процессов рекомбинации при любой концентрации дислокаций с помощью модели, предложенной Гленцером и Йорданом [14]. Во второй части рассмотрена рекомбинация горячих электронов на дислокациях в случаях, когда функция распределения определяется: 1. рассеянием на акустических фононах, 2. суммарным рассеянием на примесях и на акустических фононах, и 3. суммарным рассеянием на акустических и оптических фононах. В случаях 1 и 2 предполагается туннельное прохождение электронов через потенциальный барьер дислокации. Вычисляется возрастание подвижности, вызванное рассеянием горячих электронов на заряженных дислокациях. Приведенный анализ показывает, что время жизни горячих электронов в общем случае уменьшается, но возникновение отрицательной дифференциальной проводимости возможно только для случая 3 в определенном интервале плотности дислокаций и температуры когда туннельный переход не ограничивает скорость захвата. Приведен численный расчет для германия.

1. Introduction

The role of dislocations in determining carrier lifetime and noise in semiconductors is well known [1 to 6]. Reproducible relations between dislocation density and carrier lifetime were found by many authors [2, 6, 7] both for dislocations introduced during crystal growth or by plastic deformation. The analysis of experimental data has shown that the potential barrier around the dislocation has an important effect on the recombination mechanism [1, 8, 9]. This effect can manifest itself not only in determining the concentrations of carriers around the dislocations forming the space charge region, but also in the capture process, as it was pointed out in papers [10] and [11], if the tun-

---

[1]) Presented at the "Symposium on Crystal Defects" held in Zakopane, 1970.

nelling through the repulsive potential barrier turns out to be important. The investigations concerning the electron capture on repulsive recombination centres on the other hand have shown that the hot-electron recombination can lead to the domain instability which, in the last years, became a new field of semiconductor physics. At present, a number of recombination centres are known, both in classical semiconductors and compounds which, by capturing the hot electrons, give rise to the formation of moving domains. As for the dislocations, it seems natural to raise the question what is the capture rate or lifetime of hot electrons in the case of negatively charged dislocations, and whether it is possible to observe negative differential conductivity (NDC).

In the present work, the dislocations are considered as uniformly charged lines, and the shape of electrostatic potential will be determined by the Debye approximation [12]. Though the model of discrete charge row seems to be closer to reality [9, 13], because of mathematical simplicity, however, the former potential will be used.

In the first part of the paper the phenomenological theory of recombination on dislocations is considered, starting from paper [14], and the stationary and transient lifetimes of the model are given for arbitrary concentration of dislocations. Based on the expressions derived, the effect of high electric field on recombination is investigated under various conditions concerning the scattering mechanism determining the hot-electron distribution function. The following cases will be treated in detail: (i) scattering on acoustic phonons; (ii) combined acoustic and impurity scattering; (iii) combined optical and acoustic phonon scattering in high field (effective electron temperature approximation). For the first two cases, starting from the papers [11], [15], and [16], the electrical field dependence of the capture probability will be determined. For case (iii) the lifetime of electrons can simply be determined by substituting the lattice temperature T by the effective electron temperature T_e.

The field dependence of electron mobility due to electron scattering on the electrostatic field of the dislocation is also treated, generalizing the corresponding low-field mobility derived firstly in [17]. The numerical calculations and estimates for Ge and Si have shown that, in the first two cases, instead of NDC a deviation from Ohm's law is predicted, due to increased hot-electron capture. In these cases, with decreasing carrier concentration the potential barrier changes so that the effectiveness of the tunnelling is significantly reduced. In the third case, however, with properly chosen conditions (compensated material, background illumination, etc.), the observation of NDC in the photoconductivity seems to be possible.

2. Kinetics of the Recombination near the Equilibrium

In the present section, the kinetics of the electron–hole recombination at dislocations is investigated, starting from the model proposed by Glaenzer and Jordan [14]. The present treatment is more general compared with the previous ones [1, 8, 14] in that respect that both the stationary and transient lifetimes will be given at an arbitrary concentration of dislocations. The effect of the potential barrier on recombination was consistently taken into account in the papers by Morrison [1] and Figielski [8]. In a recent paper [14], a generalized model for recombination was proposed in which for capturing an electron it must not only have the energy $e\,\Phi$ to surmount the potential barrier, but also an additional energy $E_0 f$ to re-arrange the electrons already residing on the dislocation. The equations of the recombination kinetics can be written in the following form:

$$-\frac{dn}{dt} = N\,C_n\,[n_b\,(1-f) - n_1\,f]\,, \tag{1}$$

$$-\frac{dp}{dt} = N\,C_p\,[p_b\,f - p_1\,(1-f)]\,, \tag{2}$$

where f is the fractional electron occupation of the acceptor states. It differs from the usual Fermi-Dirac distribution function, because the energy of the dislocation states depends on the filling of these states.

Read [18] proposed the following general expression for f:

$$f = \frac{1}{1 + \exp\,[(E_2 - F_0 + E^*)/kT]}, \tag{3}$$

where $E^*(f)$ is added to the acceptor energy E_2. The actual value of E^* depends on the model accepted. Thus, for "minimum energy approximation" $E^* = e\,\Phi + E_0\,f$ where $E_0 = e^2/4\,\pi\,\varepsilon\,a$ is the electrostatic energy of adjacent occupied acceptors. In n-type material the potential barrier has the form [18]

$$e\,\Phi = E_0\,f\left[3\,\ln\frac{f}{f_c} - 1.232\right], \tag{4}$$

where $f_c = a\,[\pi\,(N_d - N_a)]^{1/3}$ is a constant which depends on the net donor density. A similar expression holds for donor-like dislocation states in p-type material.

The quantities n_b, p_b in equations (1) and (2) give the carrier concentrations in the bands at the tip of the potential barrier.

According to the model of paper [14]

$$n_b = n\,\exp\left(\frac{-E^*}{kT}\right), \qquad p_b = p\,\exp\left(\frac{+E^*}{kT}\right), \tag{5a}$$

and for the quantities n_1, p_1

$$n_1 = \frac{n_0\,(1-f_0)}{f_0\,N_c}, \qquad p_1 = \frac{p_0\,f_0}{(1-f_0)\,N_v}. \tag{5b}$$

A further relation between the variables of (1) and (2) is the conservation of charge carriers:

$$\delta n + N\,\delta f = \delta p\,, \tag{6}$$

where $\delta n = n - n_0$, $\delta p = p - p_0$, and $\delta f = f - f_0$; $N = N_D/a$ is the number of acceptor states per cm^3 and N_D the number of dislocations per cm^2.

For linearizing the above equations, according to equation (4) we can use the relation

$$E^* = E_0\,f\left[3\,\ln\frac{f}{f_c} - 0.232\right] \equiv K_1\,kT\,f\,,$$

where the quantity K_1 can be considered as a constant, since E^* is relatively insensitive to the logarithmic term. So, we can write $\delta(E^*/kT) = K_1\,\delta f$.

Let us introduce the notations $x_1 = \delta n$ and $x_2 = \delta p$, then, after properly linearizing the equation system (1), (2), for the column vector $\boldsymbol{x}$ we obtain

$$-\boldsymbol{x} = \begin{pmatrix} \alpha + \gamma_n; & -\alpha \\ -\beta; & \beta + \gamma_p \end{pmatrix} \boldsymbol{x} \equiv A\,\boldsymbol{x}\,, \tag{7}$$

where

$$\alpha = C_n n_{b0} [K_1 (1 - f_0) + f_0^{-1}] , \qquad \beta = C_p p_{b0} [K_1 f_0 + (1 - f_0)^{-1}] ,$$

$$\gamma_n = C_n N (1 - f_0) \exp \left(\frac{- E_0^*}{kT} \right), \qquad \gamma_p = C_p N f_0 \exp \left(\frac{+ E_0^*}{kT} \right).$$

Denoting by T(A) and D(A) the trace and the determinant of the matrix A, respectively, the characteristic times of the recombination kinetics are

$$\tau_{1,2}^{-1} = 2^{-1} [+ \mathrm{T}(A) \pm \sqrt{\mathrm{T}^2(A) - 4\,\mathrm{D}(A)}] .$$

If the condition $4\,\mathrm{D}(A)/\mathrm{T}^2(A) \ll 1$ is valid, the approximate roots are $\tau_1 \approx \mathrm{T}^{-1}(A)$ and $\tau_2 \approx \mathrm{T}(A)/\mathrm{D}(A)$, and finally we obtain

$$\tau_1 = \frac{1}{\alpha + \beta + \gamma_n + \gamma_p}, \tag{8}$$

$$\tau_2 = \frac{\alpha + \beta + \gamma_n + \gamma_p}{\alpha \gamma_p + \beta \gamma_n + \gamma_n \gamma_p} \tag{9}$$

which can be considered as the generalized Shockley-Read-Hall lifetime for the recombination through dislocations. For $K_1 = E_0^* = 0$, the equations (8) and (9) give the well-known formulae of the SRH model. Identifying τ_2 with the measurable carrier lifetime, some important conclusions can be drawn. In fact, for low dislocation densities the lifetime τ_2 is inversely proportional to N, the density of states associated with the dislocations, in agreement with the experimental observations of papers [2], [6], and [7]. On the other hand, for a sufficiently wide region of dislocation density, τ_2 becomes independent of N, in accordance with the experimental and theoretical results of paper [14]. It is to be noted that in paper [14], because of erroneous linearization of equations (1), (2) essentially different expressions were derived.

Let us now consider the case when the lifetime is really independent of N. Neglecting the small terms in equation (9) one obtains

$$\tau_2 = \frac{f_0 \exp \left(\frac{E_0^*}{kT} \right)}{C_n n_0 [1 + f_0 (1 - f_0) K_1]} . \tag{10}$$

Taking into account the equations (5), τ_2 can be written in a more convenient form (see equation (20) of paper [14]):

$$\tau_2 = \frac{\exp \left(\frac{E_c - E_2}{kT} \right)}{C_n N_c [K_1 f_0 + (1 - f_0)^{-1}]} . \tag{11}$$

The expression (11) shows that there are two possibilities to influence the lifetime by electrical field. In the first case, the capture parameter C_n will explicitly depend on the applied electrical field E, in the second case, it is allowed to use an effective electron temperature T_e and the lifetime will depend on the field E through the electron temperature only. In the following section we shall consider these possibilities, and also some numerical estimates will be given to demonstrate the effects which are to be expected.

3. Hot-Electron Capture on Dislocations

For the sake of definiteness we shall study the capture of hot electrons on negatively charged dislocations in an n-type material. As for the mathematical treatment of the problem we shall follow the line of paper [11] where the electron capture is considered for low fields, assuming the electron tunnelling to be the rate-limiting process.

Now, it will be supposed that this condition is met. Let the sample be placed along the x-axis. The edge dislocations of density N_D are assumed to be along the z-axis. The potential barrier in the (xy) plane can be described by the following expression [12]:

$$U(r) = \frac{\alpha_0}{e} K_0\left(\frac{r}{L_D}\right), \qquad (12)$$

where

$$\alpha_0 = \frac{f e^2}{2 \pi \varepsilon a}, \qquad L_D = \left(\frac{\varepsilon k T}{4 \pi n e^2}\right)^{1/2}. \qquad (13)$$

$K_0(x)$ is the modified Bessel function of second kind and n the electron density which leads to the screening of the dislocation potential. Using equation (12) the probability of the tunnelling was determined in [11]. In the WKB approximation the capture parameter is given in the form

$$B = \int_0^\infty v_\perp \, l(v_\perp) \, f(v_\perp) \, dv_\perp, \qquad (14)$$

where $f(v_\perp)$ is the normalized distribution function of the electrons in the (xy) plane, $l(v_\perp)$ is the capture radius which could be written as

$$\left.\begin{aligned} l(v_\perp) &= v_\perp^{\mu-2} \Psi_1(v_\perp) \, e^{-2ge^{-\gamma y}}; \qquad v_{\perp 1} < v_\perp < v_{\perp 2}, \\ l(v_\perp) &= 0; \qquad v_\perp < v_{\perp 1}, \qquad v_\perp > v_{\perp 2}, \end{aligned}\right\} \qquad (15)$$

where $y = (m v_\perp^2/2 kT)$, $\gamma = kT/\alpha_0$, and $\Psi_1(v_\perp)$ is a slowly varying function of the velocity $v_\perp$; μ is a parameter close to unity; $v_{\perp 1}$, $v_{\perp 2}$ are the lower and upper limits of velocity, respectively, prescribed by conservation laws. In the further calculations the assumption $v_{\perp 1} = 0$, $v_{\perp 2} = \infty$ will be used.

The parameter g depends on the shape of the barrier, i.e. on the extent (L_D) and on the "intensity" (α_0) of the barrier:

$$g = L_D \left(\frac{2 m \alpha_0}{\hbar^2}\right)^{1/2}. \qquad (16)$$

Equation (14) shows that for the investigation of the hot-electron recombination we need the electron distribution function at high electrical fields. In what follows, the cases listed in the Introduction will be treated separately.

3.1 Acoustic phonon scattering

It is known that if the electron–acoustic phonon interaction is dominating and the electrical field is not too high, the electron distribution function can be described by the so-called Davidov function which, under the assumptions made, represents the exact solution of the Boltzmann equation (see, e.g., [19], p. 227). For an isotropic band model, using the notation $W = m v^2/2$ one obtains

$$f(W) = N\left(1 + \frac{W}{kT \alpha}\right)^\alpha \exp\left(-\frac{W}{kT}\right), \qquad (17)$$

where the dimensionless parameter $\alpha = (3\pi/16)(\mu_l E/c_s)^2$ and μ_l is the electron mobility at low field corresponding to the acoustic phonon scattering only; c_s is the sound velocity in the crystal. The normalization constant N is given by the expression [15]

$$N = (2\pi)^{-1} (2 m kT)^{-3/2} f^{-1}(\alpha); \tag{18}$$

$$f(\alpha) = \alpha^{-\alpha} \int_0^\infty x^{1/2} (x + \alpha)^\alpha e^{-x} dx = \Gamma\left(\frac{3}{2}\right) \alpha^{1/4 - \alpha/2} \exp\left(\frac{\alpha}{2}\right) W_{\lambda,\lambda'}(\alpha),$$

where $W_{\lambda,\lambda'}$ is the Whittaker function of indices $\lambda = \alpha/2 - 1/4$; $\lambda' = \alpha/2 + 3/4$; $\Gamma(x)$ is the gamma function. We have to determine the normalized distribution function in the (xy) plane, for which the condition

$$\int_0^{2\pi} \int_0^\infty f(p, \varphi)\, p\, dp\, d\varphi = 1 \tag{19}$$

is valid. (We used here polar coordinates with notation $p = m v_\perp$.)

From the equation (17) we obtain after some calculations[2])

$$f(p, \varphi) = (2\pi m kT)^{-1} A(\alpha) (y + \alpha)^{\alpha/2 - 1/4} \exp\left(-\frac{y}{2}\right) W_{\varkappa,\varkappa}(y + \alpha), \tag{20}$$

where $W_{\varkappa,\varkappa}$ is the Whittaker function of index $\varkappa = \alpha/2 + 1/4$, $y = p^2/(2 m kT)$ is the dimensionless "transverse" energy, and

$$A(\alpha) = \alpha^{-\alpha/2 - 1/4} W_{\lambda,\lambda'}(\alpha).$$

To characterize the effect of the electrical field on the capture parameter B, the ratio $C(\alpha) = B(\alpha)/B(\alpha = 0)$ will be introduced. It is convenient to use the variable y and also to introduce the notation $\nu = (\mu - 1)/2$; then we have

$$C(\alpha) = \frac{A(\alpha) \int_0^\infty y^\nu e^{-2g e^{-\gamma y} - y} (y + \alpha)^{\alpha/2 - 1/4} W_{\varkappa,\varkappa}(y + \alpha)\, dy}{\int_0^\infty y^\nu e^{-2g e^{-\gamma y} - y}\, dy}. \tag{21}$$

Because of appearance of the function $W_{\varkappa,\varkappa}$, the analytical investigation of equation (21) is very difficult, taking into consideration also the method of steepest slope used in [11]. So, in order to illustrate the effect which is to be expected, the integrals in equation (21) were calculated numerically. In Fig. 1 the function $C(\alpha)$ is plotted for $\nu = 0.5$ and $g = 12$ varying the parameter γ characterizing the ratio of the thermal energy and the "intensity" of the barrier (α_0). From the curves one can immediately see that at low temperatures or at a relatively high value of α_0 the capture probability increases very rapidly. It is interesting to study the variation of $C(\alpha)$ at a fixed value of ν and γ but changing the parameter g. This corresponds to the physical situation when the electron concentration in the band changes instead of the filling of the dislocations (Fig. 2).

---

[2]) In the calculations the following relation was used (see [20], p. 333):

$$\int_0^\infty x^{-1/2} (x + y + \alpha)^\alpha e^{-x} dx = (y + \alpha)^{\alpha/2 - 1/4} \exp\left(\frac{y + \alpha}{2}\right) \Gamma\left(\frac{1}{2}\right) W_{\varkappa,\varkappa}(y + \alpha),$$

where $\varkappa = \alpha/2 + 1/4$.

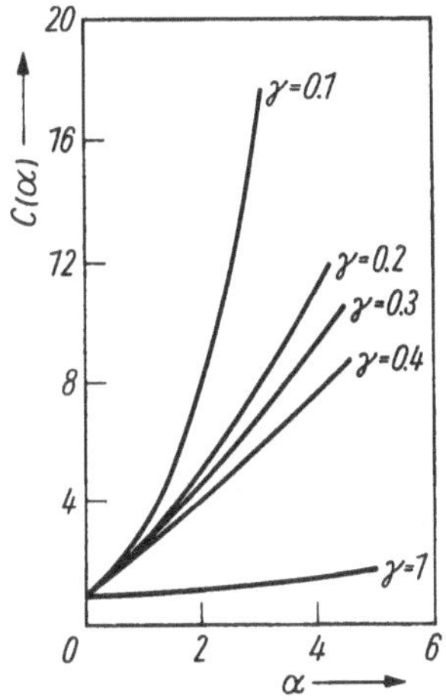

Fig. 1. The relative change of the capture parameter B. $C(\alpha) \equiv B(\alpha)/B(0)$, calculated from equation (21) using $\nu = 0.5$, $g = 12$, ($\alpha \sim E^2$)

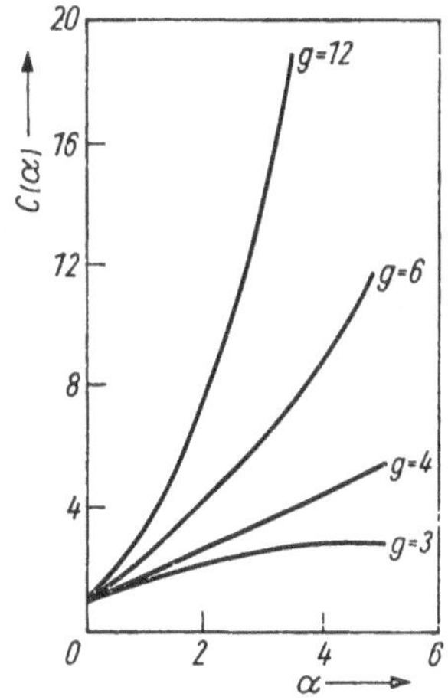

Fig. 2. The relativ change of capture parameter B using $\nu = 0.5$, $\gamma = 0.1$ in equation (21), ($\alpha \sim E^2$)

3.2 Scattering on ionized impurities

At low enough temperature the scattering of electrons on ionized impurities becomes dominant. On the other hand, to get a stationary distribution function for hot electrons, we have to take into account also the scattering on acoustic phonons, as the impurity scattering is absolutely elastic. For the case of combined acoustic phonon and impurity scattering the hot-electron distribution function was determined by Yamashita and Watanabe [21]. In the first approximation one can use the following expression [21, 16]:

$$f(W) = N' \left\{1 + \frac{1}{2}\alpha' \left(\frac{W}{kT}\right)^2 + O\left[\alpha'^2 \left(\frac{W}{kT}\right)^4\right]\right\} \exp\left(-\frac{W}{kT}\right), \tag{22}$$

where N' is the normalization constant, $\alpha' = (3\pi/16)(\mu^* E/c_s)^2$ is a dimensionless parameter with the effective mobility $\mu^* = (\frac{1}{6}\mu_l \mu_i)^{1/2}$, where μ_l and μ_i are the mobilities for the acoustic phonon and impurity scattering, respectively, but at low electrical fields. The conditions of the validity of equation (22) are $6\mu_l/\mu_i \gg 1$ and $\alpha' (W/kT)^2 \ll 1$. The normalization of $f(W)$ can simply be carried out and one obtains $N' = (1 + \frac{15}{8}\alpha')^{-1}$. The normalized distribution function in the (xy) plane can be determined, similarly to case (i). Introducing the variable $y = m v_\perp^2/(2kT)$ we obtain

$$f(p, \varphi) = N'_\perp \left[1 + \frac{\alpha'}{2}\left(y^2 + y + \frac{3}{4}\right)\right] e^{-y}, \tag{23}$$

where

$$N'_\perp = (2\pi m kT)^{-1} \left(1 + \frac{15}{8}\alpha'\right)^{-1}.$$

Now, the variation of the capture probability can be written in the form

$$C(\alpha') = \left(1 + K\frac{\alpha'}{2}\right)\left(1 + \frac{15}{8}\alpha'\right)^{-1}, \tag{24}$$

where K, independent of α', is

$$K = \frac{\int\limits_0^\infty y^\nu \left(y^2 + y + \frac{3}{4}\right) e^{-y-2ge^{-\gamma y}} dy}{\int\limits_0^\infty y^\nu e^{-y-2ge^{-\gamma y}} dy} . \tag{25}$$

In Table 1 the function $K(g, \gamma, \nu)$ is given at different values of the parameters, together with the denominator I_0 of expression (25) which is important for discussing the NDC.

Table 1

The function $K(g, \gamma, \nu)$ and its denominator I_0 defined in equation (25)

g ($\gamma = 0.1$, $\nu = 0.5$)	I_0	K	g ($\gamma = 0.1$, $\nu = 0.5$)	I_0	K
1.5	6.9×10^{-2}	8.7	6.5	4.1×10^{-5}	38.8
2.0	3.0×10^{-2}	10.0	7.0	2.4×10^{-5}	44.4
2.5	1.3×10^{-2}	11.6	7.5	1.2×10^{-5}	50.6
3.0	6.0×10^{-3}	13.4	8.0	7.1×10^{-6}	57.1
3.5	2.8×10^{-3}	15.6	8.5	4.1×10^{-6}	64.1
4.0	1.3×10^{-3}	18.2	9.0	2.5×10^{-6}	71.3
4.5	6.2×10^{-4}	21.3	9.5	1.5×10^{-6}	78.8
5.0	3.0×10^{-4}	24.9	10.0	9.4×10^{-7}	86.4
5.5	1.5×10^{-4}	29.0	10.5	6.0×10^{-7}	94.2
6.0	7.9×10^{-5}	33.6	11.0	3.9×10^{-7}	102.1
—	—	—	12.0	1.7×10^{-7}	118.1

3.3 Optical phonon scattering

It is known that the concept of electron temperature can be very useful under a set of well defined assumptions [19, 22].

Thus, in the case of combined acoustic and optical phonon interaction, the electron temperature T_e can be written as (cf. [19], p. 233)

$$kT_e = F(x_0) \frac{\hbar \omega_0}{2} + \frac{e^2 l_{op}^2 \, b \frac{m}{m_0}}{6 \, kT \left[1 + \frac{x_0 b}{2} F(x_0)\right]} E^2 . \tag{26}$$

Here $\hbar \omega_0$ is the optical phonon energy; $x_0 = \hbar \omega_0/(kT)$; $F(x_0) = (e^{x_0} + 1) \times \times (e^{x_0} - 1)^{-1}$; l_{op} is the mean free path due to the optical phonon scattering; $b = E_{1op}^2/E_1^2$, where E_{1op} and E_1 are the optical and acoustic deformation potential constants, respectively. Based on the detailed analysis by Reik and Risken [23] the above expression can be applied up to $kT_e \approx 2$ to $3 \hbar \omega_0$. This corresponds, for Ge, to fields up to 5 kV/cm. The parameters of equation (26) for Ge are: $\hbar \omega_0 = 0.037$ eV; $m/m_0 = 0.3$; $b = 0.4$; $l_{op} = 83$ Å.

4. On the Possibility of the Negative Differential Conductivity

The results of the previous section have shown that the capture parameter B increases if an external electrical field is applied to a dislocated sample. It is interesting to examine whether it is possible to observe an NDC, similarly to the case of negatively charged recombination centres. This last problem was examined by Bonch-Bruevich and Kalashnikov [16] for the cases (i) and (ii). It was shown that a negative differential conductivity can be expected only for the case (ii), but not for the case of pure acoustic phonon scattering.

Let us consider an n-type material where the current density can be written in the form $j = e\, n\, \mu\, E$. At high fields, both n and μ depend on the field strength E; thus the stationary current–voltage characteristic of the homogeneous sample is $j(E) = e\, n(E)\, \mu(E)\, E$. Thus, beside the field dependence of n one has to investigate the electron mobility in high electrical field.

Now, let us suppose that the impurity and dislocation scattering have only small influence on the determination of the distribution function, what can be expressed by equation (17). Then one can calculate the change in the mobility due to the scattering of hot electrons on the charged dislocations. It is clear by inspection that only this part of the effective mobility, $1/\mu = 1/\mu_l + 1/\mu_d$, is important, because the lattice mobility μ_l decreases with increasing electrical field.

The mobility μ_d due to scattering of electrons on the potential barrier (12) was considered both theoretically and experimentally in paper [17]. Now, the same calculation, but for the hot-electron distribution function will be given. According to the definition of mobility we can write

$$\mu_d(\alpha) = \frac{e}{kT} \frac{\int \tau_d(v_\perp)\, v_x^2\, f(v, \alpha)\, d^3\bar{v}}{\int f(v, \alpha)\, d^3\bar{v}}, \tag{27}$$

where the relaxation time for the case considered (in MKS units) is [17]

$$\tau_d(v_\perp) = \frac{8\, \varepsilon^2\, a^2\, m^2 \left(\dfrac{\hbar^2}{4\, m^2\, L_D^2} + v_\perp^2\right)^{3/2}}{N_D\, e^4\, f^2\, L_D}.$$

In the actual calculation the first term (in parentheses) can be neglected compared to the term $v_\perp^2$.

Inserting the Davidov function in this expression and carrying out the integration over variables ϑ and φ one obtains the following simple relation:

$$\mu_d(\alpha) = R(\alpha)\, \mu_d, \tag{28a}$$

where

$$\mu_d = \frac{30 \sqrt{\pi}\, \varepsilon^2\, a^2 (kT)^{3/2}}{N_D\, e^3\, f^2\, L_D\, m^{1/2}} \tag{28b}$$

is the mobility at low field, given in [17], while $R(\alpha)$ depends on the variable α only:

$$R(\alpha) = \Gamma\left(\frac{3}{2}\right) \Gamma(4)^{-1}\, J_3(\alpha)\, J_{1/2}^{-1}(\alpha), \tag{29}$$

where $J_k(\alpha) = \int_0^\infty x^k\, (1 + x/\alpha)^\alpha\, e^{-x}\, dx$. It is easy to see that $R(\alpha)$ has the following asymptotic form:

$$R(\alpha) \sim \Gamma\left(\frac{3}{2}\right) \Gamma(4)^{-1}\, \Gamma(4 + \alpha)\, \Gamma\left(\frac{3}{2} + \alpha\right)^{-1}$$

if $\alpha \ll 1$, while for $\alpha \gg 1$ we obtain

$$R(\alpha) \sim \Gamma\left(\frac{3}{2}\right) \Gamma(4)^{-1} \Gamma\left(\frac{3}{4}\right)^{-1} 2^{5/4} \alpha^{5/4} .$$

In Fig. 3 the function $R(\alpha)$ is plotted for $0 \leqq \alpha \leqq 10$. It is seen, that the mobility variation is not essential, in the range of interest. Moreover $R(\alpha)$ has a maximum at $\alpha \approx 2$, the physical interpretation of which is not clear.

In the case of combined impurity, dislocation, and lattice scattering one has to solve the Boltzmann equation to find the correct distribution function. Instead, we shall assume that the distribution function (22) can be applied also for the case when μ_{id} denote the combined impurity and dislocation scattering:

$$\frac{1}{\mu_{\mathrm{i}}} \to \frac{1}{\mu_{\mathrm{i}}} + \frac{1}{\mu_{\mathrm{d}}} \equiv \frac{1}{\mu_{\mathrm{id}}} .$$

After this remark, the field dependence of electron mobility due to dislocation can be calculated very simply, using the definition (27) and the distribution function (22):

$$\mu_{\mathrm{d}}(\alpha') = \mu_{\mathrm{d}} \frac{1 + 10\,\alpha'}{1 + \frac{15}{8}\alpha'} \equiv r(\alpha')\,\mu_{\mathrm{d}} ,$$

where, according to the above remark, the parameter α' contains besides the lattice mobility μ_{l} also the mobility μ_{id}.

Now, the possibility of NDC can be discussed. Since the case (ii) is more simple and a similar procedure is applicable for the case (i), we shall deal with the former case.

Let us assume that we have n-type material for which the condition $n \ll N f$, $N(1 - f)$ is valid; then $j(E) \sim \sigma(0)\, r(\alpha')\, C^{-1}(\alpha')\, E$. In this expression, however, the concentration dependence of the mobility (28b) was neglected, thus a more precise expression is: $j \sim \sigma(0)\, r(\alpha')\, C^{-3/2}(\alpha')\, E$ because μ_{d} is proportional to $\sqrt{n}$. Unfortunately, there is another effect to be taken into account, namely: if the carrier concentration decreases, the tunnelling parameter g increases, which leads to a drastic change in the carrier concentration, but in opposite direction. This effect can be characterized by the integral $I_0(g, \gamma, \nu)$ appearing in the denominator of the expression (25). If we assume that the parameters γ, ν remain constant, we can write

$$j(E) = \sigma(0)\, r(\alpha')\, C^{-3/2}(\alpha') \left[\frac{I_0(g)}{I_0(g_0)}\right]^{3/2} E .$$

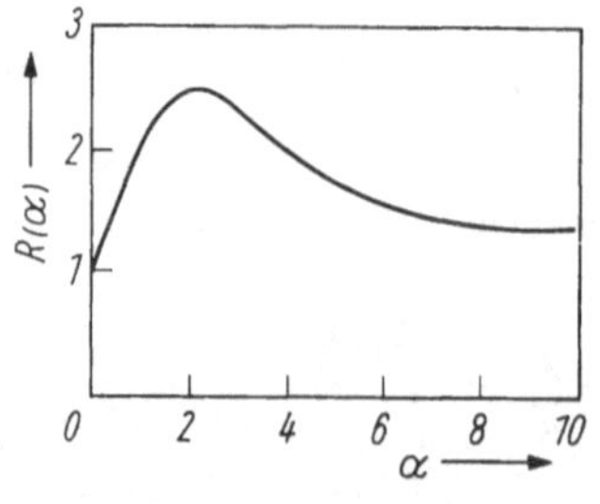

Fig. 3. Relative mobility variation $R(\alpha) \equiv \mu_{\mathrm{d}}(\alpha)/\mu_{\mathrm{d}}$ using equation (29), ($\alpha \sim E^2$)

From the analysis of this expression, we can conclude that under the conditions supposed one cannot observed NDC.

A similar conclusion can be drawn, analysing the case (i). It is interesting to note that for point-like recombination centres an NDC is possible in the case (ii) but not in the case (i), as it was demonstrated in the paper by Bonch-Bruevich and Kalashnikov [16]. It is important to emphasize that the field dependence of the mobility cannot be neglected in the case of impurity scattering, because in this case the field dependence of the capture parameter is not very strong (cf. Fig. 1 and 2 and the function (24)). In paper [16] the field dependence of mobility was neglected also in the case (ii), what can increase the value of $\alpha'_{\max.}$, where the NDC occurs, essentially.

Finally, let us examine the case (iii). If the lattice temperature is high enough, and thus the mobility is determined by the lattice scattering and at the same time, the dislocation concentration is so large that the material becomes compensated; then the carrier concentration in the band will be controlled by the barriers of dislocations. Now, the carrier lifetime is determined by expression (11). As for the electron the temperature of hot electrons can be much higher than the lattice temperature, the lifetime τ_2 will decrease, which can lead to an NDC in the stationary photoconductivity of the sample.

5. Discussion of the Results

The results of the calculations can be summarized as follows: In the first part of the paper, after properly linearizing the recombination equations the characteristic lifetimes were given at arbitrary concentration of dislocations. In Section 3, the capture of hot electrons by negatively charged dislocation lines was determined, assuming that tunnelling is the rate-limiting process. Both for acoustic and impurity scattering , the capture parameter can increase essentially with increasing electrical field. In the case of strong optical phonon scattering, the effective electron temperature rises which, under suitable conditions, can lead to the decrease of the lifetime. In Section 4, the field dependence of electron mobility was determined for cases (i) and (ii). In both cases the electron scattering on dislocations decreases, and so the mobility increases with increasing field. In our opinion this effect should be taken into consideration, also in experiments on thin silicon films, where the defect density is generally large and where a superlinear I–U characteristic has been observed in fact [24].

As for the possibility of the NDC, we concluded that in cases (i) and (ii) one can hardly expect NDC, while for case (iii), under proper conditions, such effect is quite likely.

As for the assumptions in the calculations, two points need further discussion: the applicability of potential (12) and the validity of hot–electron distribution function (22) in the presence of dislocations. As for the former problem, it seems interesting to calculate the capture parameters using the potential of a discrete charge row [9], which is bounded also for $r \to 0$. On the other hand, to give the correct hot–electron distribution function in the presence of dislocations, the Boltzmann equation has to be solved. The experimental and theoretical works [9] and [11] show that the tunnelling of electrons through the potential barrier can take place only at low temperatures. Our calculations predict for this case a large variation of capture parameter both for lattice and impurity scattering of electrons. Similarly for the case (iii) the lifetime can change markedly if the capture is barrier-limited, because in this case the lifetime is very sensitive to the electron temperature T_e (see equation (11)). Based on the works [8] and [14], both for Ge and Si one can choose an appropriate range of temperature where τ_2 decreases with increasing temperature.

The results of the calculation can be checked experimentally. It seems to be convenient to measure the mobility variation of hot electrons. Such experiments are in progress in our laboratory, using plastically deformed n-type Ge at temperatures where the scattering of electrons on dislocations is not negligible. In paper [3] the electron mobility did not show any peculiarity up to 3 kV/cm. This is, however, not surprising since the measurements were made at room temperature and according to paper [17] the effect of dislocations on carrier mobility begins below room temperature. It is possible also to look for the NDC in the photocurrent of a highly compensated dislocated sample, but it seems easier to measure the lifetime of hot electrons. These possibilities, including also the dislocations acting as surface recombination centres [25], can give new informations on the capture mechanism and on the structure of dislocations.

References

[1] S. R. MORRISON, Phys. Rev. **104**, 619 (1956).
[2] G. K. WERTHEIM and G. L. PEARSON, Phys. Rev. **107**, 694 (1957).
[3] J. B. ARTHUR, A. F. GIBSON, J. W. GRANVILLE, and E. G. S. PAIGE, Phil. Mag. **3**, 950 (1958).
[4] J. J. BROPHY, J. appl. Phys. **27**, 1383 (1956).
[5] A. D. KURTZ, S. A. KULIN, and B. L. AVERBACH, Phys. Rev. **101**, 1285 (1956).
[6] J. OKADA, J. Phys. Soc. Japan **12**, 1338 (1957).
[7] H. LEMKE, phys. stat. sol. **12**, 125 (1965).
[8] T. FIGIELSKI, phys. stat. sol. **6**, 429 (1964).
[9] T. FIGIELSKI, phys. stat. sol. **9**, 555 (1965).
[10] V. L. BONCH-BRUEVICH, Fiz. tverd. Tela (Sbornik II) 182 (1959).
[11] YU. V. GULYAEV, Fiz. tverd. Tela **4**, 1285 (1962).
[12] V. L. BONCH-BRUEVICH and V. B. GLASKO, Fiz. tverd. Tela **3**, 36 (1961).
[13] Z. GOŁACKI, T. FIGIELSKI, and M. JASTRZĘBSKA, phys. stat. sol. **11**, K35 (1965).
[14] R. H. GLAENZER and A. G. JORDAN, Solid State Electronics **12**, 247 (1969).
[15] V. L. BONCH-BRUEVICH, Fiz. tverd. Tela **6**, 2047 (1964).
[16] V. L. BONCH-BRUEVICH and S. G. KALASHNIKOV, Fiz. tverd. Tela **7**, 750 (1965).
[17] B. PÖDÖR, Acta phys. Hungar. **23**, 393 (1967).
[18] W. T. READ, Phil. Mag. J. Sci. **45**, 775 (1954).
[19] E. M. CONWELL, High Field Transport in Semiconductors, Academic Press, New York/London 1967.
[20] I. S. GRADSHTEIN and I. M. RIZHIK, Tablitsy integralov, summ Izd. Fiz.-Mat. Lit., Moscow 1963.
[21] J. YAMASHITA and M. WATANABE, Progr. theor. Phys. (Kyoto) **12**, 443 (1954).
[22] R. STRATTON, Proc. Phys. Soc. **A242**, 355 (1957).
[23] H. G. REIK and H. RISKEN, Phys. Rev. **124**, 777 (1961).
[24] E. C. ROSS and G. WARFIELD, J. appl. Phys. **41**, 2657 (1970).
[25] M. PAJDZINSKA, A. MORAWSKI, and J. LAGOWSKI, Surface Sci. **18**, 446 (1969).

(Received August 24, 1971)

phys. stat. sol. (a) 8, 423 (1971)

Subject classification: 10.1; 21; 21.3

Sektion Metallurgie und Werkstofftechnik, Bergakademie Freiberg

Zur Gleitung und Zwillingsbildung von Magnesium- und Magnesium–Kadmium-Vielkristallen bei Raumtemperatur und 77 °K[1])

Von

B. Reh und G. Hötzsch

An Hand von Oberflächenuntersuchungen röntgenographisch orientierter Einzelkristallite wurden die Verformungsvorgänge in vielkristallinem Magnesium und Magnesium mit 10 At% Kadmium untersucht. Die plastische Verformung beider Materialien erfolgt bei 77 °K und RT hauptsächlich durch Basisgleitung und $\{10\bar{1}2\}$-Zwillingsbildung.

The deformation mechanisms of polycrystalline magnesium and magnesium with 10 at% cadmium were investigated using metallographic surface observations. The orientation of individual grains of these materials were determined by a microbeam Laue X-ray technique. Basal slip and $\{10\bar{1}2\}$ twinning were found to be the main mechanisms of plastic deformation at 77 °K and room temperature.

1. Einleitung

In den letzten Jahren ist das Interesse an der Erforschung der Verfestigung der Vielkristalle stärker in den Vordergrund getreten. Das betrifft insbesondere die k.f.z.-Metalle und -Legierungen, die sich durch eine große Zahl gleichwertiger Deformationssysteme verformen können. Demgegenüber existieren wesentlich weniger systematische Arbeiten über die Verfestigung hexagonaler Vielkristalle, bei denen die Geometrie der Verformung sehr komplex ist.

Zur plastischen Verformung eines Vielkristalls sind nach von Mises fünf unabhängige Gleitsysteme erforderlich [1]. In Tabelle 1 sind die im hexagonalen Gitter möglichen Gleitsysteme zusammengestellt [2]. Die Systeme 1 bis 3 besitzen jeweils die Gleitrichtung $\langle 11\bar{2}0\rangle$.

Es ist zu erkennen, daß die Systeme 1 und 2 zusammen dem System 3 äquivalent sind, d. h. vier unabhängige Gleitsysteme ergeben und somit das von Mises-Kriterium nicht erfüllen, da eine Längenänderung parallel zur c-Achse nicht möglich ist. Lediglich die Betätigung des Systems 4 mit der Gleitrichtung $\langle 11\bar{2}3\rangle$ würde eine Vielkristalldeformation im Sinne des von Mises-Kriteriums ermöglichen.

Die Betrachtung der in Magnesium-Vielkristallen beobachteten Gleitsysteme (Tabelle 2 in [3]) zeigt jedoch, daß nur das Basisgleitsystem, das Prismengleitsystem I. Art und das Pyramidengleitsystem I. Art, 1. Ordnung im interessierenden Temperaturbereich gefunden wurden, wobei noch die Kristallorientierung und der Legierungsgehalt eine Rolle spielen.

Die plastische Deformation von Magnesium-Vielkristallen ist also durch Gleitung allein nicht möglich, vielmehr müssen Zwillingsbildung und Knickung sowie auch Korngrenzengleitung [4, 7] mit zur Anpassung der Verformung an den Korngrenzen beitragen. Dieser Sachverhalt läßt die besondere Problematik der Verformung hexagonaler Vielkristalle erkennen.

[1]) Vortrag gehalten auf der 5. Jahrestagung der DVK in Leipzig 1970.

Tabelle 1

Gleitsysteme im hexagonalen Gitter [2]

Gleit-system	Burgers-Vektor	Gleit-richtung	Gleitebene	Anzahl der Gleitsysteme insges.	unabh.
1	***a***	⟨11$\bar{2}$0⟩	Basisebene (0001)	3	2
2	***a***	⟨11$\bar{2}$0⟩	Prismenebene I. Art {10$\bar{1}$0}	3	2
3	***a***	⟨11$\bar{2}$0⟩	Pyramidenebene I. Art, 1. Ordnung {10$\bar{1}$1}	6	4
4	***c*** + ***a***	⟨11$\bar{2}$3⟩	Pyramidenebene II. Art, 2. Ordnung {11$\bar{2}$2}	6	5
5	***c***	⟨0001⟩	Prismenebene I. Art {10$\bar{1}$0}	3	2
6	***c***	⟨0001⟩	Prismenebene II. Art {11$\bar{2}$0}	3	2

In Tabelle 3 sind die Zwillingssysteme einiger hexagonaler Metalle mit ihren kristallographischen Elementen dargestellt. Bei Magnesium erfolgt die mechanische Zwillingsbildung vorwiegend auf den Ebenen {10$\bar{1}$2}. Damit ergibt sich eine Dehnungskomponente parallel zur c-Achse, wodurch eine Vielkristalldeformation in Verbindung mit den experimentell gefundenen Gleitsystemen möglich wird.

Bisher wurde die parabolische Verfestigungskurve als typisch für den Verfestigungsverlauf von Magnesium angesehen [4 bis 6]. Das liegt einfach daran, daß die Kristallorientierungen von Proben aus einem Blech bzw. Draht ähnlich sind in bezug auf die Beanspruchungsrichtung. Das Bild ändert sich grundlegend, wenn die Proben z. B. mit ihrer Zugachse in Radialrichtung aus einem Preßstrang entnommen werden. In Fig. 1 sind die Verfestigungskurven von Längs- und Radialproben von stranggepreßtem Magnesium dargestellt [3]. Die Spannungs–Dehnungs-Kurve der Längsproben besitzt die typische para-

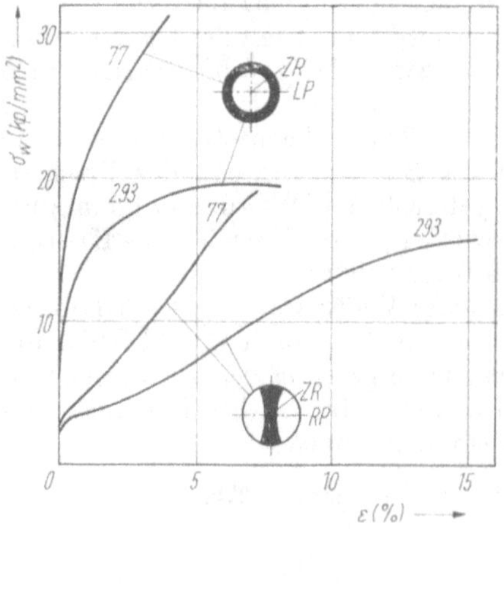

Fig. 1. Verfestigungskurven von vielkristallinem Magnesium sowie Basispolverteilungen von Längsproben (LP) und Radialproben (RP); ZR Zugrichtung

Tabelle 2

Experimentell beobachtete Gleitsysteme einiger hexagonaler Metalle (ref. in [3]); χ Winkel zwischen Basisebene und Verformungsrichtung

Metall	Gleitsystem	Verformungsbedingungen Temperatur	Korngröße	Textur	Anmerkungen
Kadmium	$(0001)\ [1\bar{2}10]$	Zug, 77 bis 367 °K	20 bis 200 μm	$\chi = 0°$ bis 40°	bei allen Temp. domin.
	$(11\bar{2}2)\ [\bar{1}123]$				bevorzugt bei tiefen Temp., kleinem Winkel χ in Korngrenzennähe
Magnesium	$(0001)\ [1\bar{2}10]$	Zug, 77 bis 553 °K	50 μm bis 1 mm	$\chi = 0°$ bis 15°	domin. bei allen Temp.
	$(10\bar{1}0)\ [1\bar{2}10]$				bevorzugt in Korngrenzennähe und bei Temp. unterhalb 293 °K
	$(10\bar{1}1)\ [1\bar{2}10]$				ab 473 °K auftretend
Zirkonium	$(10\bar{1}0)\ [1\bar{2}10]$	77 bis 1075 °K	Einkristall; Vielkristall ohne Angabe	$\chi = 0°$ bis 60° / $\chi = 0°$ bis 20°	domin. bei allen Temp.
	$(10\bar{1}1)\ [1\bar{2}10]$				auf Grund von Texturuntersuchungen an Blech
	$(0001)\ [1\bar{2}10]$				elektronenmikroskopisch bei sehr reinem Zirkonium als Nebengleitsystem

bolische Form, während die der Radialproben drei Bereiche mit einem linearen Verfestigungsanstieg aufweist. Zur Interpretation dieses unterschiedlichen Verfestigungsverlaufes ist die Kenntnis der Deformationsmechanismen von primärer Bedeutung. Aus diesem Grunde wurden mit Hilfe lichtoptischer Oberflächenuntersuchungen die Gleit- und Zwillingssysteme von Magnesium- und Magnesium-Kadmium-Vielkristallen bestimmt.

Vergleichende Untersuchungen wurden bisher nur von Hauser et al. [4, 7] an grobkörnigem Material (ca. 1 mm Korndurchmesser) ohne genügende Beachtung des Textureinflusses durchgeführt.

2. Experimente und Auswertung

Als Probenmaterial diente stranggepreßtes technisch reines Magnesium sowie Magnesium mit 10 At% Cd. Daraus wurden Flachzugproben mit 20 mm Meßlänge und einem Querschnitt von $3 \times 2\ mm^2$ als Längsproben parallel und

Tabelle 3

Kristallographische Elemente der wichtigsten Zwillingssysteme einiger hexagonaler Metalle (ref. in [3])

Metall	Zwillings-ebene (1. Kreis-schnittebene) (K_1)	Richtung der Schiebung (η_1)	2. Kreis-schnitt-ebene (K_2)	Richtung der Schnittlinie der Schie-bungsebene mit K_2 (η_2)	Ebene der Schiebung senkrecht zu K_1 und K_2	Größe der Schiebung
Cd	$\{10\bar{1}2\}$	$\langle 10\bar{1}\bar{1}\rangle$	$\{\bar{1}012\}$	$\langle 10\bar{1}1\rangle$	$\{1\bar{2}10\}$	0,171
Zn						0,139
Mg						0,131
Ti						0,167
Be						0,199
Ti	$\{11\bar{2}1\}$	$\langle 11\bar{2}\bar{6}\rangle$	(0001)	$\langle 11\bar{2}0\rangle$	$\{\bar{1}100\}$	0,638
	$\{11\bar{2}2\}$	$\langle 11\bar{2}\bar{3}\rangle$	$\{11\bar{2}4\}$	$\langle 22\bar{4}3\rangle$	$\{\bar{1}100\}$	0,225

als Radialproben senkrecht zur Preßrichtung dem Strang entnommen. Die Längsproben besitzen die Ringfasertextur der Preßstange, d. h., die Basisnormalen streuen $\pm 20°$ um den Äquator der Lagenkugel (siehe Fig. 1). Die Basisnormalen der Radialproben liegen im Bereich von 0° bis 90° zur Zugrichtung.

Mit Hilfe eines Feinstrahl-Laue-Verfahrens wurden Einzelkristallite des Probenmaterials, das einen mittleren Korndurchmesser von 200 μm besaß, röntgenographisch orientiert. Die gesamte Anordnung bestand aus dem Blendenhalter mit der Glaskapillare, die als Totalreflexionskollimator zur Erzeugung eines Primärstrahles von ca. 80 bis 100 μm ∅ diente, sowie einer konventionellen Laue-Kammer und dem Probenhalter.

Die polierten Proben wurden im dynamischen Zugversuch plastisch verformt und danach die Oberfläche im Interferenzkontrast eines Reichert-Mikroskops MeF untersucht. Die Indizierung der von kristallographischen Ebenen stammenden Oberflächenspuren wurde mittels stereographischer Ein- und Zweiflächenanalyse durchgeführt. Ein Beispiel für eine Zweiflächenanalyse zeigt Fig. 2.

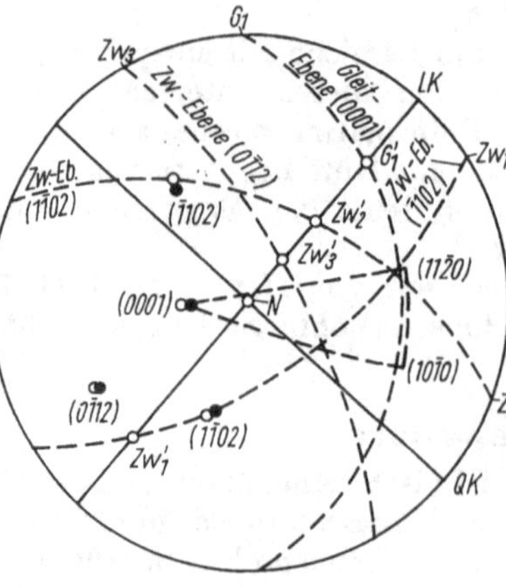

Fig. 2. Stereographische Zweiflächenanalyse; Magnesium-Längsprobe

Es wurden hierbei die Gleitebene (0001) sowie drei Zwillingsebenen (0$\bar{1}$12), (1$\bar{1}$02) und ($\bar{1}$102) gefunden. Die vollen Kreise stellen die röntgenographisch ermittelten Flächenpole dar. Ihre nur unwesentliche Abweichung von der Lage der stereographisch bestimmten Flächenpole demonstriert die gute Reproduzierbarkeit der Methode.

3. Ergebnisse und Diskussion

3.1 Basisgleitung

Als dominierendes Gleitsystem bei der plastischen Verformung der bei 77 und 293 °K untersuchten Proben tritt das Basissystem in Erscheinung. Dieser Sachverhalt ist auch nach theoretischen Vorstellungen zu erwarten [2, 13 bis 15]. Die Gleitbänder des Basissystems zeichnen sich dadurch aus, daß sie besonders geradlinig ausgebildet sind (Fig. 3 und 4). Vergleicht man den Einfluß von Textur und Legierungsgehalt bei 293 °K, so zeichnet sich ab, daß die Radial- gegenüber den Längsproben und die Magnesium–Kadmium- gegenüber den Magnesiumproben eine ausgiebigere Basisgleitung aufweisen.

3.2 Prismengleitung

Das Prismensystem I. Art ergab sich als wichtigstes Nichtbasisgleitsystem. Es kann in zwei Erscheinungsformen auftreten [7]:

1. Als „unabhängiges" Prismengleitsystem, wobei Richtung und Betrag der Gleitung vom Hauptgleitsystem (0001) unabhängig sind;

2. als „pencil slip", d. h., es tritt gleichzeitig Gleitung auf Prismen- und Basisebene in der gleichen Richtung um den gleichen Betrag auf.

Charakteristisch für die unabhängige Prismengleitung sind gewellte Gleitbänder, die je nach Lage der Gleitrichtung zur Kornoberfläche mehr oder weniger grob erscheinen (Fig. 4). In Fig. 5 ist die relative Häufigkeit des Prismensystems dargestellt. Das Säulendiagramm zeigt deutlich, daß die Prismengleitung in Magnesium bei tiefen Temperaturen häufiger auftritt. Diese

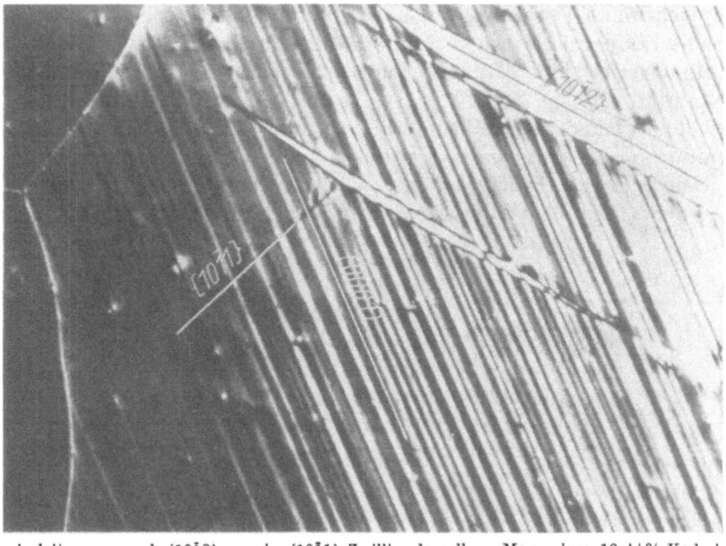

Fig. 3. Basisgleitspuren und {10$\bar{1}$2}- sowie {10$\bar{1}$1}-Zwillingslamellen; Magnesium–10 At% Kadmium, Radialprobe, T = 77 °K, ε = 0,3% (250×)

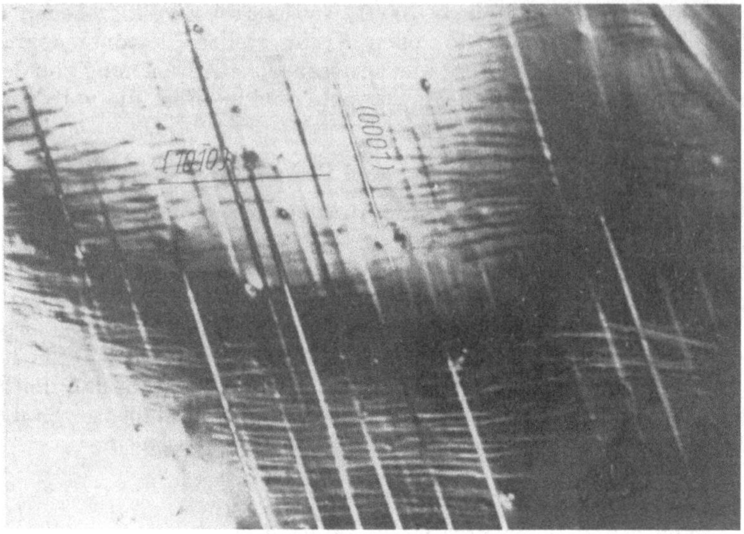

Fig. 4. Spuren von Basis- und Prismengleitung (gewellt); Magnesium–10 At% Kadmium, Längsprobe, $T = 293$ °K, $\varepsilon = 6{,}9\%$ (250 ×)

auch in [7] beschriebene Tendenz wird mit lokalen Spannungskonzentrationen in Verbindung gebracht, da relativ hohe Schubspannungen zur Aktivierung dieses Gleitsystems benötigt werden [8 bis 11]. Tatsächlich wurde die unabhängige Prismengleitung meist nur in Kornecken und in Korngrenzennähe beobachtet. Es wäre außerdem denkbar, daß die mit abnehmender Temperatur sich verringernde Korngrenzengleitung durch diesen Mechanismus ersetzt wird.

„Pencil slip“ (Fig. 6), die zweite Form der Prismengleitung, äußert sich meist in treppenförmig verlaufenden, kräftigen Gleitbändern, die oft auch einen makroskopisch-nichtkristallographischen Verlauf vortäuschen. Bei geeigneter Vergrößerung können aber die einzelnen Prismen- und Basissegmente des Gleitbandes aufgelöst werden. Zum Mechanismus dieser Gleiterscheinung ist sehr wenig bekannt. Da sie bei Einkristallen selten auftritt, kann es sich um einen typischen Vielkristalleffekt handeln. “Pencil slip“ ist für die Tieftemperaturverformung der Magnesium–Kadmium-Längsproben von erheblicher Bedeutung. Die relative Häufigkeit ist bei tiefen Temperaturen und bei Längsproben größer (Fig. 7). Zulegieren von Kadmium begünstigt „pencil slip“ ebenfalls.

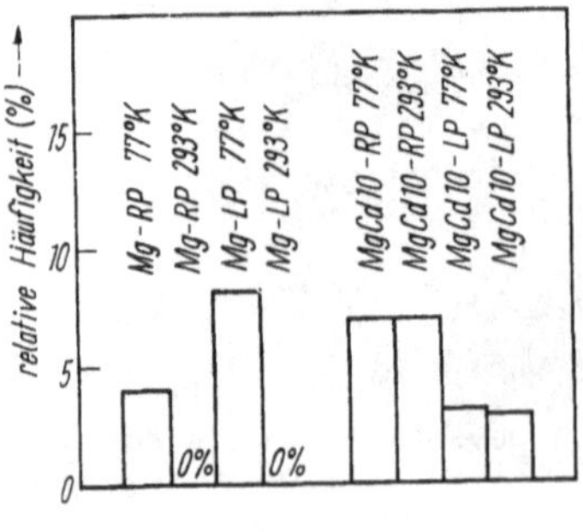

Fig. 5. Relative Häufigkeit der Prismengleitung $\{10\bar{1}0\}$ $\langle 11\bar{2}0\rangle$

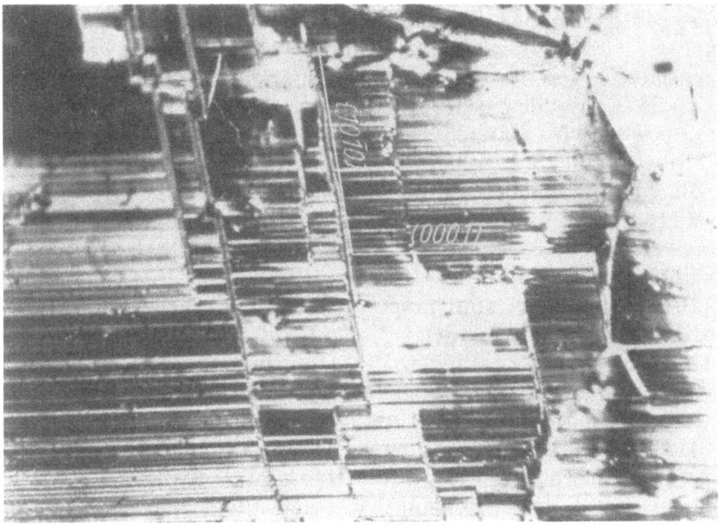

Fig. 6. Spuren von „pencil slip“ und Basisgleitung; Magnesium–10 At% Kadmium, Längsprobe, $T = 77\,°K$, $\varepsilon = 1{,}3\%$ (250 ×)

Bei den Verformungsexperimenten waren in den Proben, die „pencil slip“ zeigten, deutliche Geräusche zu hören, die mit Zacken im Kraft–Verlängerungs-Diagramm verbunden waren. Eine Verwechslung mit mechanischer Zwillingsbildung ist kaum möglich, da das Diagramm bei alleinigem Auftreten von Zwillingen und Basisgleitung keine Zacken aufweist. Deshalb sollte diese Beobachtung den gekoppelten Gleitvorgängen auf Prismen- und Basisebene zugeschrieben werden. Als weitere Gleitebenen wurden — allerdings in geringem Maße — Pyramidenebenen I. Art, 1. Ordnung und II. Art, 2. Ordnung gefunden.

3.3 Zwillingsbildung

Die untersuchten Radialproben zeigten eine recht erhebliche mechanische Zwillingsbildung, da diese bei der Erfüllung der Kompatibilitätsbedingungen an den Korngrenzen eine wesentliche Rolle spielt.

Das Auftreten der Zwillingssysteme in einem bestimmten Kristallgitter hängt wie bei der Gleitung von energetischen und geometrischen Kriterien ab. Es

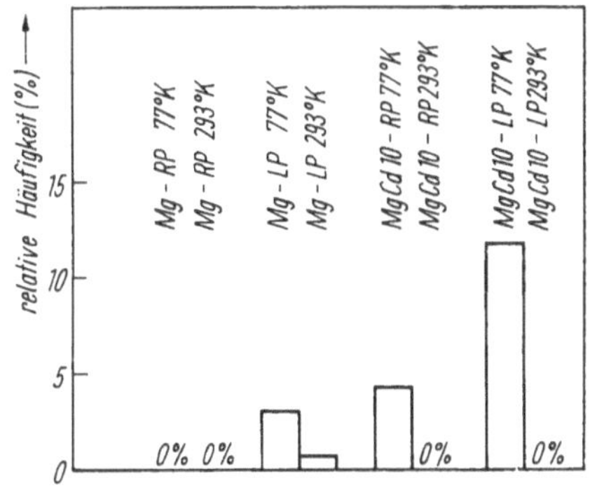

Fig. 7. Relative Häufigkeit von „pencil slip“ (0001) {10$\bar{1}$0}

sollte im wesentlichen das System beobachtet werden, das den geringsten Schiebungsbetrag und den einfachsten „shuffle"-Mechanismus besitzt [16]. Darüber hinaus spielt wegen der Orientierungsabhängigkeit der mechanischen Zwillingsbildung die Kristallorientierung bzw. Textur eine entscheidende Rolle.

Als dominierend wird danach für Magnesium das Zwillingssystem {10Ī2} ⟨Ī011⟩ erwartet, was auch experimentell bestätigt wird [4 bis 7, 16 bis 20]. Im Oberflächenbild erscheinen die Zwillinge dieses Systems als linsenförmige Lamellen (Fig. 3), wobei die relativ große Breite auf die kleine Schiebung des Systems hinweist. Fig. 8 zeigt, daß die Magnesium-Radialproben die meisten {10Ī2}-Zwillinge aufweisen.

Außer {10Ī2}-Zwillingen sind noch {10Ī1}- (Fig. 3), {10Ī3}-, {11$\bar{2}$2}- und {11$\bar{2}$4}-Zwillinge gefunden worden. Nach Reed-Hill [21] handelt es sich bei den {10Ī1}- und {10Ī3}-Zwillingen um Doppelzwillinge. Fig. 9 enthält alle beobachteten Gleit- und Zwillingsebenen der Magnesium-Radialproben für beide Versuchstemperaturen. Dabei wird die dominierende Rolle von Basisgleitung und {10Ī2}-Zwillingsbildung deutlich. Die im wesentlichen gleiche Aussage trifft auch für die Magnesium–Kadmium-Radialproben zu.

Die Kenntnis der Deformationsmechanismen allein genügt jedoch nicht zur Interpretation der unterschiedlichen Verläufe der Spannungs–Dehnungs-Kurven der beiden Probenarten.

Da die mechanische Zwillingsbildung in diesem Zusammenhang eine entscheidende Rolle zu spielen scheint, wurde der Beitrag der Zwillingsbildung zur Gesamtdeformation quantitativ aus der Flächendichte der Zwillinge ermittelt. Aus dieser Größe konnte unter Berücksichtigung eines mittleren Orientierungsfaktors die Zwillingsdehnung berechnet werden [22].

In Fig. 10 sind die Anteile der Zwillingsdehnung an der plastischen Gesamtdehnung $\varepsilon_z/\varepsilon_{ges.}$ über der Gesamtdehnung aufgetragen und den Verfestigungskurven der Radial- und Längsproben von Magnesium–Kadmium zugeordnet worden. Der Quotient $\varepsilon_z/\varepsilon_{ges.}$ verläuft bei der Radialprobe über ein Maximum, das annähernd mit dem Beginn des linearen Verfestigungsanstieges korrespondiert. Ähnliche Verhältnisse treffen wir auch bei Kadmium-Vielkristallen für den Fall an, daß intensive Zwillingsbildung die Verformung einleitet [23, 24].

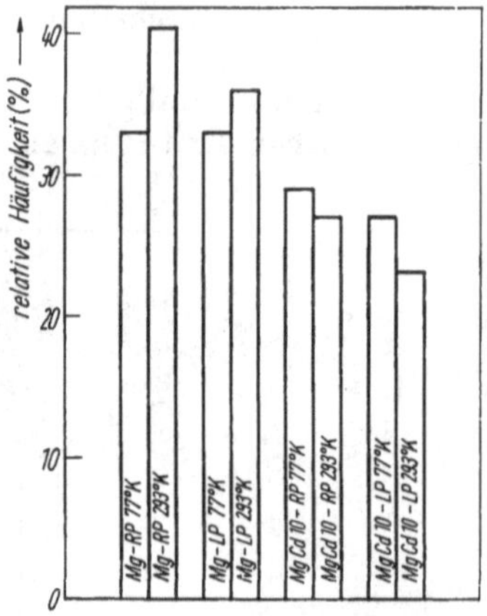

Fig. 8. Relative Häufigkeit des {10Ī2} ⟨Ī011⟩-Zwillingssystems

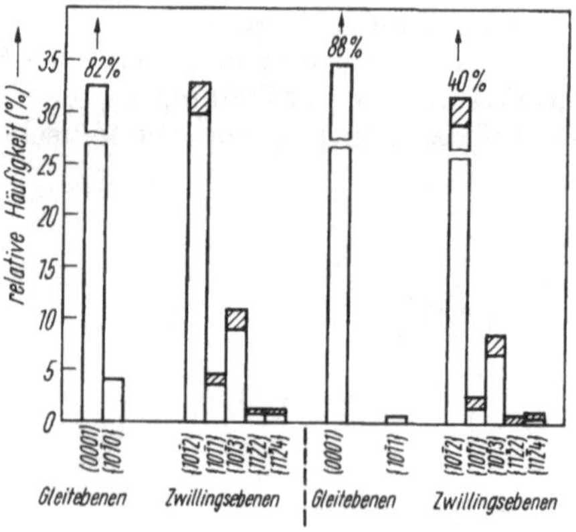

Fig. 9. Relative Häufigkeit der beobachteten Gleit- und Zwillingsebenen von Magnesium-Radialproben; schraffiert — unsichere Indizierungen; links bei 77 °K, rechts bei 293 °K

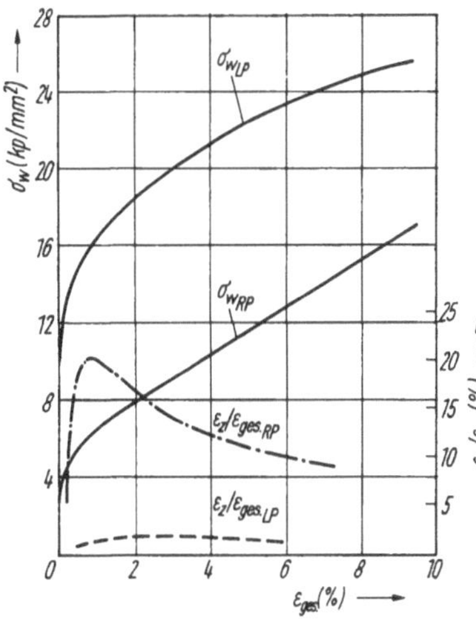

Fig. 10. Verfestigungskurven und Anteile der Zwillingsdehnung $\varepsilon_z/\varepsilon_{ges.}$ von Magnesium–10 At% Kadmium bei Raumtemperatur (293 °K)

Weiterhin ist es interessant festzustellen, daß der Anteil der Zwillingsdehnung bei den Längsproben eine Größenordnung kleiner als bei den Radialproben ist. Dieser Befund wird auf Grund der Orientierungsabhängigkeit der Zwillingsbildung und der vorliegenden Ringfasertextur auch für das bevorzugte System $\{10\bar{1}2\}$ erwartet.

4. Zusammenfassung

Die plastische Verformung von Magnesium- und Magnesium–10 At% Kadmium-Vielkristallen erfolgt bei 77 und 293 °K hauptsächlich durch Basisgleitung und $\{10\bar{1}2\}$-Zwillingsbildung. Weitere Deformationsmechanismen sind: Prismengleitung, „pencil slip", Pyramidengleitung auf $\{10\bar{1}1\}$- und $\{11\bar{2}2\}$- sowie Zwillingsbildung auf $\{10\bar{1}1\}$-, $\{10\bar{1}3\}$-, $\{11\bar{2}2\}$- und $\{11\bar{2}4\}$-Ebenen.

Die unterschiedliche Form der Verfestigungskurven von Radial- und Längsproben kann neben der besseren Möglichkeit der Radialproben zur Basisgleitung auf die erhebliche Zwillingsbildung in den Radialproben zurückgeführt werden.

Die Verfasser möchten Herrn Prof. Dr. Günther für das fördernde Interesse danken, das er der Arbeit stets entgegengebracht hat.

Literatur

[1] R. von Mises, Z. angew. Math. Mech. **8**, 161 (1928).
[2] P. G. Partridge, Metals Rev. **118**, 169 (1967).
[3] P. Männel, Dissertation, Bergakademie Freiberg, 1970.
[4] F. E. Hauser, C. D. Starr, L. Tietz und J. E. Dorn, Trans. ASM **47**, 102 (1955).
[5] F. E. Hauser, P. R. Landon und J. E. Dorn, Trans. ASM **50**, 856 (1958).
[6] C. S. Roberts, Magnesium and Its Alloys, Wiley, 1960.
[7] F. E. Hauser, P. R. Landon und J. E. Dorn, Trans. ASM **48**, 986 (1956).
[8] P. W. Flynn, J. Mote und J. E. Dorn, Trans. MS AIME **221**, 1148 (1961).
[9] H. Yoshinaga und R. Horiuchi, Trans. Japan Inst. Metals **5**, 14 (1963).
[10] A. Ahmadieh, J. Mitchell und J. E. Dorn, Trans. MS AIME **233**, 1130 (1965).
[11] A. Akhtar, Ph.D. Thesis, University of British Columbia, 1968.
[12] C. S. Barrett und T. B. Massalski, Structure of Metals, McGraw-Hill Publ. Co., 1966.
[13] M. H. Yoo und C. T. Wei, J. appl. Phys. **38**, 4317 (1967).
[14] R. B. Roy, Phil. Mag. **15**, 477 (1967).
[15] D. Rühlicke, Metall **23**, 130 (1969).
[16] J. W. Christian, The Theory of Transformations in Metals and Alloys, Pergamon Press, 1965.

[17] E. Schiebold und G. Siebel, Z. Phys. **69**, 458 (1931).
[18] E. Schmid, Z. Elektrochem. **37**, 450 (1931).
[19] C. H. Mathewson und A. J. Philips, AIME Tech. Publ. **53**, 445 (1928).
[20] P. A. Bakarian und C. H. Mathewson, Trans. AIME **152**, 226 (1943).
[21] R. E. Reed-Hill, Trans. MS AIME **218**, 554 (1960).
[22] B. Reh, unveröffentlicht.
[23] P. Männel und G. Hötzsch, Metall **24**, 943 (1970).
[24] N. R. Risebrough und E. Teghtsoonian, Canad. J. Phys. **45**, 591 (1967).

(Received September 27, 1971)

phys. stat. sol. (a) 8, 433 (1971)

Subject classification: 7 and 14.4.1; 22.1.3

Groupe de Physique des Solides, Ecole Normale Supérieure, Paris

Study of the Acoustoelectric Amplification in Tellurium with Epitaxially Grown Selenium Transducers

By

J. BARBOT, M. JOVENIAUX, and J. M. THUILLIER

The amplification of longitudinal acoustic waves in tellurium at 77 °K has been investigated in the frequency range 10 to 500 MHz. Both electric current and acoustic propagation were along the binary axis. Selenium epitaxial transducers were used to avoid bonding problems. The measured insertion losses agreed fairly well with theoretical calculations. The attenuation versus drift field is interpreted for low drift fields using the classical theory of Hutson and White. With respect to the known constants, a gain is observed higher than expected. This casts some doubt on the values of the electro-mechanical coupling constant and the electrical conductivity at high frequencies.

On a étudié l'amplification des ondes acoustiques longitudinales dans le tellure à 77 °K, pour la gamme des fréquences s'étendant de 10 à 500 MHz. Le courant électrique et le vecteur d'onde acoustique étaient orientés selon un axe binaire. Pour s'affranchir des problèmes de collage, on a utilisé des transducteurs de sélénium épitaxiés, dont les pertes d'insertion mesurées étaient en bon accord avec le calcul théorique. L'atténuation en fonction du champ électrique a été interprétée aux faibles champs à l'aide de la théorie classique de Hutson et White. Par rapport aux constantes admises, on observe un gain supérieur à ce qui est attendu. Ceci nous amène à douter des valeurs du coefficient de couplage et de la conductivité électrique à haute fréquence.

1. Introduction

The amplification of bulk acoustic waves in tellurium was studied at 77 °K by Ishiguro and Tanaka [1] along the Y-axis at 45 MHz, and more recently by Arlt and Reiber [2] along the Z-axis at 30 MHz. These authors used quartz or ceramic platelets for the acoustic excitation. The low conversion efficiency and small bandwidth of these bonded transducers are serious limitations in the experiments. In the present study, we used epitaxial selenium transducers, as described by Shiosaki et al. [3], to avoid bonding difficulties. We investigated the acoustic interaction of holes in a wide frequency range at 77 °K along the binary axis. For this direction the phase and the group velocity vectors of longitudinal waves are collinear.

2. Experimental Methods

The samples are parallelepipeds, the transverse dimensions are 4×4 mm^2 and the length ranging from 6 to 10 mm is along the binary axis. Both end faces are carefully polished and etched. The selenium transducers have a diameter of 1 mm and a thickness ranging from 2 to 30 μm. They are epitaxially grown in a vacuum of 10^{-6} Torr, the substrate temperature being regulated at 130 °C. Gold contacts are finally evaporated on the deposited selenium and on the surrounding tellurium end face (Fig. 1). As these contacts are not made on the whole surface of the ends of the bar, a voltage drop can exist in the active region

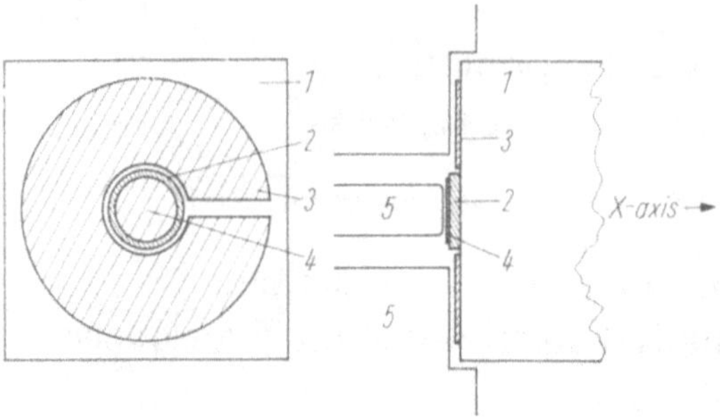

Fig. 1. End view and longitudinal cut of the transducer structure. 1 Tellurium bar, 2 Selenium transducer, 3 Gold ring contact on tellurium, 4 Gold centre contact on the transducer, 5 Coaxial line for r.f. excitation or detection

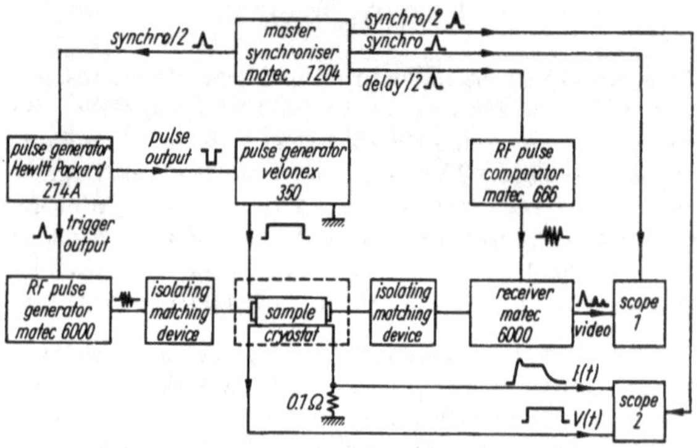

Fig. 2. Experimental set-up for attenuation measurements versus drift field

under the transducer. So we measured the voltage between the gold contact on tellurium and the centre contact on selenium with a high impedance probe, when the drift field is applied to the ends of the bar. This voltage drop was found to be only a few per cent of the total voltage, so the resulting disturbance is not serious.

The experimental set-up is classical, the main apparatus being a "Matec" RF generator and receiver for acoustic excitation and detection, and a "Velonex" pulse generator for the drift voltage (Fig. 2). Care was taken to maintain the characteristic impedance of the r.f. coaxial line up to the transducer.

3. Results and Discussion

Each transducer has an untuned conversion efficiency of 17 dB at optimal conditions as compared with the computed value of 14 dB (Fig. 3). The relative bandwidth is more than 100%, but measurements are still possible in a larger frequency range. Typical results for the variations of attenuation with drift field are shown on Fig. 4. In this case the measurements could be performed

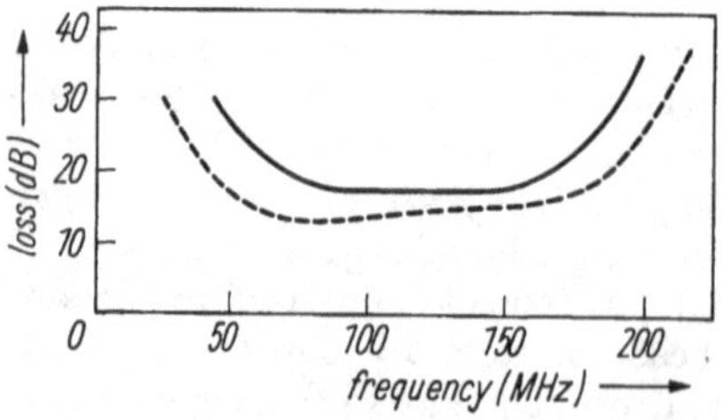

Fig. 3. Experimental (solid curve) and theoretical (dashed curve) insertion loss for one transducer. $\varnothing = 0.9$ mm, thickness $= 8\ \mu m$, $Z_0 = 25\ \Omega$

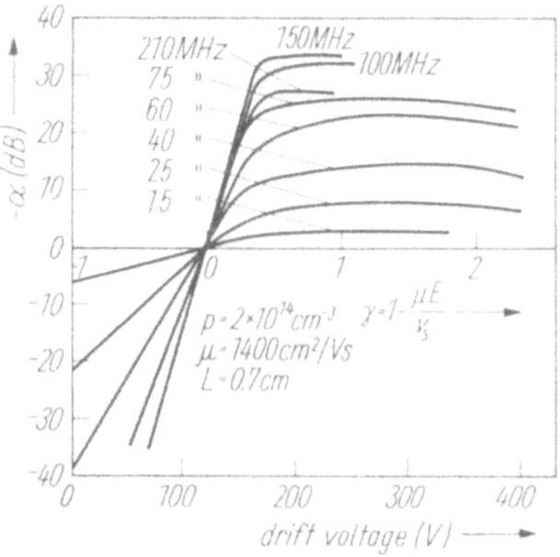

Fig. 4. Variation of acoustic attenuation versus drift voltage (sample X 104)

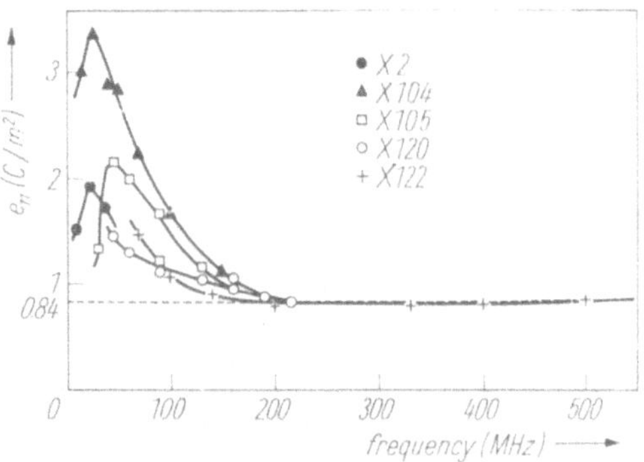

Fig. 5. Variations of e_{11} with frequency for various samples

between 15 and 210 MHz on the same sample. We first notice that the linear region at fields higher than the critical one E_c, for which the carriers become supersonic, is very narrow. Strong saturation effects occur for $E > 1.2\,E_c$, and are associated with a noise build-up and non-ohmic I versus U characteristics. This arises from the very important amplification expected in the GHz range.

We attempted an interpretation of our results in the so-called linear region. According to Hutson and White [4, 5], the attenuation coefficient is

$$\alpha = \frac{K^2\,\omega_c}{2\,v_s}\,\frac{\gamma}{\gamma^2 + \left(\frac{\omega_c}{\omega} + \frac{\omega}{\omega_D}\right)^2}$$

with the notations of these authors. In our case, ω is smaller than the maximum gain frequency $\omega_M = (\omega_c\,\omega_D)^{1/2} \approx 10^{10}$ Hz, and γ is relatively small. Then α can be simplified as

$$\alpha = \frac{K^2\,\omega^2\,\gamma}{2\,v_s\,\omega_c} = \frac{e_{11}^2\,\gamma\,\omega^2}{2\,\varrho\,v_s^3\,\sigma},$$

where ϱ is the specific mass, and σ the electric conductivity. An important point is that the dielectric constant ε_1 is not used in this expression and all the constants are rather well-known, the only uncertainty being e_{11}. Therefore we computed e_{11} as a function of frequency using the preceding expression (Fig. 5). The low frequency value $e_{11} = 0.84$ C/m^2 was measured at about 30 kHz by Quentin and Thuillier [6]. At higher frequencies, in the range from 10^7 to 10^8 Hz, we observe an important increase of the coefficient e_{11}. This effect varies from one sample to another. In other words, in this anomalous region, the gain is higher than expected from the known piezoelectric constant at low frequencies.

4. Conclusion

The use of selenium epitaxial transducers permits the investigation of the acoustoelectric effects in tellurium in a wide frequency range. The results are not consistent with the Hutson-White theory, even in the linear region if one uses the low frequency value of e_{11}. This is probably due to interactions with crystal defects. Measurements of the frequency dependence of the conductivity and dielectric constants of tellurium are under way, in order to obtain more information about these phenomena.

References

[1] T. Ishiguro and T. Tanaka, Conf. Phys. Semicond., Kyoto 1966; J. Phys. Soc. Japan, Suppl. **21**, 489 (1966).
[2] G. Arlt and K. Reiber, Solid State Commun. **7**, 1155 (1969).
[3] T. Shiosaki, A. Kawabata, and T. Tanaka, Japan. J. appl. Phys. **9**, 631 (1970).
[4] A. R. Hutson and D. L. White, J. appl. Phys. **33**, 40 (1962).
[5] D. L. White, J. appl. Phys. **33**, 2547 (1962).
[6] G. Quentin and J. M. Thuillier, Solid State Commun. **2**, 115 (1964).

(Received September 27, 1971)

phys. stat. sol. (a) 8, 437 (1971)

Subject classification: 20.3; 13.1; 22.2.4

Laboratoire de Physique des Solides, C.N.R.S., Bellevue

Photoluminescence of Undoped $Ga_xIn_{1-x}P$ Alloys

By

J. CHEVALLIER and A. LAUGIER

Changes in photoluminescence spectra at low temperatures (2 to 20 °K) are reported of non-intentionally doped $Ga_xIn_{1-x}P$ solution-grown alloys in the whole range of compositions. Below the Γ_{15}–X_1 band cross-over composition ($x = 0.65 \pm 0.03$), the spectra are similar to those of InP, with a tendency to shift to lower energies relatively to the bandgap. For larger values of x, some lines are observed which differ from those of GaP. The value of Γ_{15}–X_1 gap in InP is deduced to be (2.30 ± 0.04) eV at 4.2 °K. A pair transition line is observed throughout the composition range, with discontinuities of the half-width and rate of excitation shift at the cross-over composition.

Les spectres de photoluminescence d'alliages non dopés $Ga_xIn_{1-x}P$ obtenus par croissance en solution ont été étudiés entre 2 et 20 °K dans toute la gamme de composition. Pour x inférieur à $0,65 \pm 0,03$, composition de croisement des bandes Γ_{15} et X_1, les spectres sont semblables à ceux de InP avec une tendance des énergies des raies à décroître par rapport à la largeur de bande interdite. Pour x plus élevé, on observe des raies qui diffèrent de celles de GaP. On déduit pour la transition Γ_{15}–X_1 de InP une énergie de $2,30 \pm 0,04$ eV à 4,2 °K. Une transition de paires a été observée de façon continue dans toute la gamme de compositions, avec des discontinuités de sa largeur à mihauteur et de son taux de déplacement en fonction de l'intensité d'excitation à la composition de croisement.

1. Introduction

In $Ga_xIn_{1-x}P$ alloys direct transitions are possible with photon energies corresponding to the highest visual sensitivity range of the human eye. This material appears to be very attractive in view of electroluminescent devices. By means of photoluminescence measurements we have studied the changes of the electronic structure for undoped alloys when x passes through the "cross-over" composition where the energies of direct and indirect transitions become equal.

2. Experimental

A new solution growth process was recently described for these alloys [1, 2]. An indium bath is continuously fed by phosphorus coming from vapour and by gallium from solid gallium phosphide. Crystals have a good homogeneity in composition determined with an electron microprobe. We used commercial 6N purity elements (indium shot and gallium from Kock-Light, phosphorus from M.C.P.). In the grown crystals we found residual impurities. Table 1 gives the main impurity contents from visual mass spectrographic analysis. These impurities seem to be either present in the indium shot or added during the process of crystal preparation. Relatively high quantities of carbon and oxygen are to be noticed.

All the alloys are n-type at 300 °K in the range $0 < x < 0.50$ (e.g. for $x = 0.34$ at 300 °K, $n = 2.5 \times 10^{16}$ cm^{-3} and $\mu_n = 360$ cm^2/Vs).

The samples were immersed in a cryogenic liquid. Their luminescence was excited with radiations emitted from a mercury vapour lamp (200 W). Filters allowed to obtain adequate exciting wavelengths. The emitted radiations were diffracted by a grating spectrometer with 30 Å/mm dispersion. An AS-20 or S-1 photomultiplier was used as detection system.

Table 1

Mass spectrographic analysis of $Ga_{0.36}In_{0.64}P$. Values are in atomic p.p.m.

C	27	Al	0.8	Ca	2.7
N	0.8	Si	0.3	Fe	0.3
O	27	S	0.3	Mn	0.09
F	2.7	Ce	0.25	Zn	0.18
Mg	0.5	K	9	As	0.01

3. Direct Band Gap Materials

Solution grown InP crystals, prepared in the same way as described in [3], are used as references for this study. They are n-type from 300 to 4.2 °K. Hall effect measurements give electron concentrations of 6.5×10^{16} and 5×10^{16} cm^{-3} and mobilities of 2150 and 7100 cm^2/Vs at 300 and 77 °K respectively. These values indicate that the crystals are partially compensated. Fig. 1 shows a typical photoluminescence spectrum at 4.2 °K. It is similar to those reported by Leite [4]. The peak at 1.417 eV (a, b on Fig. 1) is unresolved. It has been shown [5, 6] that, under particular conditions of doping and excitation intensity, it can be resolved into four peaks which are assigned to 1. the excited level of a free exciton, 2. a free exciton (a-line), 3. a bound exciton and 4. the transition between a donor level and the valence band (b-line). The c-line at 1.376 eV was identified by Heim [7] as a donor–acceptor pair transition; the two impurities involved are not yet known. Its phonon replica with L.O. phonon explains the d-line at 1.331 eV.

The spectra of alloys with composition in the range $0 < x < 0.26$ are very similar to those of InP (Fig. 1). The log–log plot of photoluminescence intensity versus excitation intensity (Fig. 2) shows, for InP and such alloys, identical

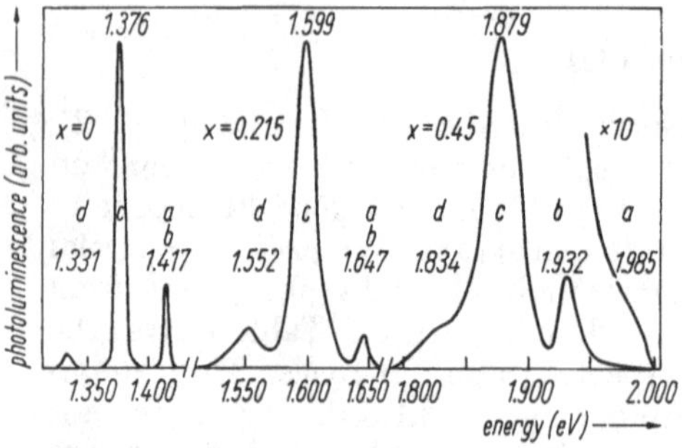

Fig. 1. 4.2 °K photoluminescence spectra for InP and two direct band gap $Ga_xIn_{1-x}P$ alloys with $x = 0.215$ and 0.45. Spectra are normalized but not corrected for photomultiplier and analyser characteristics

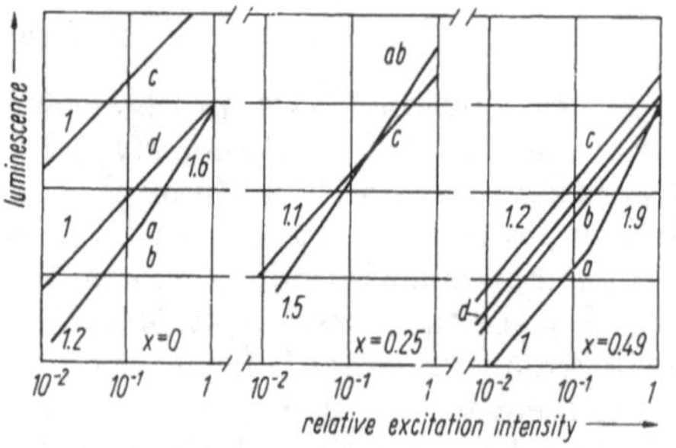

Fig. 2. Log–log plot of relative photoluminescence intensity vs. excitation intensity at 4.2 °K for InP and two direct band gap alloys $x = 0.25$ and $x = 0.49$. Slopes of curves are indicated for each line a, b, c, d

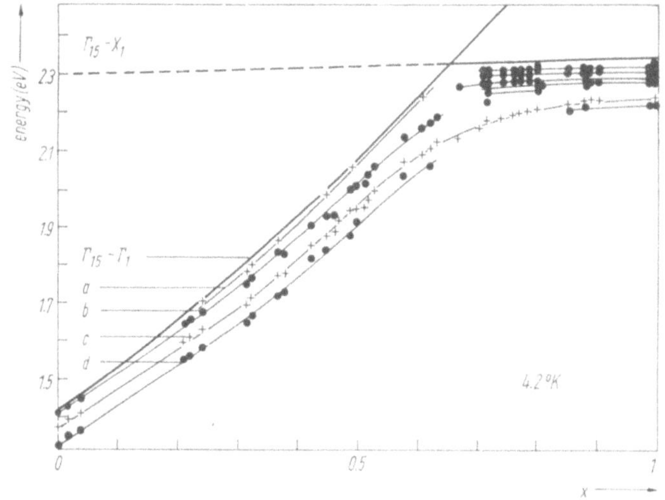

Fig. 3. Band structure and photoluminescence lines of $Ga_xIn_{1-x}P$ alloys

slopes for comparable peaks. When the excitation intensity is increased, the maximum of c-line shifts to higher energies with a rate of 1 or 2 meV per decade. This shift is characteristic of a pair transition. The d-line is the phonon replica of this transition.

When x increases to higher values, higher energy transitions generally appear in the spectra as shown on Fig. 1 (a-line). Their energies are near those of the band gap at 4.2 °K as discussed later. As seen on Fig. 2, these a-transitions are superlinear with a slope 1.9, whereas b-transitions have a slope 1. This means that the peak of highest energy in InP and In-rich alloys is now resolved at least into two lines. The a-peaks seem to be free exciton or shallow level-to-band transitions.

The b-transition is observed up to $x = 0.64$ as seen on Fig. 3. It is assumed that a transition between a non-identified donor level bound to the Γ_1 conduction band and the Γ_{15} valence band is the origin of this recombination. This

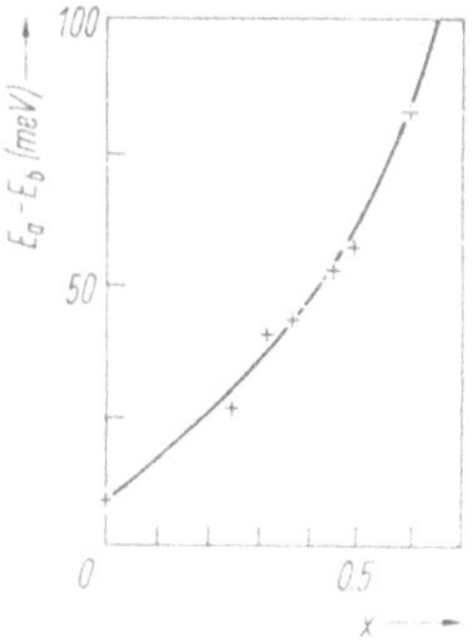

Fig. 4. Ionization energy (E_a–E_b) vs. composition at 4.2 °K

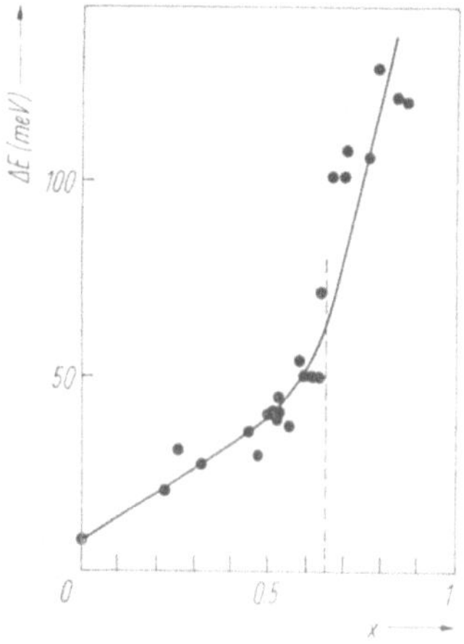

Fig. 5. Half-width of c-transition vs. composition at 4.2 °K. Dashed line indicates the cross-over

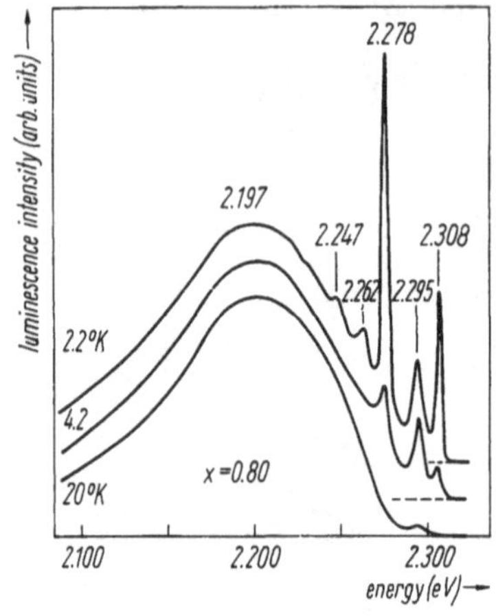

Fig. 6. Typical photoluminescence spectrum for an indirect band gap alloy

hypothesis seems reasonable since the ionization energy E_a–E_b of the defect is strongly modified when x increases to the cross-over composition (Fig. 4). This would not be the case for an acceptor.

Up to the cross-over, the c-lines have the same characteristics as in the In-rich compositions. However, their half-width increases with x as shown on Fig. 5.

4. Indirect Band Gap Materials

For $x > 0.66$, the c-transition is still present but its half-width is larger (Fig. 5). Its maximum shifts to higher energies when the excitation intensity increases, at a rate of 7 meV per decade. The discontinuities of half-width variation and shift rate are certainly related to the occurrence of the band cross-over.

For compositions $0.66 < x < 0.88$, three peaks appear at energies higher than the c-lines (Fig. 6). Two of them depend on temperature almost with the same ratio for 1.5 and 20 °K, and seem to have the same origin. The third line is still present at 20 °K. We notice that the energies of these lines vary very slowly with composition.

5. Discussion

On Fig. 3, the direct band-gap Γ_{15}–Γ_1 is reported according to the values of Hakki et al. [8], Jones and Lettington [9], Rodot et al. [10] determined at 300 °K from pressure, thermoreflectance and optical absorption measurements, respectively. We have corrected them with additive terms, due to temperature variation, estimated from InP and GaP. As seen on Fig. 3, the a-peak energies well agree with these values of band gap.

The indirect band gaps Γ_{15}–X_1 versus composition are estimated, assuming a linear variation from the GaP value with the same slope as that of the highest lines of photoluminescence spectra.

The intersection of this line with the direct band gap line gives the values $E_g = 2.33 \pm 0.02$ eV and $x = 0.65 \pm 0.03$. The extrapolation of the straight line Γ_{15}–X_1 to $x = 0$ gives the value 2.30 ± 0.04 eV. This value is larger than that of Logan [11].

6. Conclusion

In conclusion, the luminescence spectra of direct band gap alloys are very similar to those of InP. However, the transitions are shifted to lower energies relatively to the band gap. This agrees with Mabbitt's results [12]. In indirect gap materials, the spectra differ from those of GaP and we cannot yet give a complete and sure interpretation, except for a pair line which has been observed continuously in the whole range $0 < x < 1$. Studies in the range $0.9 < x < 1$ may allow to state more precisely the nature of transitions.

References

[1] A. LAUGIER and J. CHEVALLIER, phys. stat. sol. (a) **7**, 427 (1971).
[2] J. CHEVALLIER and H. RODOT, C. R. Acad. Sci. (France) B **271**, 1037 (1970).
[3] H. RODOT, A. HRUBY, and M. SCHNEIDER, J. Crystal Growth **3**, 305 (1968).
[4] R. C. C. LEITE, Phys. Rev. **157**, 672 (1967).
[5] O. RÖDER, U. HEIM, and M. H. PILKUHN, J. Phys. Chem. Solids **31**, 2625 (1970).
[6] U. HEIM, O. RÖDER, H. J. QUEISSER, and M. H. PILKUHN, J. Lumin. **2**, 542 (1970).
[7] U. HEIM, Solid State Commun. **7**, 445 (1969).
[8] B. W. HAKKI, A. JAYARAMAN, and C. K. KIM, J. appl. Phys. **41**, 5291 (1970).
[9] O. JONES and A. LETTINGTON, J. Phys. C **4**, 1534 (1971).
[10] H. RODOT, J. HORAK, G. ROUY, and J. BOURNEIX, C. R. Acad. Sci. (France) B **269**, 381 (1969).
[11] R. A. LOGAN, P. J. DEAN, H. G. WHITE, and W. WILGMANN, J. appl. Phys. **42**, 2328 (1971).
[12] A. W. MABBITT, Solid State Commun. **9**, 245 (1971).

(Received September 27, 1971)

phys. stat. sol. (a) 8, 443 (1971)

Subject classification: 20.1; 12; 22.1.2

Van der Waals Laboratory, Amsterdam

The Pressure Coefficient of the Infra-Red Refractive Index of Silicon

By

R. VETTER[1])

From the pressure shift of the interference fringes obtained with a plane parallel silicon single crystal of 0.5 mm thickness the pressure coefficient of the refractive index in the near infra-red has been found to be $-0.14\times10^{-6}\ \text{bar}^{-1}\pm12\%$.

A partir du déplacement à haute pression des franges d'interférences d'un échantillon de Si, plan-parallèle et d'une épaisseur de 0,5 mm, on a déterminé le coefficient de pression de l'indice de réfraction dans l'infrarouge proche; on a trouvé: $-0{,}14\times10^{-6}\ \text{bar}^{-1}\pm12\%$.

1. Introduction

With a high resolution infra-red spectrometer it is possible to obtain accurate pressure shifts of the multiple reflection interference spectra of relatively thick semiconductor specimens thereby reducing the error of the polishing damaged surface layer. This method has been successfully applied to germanium [1] and it is now extended to silicon. As is generally known constructive interference in a plane parallel slab of thickness t will occur at a wavenumber

$$\sigma=\frac{m}{2\,t\,n} \tag{1}$$

when an integer number of waves, m, exactly fills the optical path $2\,t\,n$ (n being the refractive index) that has been traversed by a beam reflected between back and front plane. With a silicon specimen of a few microns thick the interference fringes are well resolved by a prism spectrometer but when the thickness is of the order of 1 mm the fringe length

$$\Delta\sigma=\frac{1}{2\,t\,n} \tag{2}$$

becomes of the order of 1 cm^{-1} and a grating spectrometer must be used. From the pressure shift $\mathrm{d}\sigma$ of an interference maximum of constant m the coefficient of the refractive index is found by differentiating (1) with respect to the pressure p:

$$-\frac{1}{n}\frac{\partial n}{\partial p}=\frac{1}{\sigma}\frac{\partial\sigma}{\partial p}+\frac{1}{t}\frac{\partial t}{\partial p}. \tag{3}$$

The last term represents the compressibility correction. The pressure coefficient of the refractive index of common semiconductors is of the order of $10^{-6}\ \text{bar}^{-1}$ so that either very high pressures or a high spectrometric accuracy are required.

[1]) Present address: Laboratorium voor Metaalkunde, Technische Hogeschool, Delft, The Netherlands.

With a high-resolution grating spectrometer the second requirement is fulfilled. This was not the case in an earlier determination of the shift of the interference fringes of silicon by Cardona [2] where no measurable effect was detected below 2000 bar.

2. Experimental

The p-type silicon specimen is prepared from a pure single crystal having a resistivity of about 400 Ωcm. The thickness is (482.5 ± 0.5) μm. It has been polished on both sides with diamond abrasive, in the finishing stage with 1 μm grit size. Flatness and parallelism are better than 1 μm per cm. A region of about 5 mm dia. is exposed to the infra-red beam. As an interferometer it has a low finesse; nevertheless the interference fringes clearly appear as a 10% modulation of the transmittance. The specimen is freely suspended in a holder inside the pressure vessel, which is provided with quartz windows. The pressure fluid is nitrogen gas. Pressures were read from a calibrated Bourdon manometer.

The spectrometer, equipped with a Bausch & Lomb grating of 120000 lines, is used in first and second order covering a wavelength range of about 1 to 2.7 μm. A potassium dihydrogen phosphate filter and a germanium filter, having a cut-off wavelength of 1.8 μm and a cut-on wavelength of 1.6 μm, are used as order selection filters. The infra-red spectra are measured in terms of a simultaneously recorded auxiliary channel spectrum generated by a Fabry-Perot interferometer in the fourth order of the grating. The channels are spaced at 0.90645 cm^{-1} when expressed in the first order infra-red scale. The channel spectrum can be converted to an absolute wavenumber scale by some neon calibration emission lines. The silicon interference maxima lie about 2.8 cm^{-1} apart, depending on the spectral value of the refractive index. Without an order filter the first- and second-order interference spectra have fringe lengths in the ratio of nearly 2:1. They form beat patterns (Fig. 1) which can be used for a relative pressure shift determination.

The interference spectra of silicon were measured at atmospheric pressure and about 1000 bar in three regions of the wavelength range. The room temperature was regulated at 294 K within a few deci-deg and the specimen temperature was measured at the wall of the pressure vessel.

3. Results

The pressure shift was found to be about 2 cm^{-1} at 1000 bar in the first order of the grating and 4 cm^{-1} in the second order. This means that the shift on

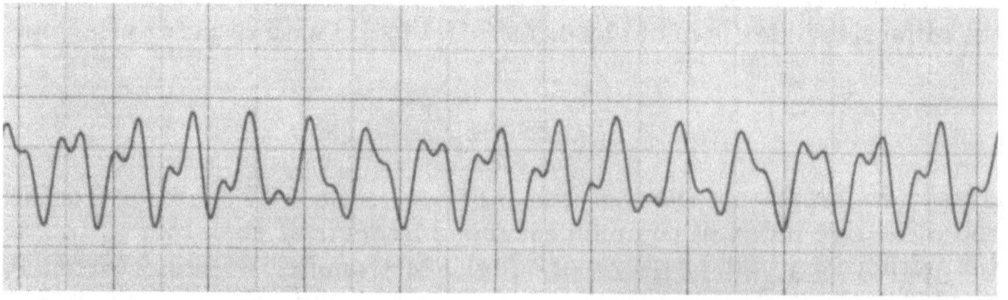

Fig. 1. Beat patterns resulting from the superposition of large first-order interference fringes in the vicinity of 4500 cm^{-1} upon small second-order fringes near 9000 cm^{-1}

recorder paper is about the same for both orders because the linear dispersion or rather the "plate factor" (increase in wavenumber per cm of recorder chart) stands also in the ratio 1:2. In order to check the relative shift, a measurement of the beat pattern was also made in addition to the first-order spectrum near 4000 cm^{-1} and the second-order spectra near 8000 cm^{-1} and 9150 cm^{-1}. It became obvious that the spectral pressure shift is much more constant for silicon than for germanium [1] although 9150 cm^{-1} is well inside the indirect absorption region, which starts at about 8500 cm^{-1} ($\hat{=}$ 1.05 eV). It was thus reasonable to determine in each region the mean pressure shift of 20 to 50 interference maxima and to divide this by the mean wavenumber of that region and by the pressure difference of 963 bar in order to compute $\Delta\sigma/(\sigma\,\Delta p)$. The results are shown in Table 1. The values are seen to be constant to $\Delta\sigma/(\sigma\,\Delta p) =$

Table 1
Pressure shift of silicon interference fringes

wave-number (cm^{-1})	$\Delta\sigma/(\sigma\,\Delta p)$ (bar^{-1})	standard deviation (%)
4000	0.49×10^{-6}	4
8000	0.495×10^{-6}	3
9150	0.485×10^{-6}	4

$= 0.49 \times 10^{-6}$ bar^{-1} within the experimental error of about 3%. Almost the whole error contribution comes from the fluctuations in the position difference of the maxima, $\Delta\sigma$, individual readings of which are uncertain by about 0.1 cm^{-1} in 2 cm^{-1} . From $\Delta\sigma/(\sigma\,\Delta p)$ the pressure coefficient of the refractive index is, in accordance with (3), obtained by subtraction of the linear compressibility, $\varkappa/3$. The compressibility determined with ultrasonic wave propagation in high-purity, p-type single crystal silicon is [3] 1.0216×10^{-6} bar^{-1}. So $\varkappa/3$ is taken as 0.34×10^{-6} bar^{-1}. It is clear that the compressibility is responsible for the larger fraction of the measured effect, and so a correction for room temperature fluctuations in $\Delta\sigma$ should be included. From (1) it follows that

$$\left(-\frac{\partial n}{n\,\partial p}\right)_T = \frac{\Delta\sigma}{\sigma\,\Delta p} - \frac{\varkappa}{3} + \left\{\frac{1}{t}\left(\frac{\partial t}{\partial T}\right)_p + \frac{1}{n}\left(\frac{\partial n}{\partial T}\right)_p\right\}\frac{\Delta T}{\Delta p}. \tag{4}$$

A temperature difference of -0.3 K was noticed for the high-pressure series with respect to the atmospheric one. The linear expansion coefficient of silicon is 2.33×10^{-6} deg^{-1} [4], and the temperature coefficient of its refractive index is $(39 \pm 4) \times 10^{-6}$ deg^{-1}, according to Cardona et al. [5]. This contributes $(-0.013 \pm 0.001) \times 10^{-6}$ bar^{-1} to (4). So finally the pressure coefficient of the infra-red refractive index of silicon is obtained as $-\partial n/(n\,\partial p) = 0.14 \times \times 10^{-6}$ $bar^{-1} \pm 12\%$, where the errors in $\Delta\sigma$, in Δp $(= (963 \pm 3)$ bar$)$, and in the temperature coefficient of n have been included.

4. Discussion

The pressure shift of the fringes is very small, but nevertheless of the order of one fringe length at 1000 bar. An incorrect assignment of corresponding fringes at 1 and 964 bar would cause a big error. Tentatively, the pressure shift was also calculated with the assumption that a certain 1 bar fringe would correspond to a next higher or lower fringe at 964 bar, and the resulting coefficients were used to predict the shift over 477 bar. These predictions were compared with the possible shifts of an actual measurement over 477 bar, from which the final assignment was found unambiguously.

The resistivity of the specimen material is not as high as it is possible to obtain for silicon, but a value of 400 Ωcm would lead to a free-carrier absorption that is low compared with the analogous one for germanium of highest purity, because the absorptivity is inversely proportional to resistivity. It is expected to have no measurable influence on the optical constants.

Cardona et al. [5] have reported a pressure coefficient for silicon $-\partial n/(n\ \partial p) = 0.3 \times 10^{-6}\ \mathrm{kp^{-1}\ cm^2}$ with an uncertainty of $\pm 67\%$. This value is twice as large as the present one, although there is agreement inside the error limits.

The present value of $\partial n/(n\ \partial p)$ also agrees with half the coefficient of the dielectric constant at radio frequencies of Cardona et al. [6]: $-\partial \varepsilon/(2\ \varepsilon\ \partial p) = (0.2 \pm 0.05) \times 10^{-6}\ \mathrm{bar^{-1}}$.

In comparison with the pressure coefficient of the infra-red refractive index of germanium, $-\partial n/(n\ \partial p) = 1.70 \times 10^{-6}\ \mathrm{bar^{-1}}$ [1], it is remarkable that the silicon coefficient is smaller by about one order of magnitude. This fact, however, is supported by theory. Brust and Liu [7] have performed theoretical calculations of the pressure coefficient of the static dielectric constant based on a Kramers-Kronig integral of the imaginary part $\varepsilon_2(E)$ of the dielectric function, and they also find that the coefficient for silicon is less than that for germanium by about an order of magnitude. This is due to the smaller pressure shift along the energy (E) scale of the corresponding peaks in $\varepsilon_2(E)$. In addition these peaks lie at higher energies and therefore contribute less to the integral on account of the smaller weighting function. Rather fortuitous there is exact agreement between the theoretical value of Brust and Liu $\partial \varepsilon/(2\ \varepsilon\ \partial p) = -3.4/(2 \times 12) \times 10^{-6}\ \mathrm{kp^{-1}\ cm^2} = -0.14 \times 10^{-6}\ \mathrm{kp^{-1}\ cm^2}$ and the present value of $\partial n/(n\ \partial p)$. (Here ε of silicon is taken to be equal to 12.)

In the second place there is the remarkable flatness of the spectrum of the silicon coefficient while for germanium an increase by a factor of two is found over the region 0.5 to 0.775 eV. This may be caused by the larger distance from the last measuring point ($9150\ \mathrm{cm^{-1}} \triangleq 1.12\ \mathrm{eV}$) to the direct gap of silicon (3.5 eV) as opposed to the germanium case (direct gap at 0.805 eV). A small spectral dependence of the silicon coefficient was revealed by a kind of null method. If the coefficient were perfectly constant there would be no change in the beat patterns of combined first- and second-order interference spectra (Fig. 1) with pressure, as $\Delta\sigma$ and the plate factor would be exactly in proportion. Such a pattern was investigated for the combination 4500 and 9000 $\mathrm{cm^{-1}}$ at 1, 500, and 1000 bar. A slightly larger shift of the small 9000 $\mathrm{cm^{-1}}$ fringes with respect to the 4500 $\mathrm{cm^{-1}}$ fringes would result in a comparatively large change in the beats. Indeed a small change could be detected at the W-like structures, where a maximum of the one system practically coincides with a minimum of the other system. Some of the W's first sag to the left and at high pressures progressively more to the right-hand side, indicating an increased shift of the 9000 $\mathrm{cm^{-1}}$ pattern relative to the other one by nearly 8%. This result should be regarded as an order of magnitude estimate only. It does not conflict with the measurements at these wavenumbers, which are both uncertain by about 4% (Table 1). An increase of 8% in $\Delta\sigma/\sigma$ would result in an increase of 24% in the spectrum of $-\partial n/(n\ \partial p)$.

In 1964 Gibbs and Hill [8] noticed an empirical regularity in the volume coefficient of the microscopic polarizability, calculated for C, Si, and Ge and partly based on data of Cardona et al. [5]. With the new data on the infra-red refractive index this regularity is much less apparent. This also casts some doubts on the existence of a simple relation between atomic volume and the calculated polarizability as given by Czaja and Gränacher [9]. These rules can be said merely to predict the right order of magnitude.

5. Conclusions

By means of high resolution infra-red spectroscopy the accuracy of measuring the pressure coefficient of the refractive index of silicon could be increased by one order of magnitude with respect to earlier determinations. The results yield a value of $\partial n/(n\ \partial p) = -0.14 \times 10^{-6}\ \text{bar}^{-1} \pm 12\%$, and a spectrum that is relatively flat although an increase of the absolute value by about 24% near 9000 cm^{-1} ($\hat{=}$ 1.1 eV) seems probable. The observed relative magnitude of the silicon coefficient with respect to the germanium coefficient is supported by a theoretical estimate of Brust and Liu. Some semi-empirical rules about the atomic polarizability do not stand the test with the present more precise value of silicon.

Acknowledgement

This work is part of a research program of the Foundation for Fundamental Research on Matter (F.O.M.), which is supported by the Netherlands Organisation for the Advancement of Pure Research (Z.W.O.).

References

[1] N. J. Trappeniers, R. Vetter, and H. A. R. de Bruin, Physica (Amsterdam) **45**, 619 (1970).
[2] M. Cardona, Dielectric Constant of Germanium and Silicon as a Function of Volume, Thesis, Harvard University, Cambridge, Massachusetts, 1959.
[3] H. J. McSkimin and P. Andreatch, Jr., J. appl. Phys. **35**, 2161 (1964).
[4] A. S. Bhalla and E. W. White, phys. stat. sol. (a) **5**, K51 (1971).
[5] M. Cardona, W. Paul, and H. Brooks, J. Phys. Chem. Solids **8**, 204 (1959).
[6] M. Cardona, W. Paul, and H. Brooks, Solid State Physics in Electronics and Telecommunications, Vol. 1, Eds. M. Desirant and J. L. Michiels, Academic Press, New York 1960 (p. 206).
[7] D. Brust and L. Liu, Phys. Rev. **154**, 647 (1967).
[8] D. F. Gibbs and G. J. Hill, Phil. Mag. **9**, 367 (1964).
[9] W. Czaja and I. Gränacher, Helv. phys. Acta **36**, 1073 (1963).

(Received July 16, 1971)

phys. stat. sol. (a) 8, 449 (1971)

Subject classification: 10.1; 21.1

Department of Physics, University of Alberta, Edmonton

On the Variation of Stacking Fault Contrast with Crystal Thickness

By

S. S. SHEININ and K. Z. BOTROS

An investigation of the dependence of stacking fault contrast on crystal thickness in f.c.c. cobalt has shown that the effects of systematic reflections can change markedly with crystal thickness and defect depth. When the (111) set of systematic reflections was excited, the results showed that good agreement was obtained between the two beam theory and experiment for crystals whose thicknesses were in the range five to nine extinction distances. In thicker crystals, however, agreement between theory and experiment could only be obtained by taking the effects of systematic reflections into account. That defect depth is also a factor in determining whether or not systematic reflections play an important role in image contrast was demonstrated by the fact that the fringes at the edges of faults in thick crystals are two beam in character while the contrast of fringes at the centre of the fault can only be predicted by taking systematic reflections into account. Finally, it has been shown that the effect of the (111) systematic reflections, which is to reduce fringe contrast, can be avoided when the (220) systematic set is excited thus giving rise to strong contrast in much thicker crystals than would otherwise be possible.

Eine Untersuchung der Abhängigkeit des Stapelfehlerkontrasts von der Kristalldicke in k.f.z. Kobalt hat gezeigt, daß sich die Einflüsse systematischer Reflexionen merklich mit der Kristalldicke und Defekttiefe ändern können. Im Falle der Anregung des (111)-Satzes der systematischen Reflexionen zeigten die Ergebnisse, daß gute Übereinstimmung zwischen der Zweistrahltheorie und dem Experiment für Kristalle, deren Dicke im Bereich zwischen fünf und neun Extinktionsabständen liegt, erhalten wurde. In dickeren Kristallen konnte jedoch Übereinstimmung zwischen Theorie und Experiment nur durch Berücksichtigung der Einflüsse systematischer Reflexionen erreicht werden. Daß die Defekttiefe ebenfalls einen Faktor bei der Feststellung darstellt, ob systematische Reflexionen eine bedeutende Rolle im Bildkontrast spielen oder nicht, wurde durch die Tatsache gezeigt, daß die Streifen an den Kanten der Stapelfehler in dicken Kristallen Zweistrahlcharakter besitzen, während der Streifenkontrast im Zentrum des Fehlers nur durch Berücksichtigung systematischer Reflexionen vorhergesagt werden kann. Es wurde ebenfalls gezeigt, daß die Einflüsse von systematischen (111)-Reflexionen, die den Streifenkontrast reduzieren, dadurch vermieden werden können, daß der systematische (220)-Satz angeregt wird und damit Anlaß zu starken Kontrasten in viel dickeren Kristallen, als es auf andere Weise möglich wäre, gegeben wird.

1. Introduction

The two beam approximation has been used extensively to describe the contrast of stacking faults at low accelerating voltages. Whelan and Hirsch [1], for example, showed that when absorption is neglected, the profile of a stacking fault inclined to the surface of a thin foil consists of fringes which have a periodicity of one-half the extinction distance. Hashimoto et al. [2, 3] examined the contrast to be expected in thick crystals by taking the effects of anomalous absorption into account. They showed that the fringe spacing in images of $\alpha = \pm 2\pi/3$ faults increased to one extinction distance near the edges of the fault while near the centre of the fault weak fringe contrast resulted. Gevers et al. [4] have also studied images of stacking faults in thick crystals. They showed that the fringe spacing near the centre of the fault could vary from a half to one extinction distance, depending on the thickness of the crystal. The effects of systematic reflections on stacking fault contrast in gold have been studied by Humphries et al. [5] who found that for a crystal six extinction distances thick, a marked increase in contrast resulted when the effects of systematic reflections were taken into account.

A question which arises when considering the possible effects of systematic reflections on image contrast is that of the possible factors which determine whether or not these effects will be significant. It has been commonly assumed, on the basis of theoretical calculations performed in the past [5 to 8], that these factors include the diffraction conditions, i.e. crystals orientation (or deviation from the Bragg condition) and accelerating voltage, as well as the material under investigation, the particular row of systematic reflections excited and whether the image is bright or dark field. Extinction contour calculations performed by the authors [9] have indicated that crystal thickness may also be a factor which could influence the role systematic reflections play in image contrast. The present investigation of the variation of stacking fault contrast with crystal thickness was undertaken with a view to exploring this possibility. The investigation has been carried out by obtaining experimental stacking fault profiles at values of crystal thickness in the approximate range of from five to fifteen extinction distances and comparing the results with computer calculations of image contrast based on the two- and multi-beam theories. It might be noted that in order to carry out such an investigation, the values of the anomalous absorption parameters must be known with a reasonable degree of precision. To this end, an experimental method for determining the parameters from the stacking fault profile itself has been used.

2. Theoretical Calculations and Experimental Procedure

Many beam theoretical calculations of stacking fault contrast were performed by using the scattering matrix approach of Howie and Whelan [10] as applied to a deformed crystal. The absorption parameters used were determined from the experimental variation of fault contrast with thickness (see Section 3.1) and the scattering factors used were those of Smith and Burge [11]. The Fourier coefficients of the lattice potential were Debye-Waller corrected in all cases.

Experimental observations of stacking fault contrast in the bright field as a function of crystal thickness were carried out using wedge crystals of f.c.c. cobalt (see Fig. 1). Experimental stacking fault profiles were obtained at a given value of thickness by simply taking densitometer traces in a direction parallel to the thickness extinction contours. The traces were taken as nearly as was possible at values of crystal thickness corresponding to maxima or minima in the directly transmitted intensity, i.e. at values of crystal thickness corresponding to n and $(n + 1/2)$ extinction distances where n is an integer. All micrographs were recorded with the (111) reflection in f.c.c. cobalt in the exact Bragg condition and the crystal was oriented so that the effects of non-systematic reflections were minimized. The results obtained have been divided into two ranges of thickness, referred to as moderately thick and thick respectively. Moderately thick crystals were taken to be those with thicknesses in the range of from five to nine extinction distances while thick crystals were those with thicknesses greater than nine extinction distances.

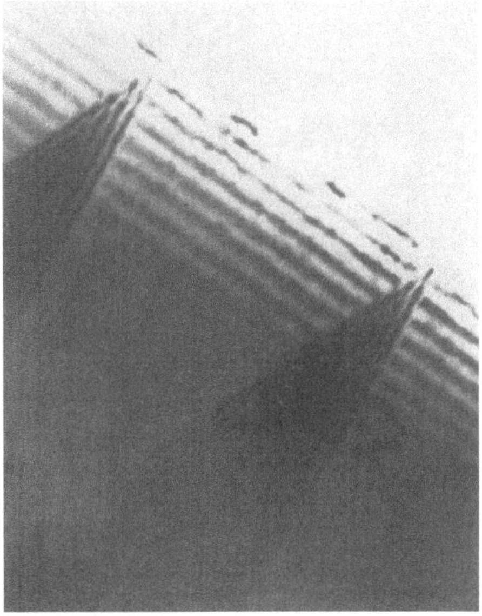

Fig. 1. Micrograph showing bright-field image of stacking faults in a wedge crystal of f.c.c. cobalt. Experimental profiles such as those in Fig. 3 and 5 were obtained by taking a microdensitometer trace in a direction parallel to the extinction contours

An important question arises when carrying out multi-beam calculations of stacking fault contrast which are to be compared with the results of either two-beam calculations or experiment and that is at what values of crystal thickness should the multi-beam calculations be performed. The approach adopted by the authors was to simply determine the positions of the appropriate intensity maxima or minima from a multi-beam extinction contour plot and to perform stacking fault calculations at these thicknesses. This method is completely analogous to the way in which stacking fault profiles are obtained in the two-beam case when crystal thickness is given in terms of a two-beam extinction distance, as well as being consistent with the method outlined above for obtaining experimental profiles. It might be noted that defining a multi-beam extinction distance, in analogy with the two-beam case, as being equal to $1/(\gamma^1 - \gamma^2)$, where the γ's are eigenvalues corresponding to the two uppermost branches of the dispersion surface, is not necessarily equivalent to the method used by the authors since the values of $1/(\gamma^1 - \gamma^2)$ can, under certain circumstances, differ appreciably from the values of the peak to peak spacing.

3. Results

3.1 Determination of absorption parameters

The first problem was to determine experimental values of the anomalous absorption parameters. The manner in which this was done is illustrated in Fig. 2 and 3. Fig. 2a, b, c shows sets of theoretical bright-field profiles at crystal thicknesses of 3.5, 5.5 and 7.5 extinction distances respectively and for different sets of absorption parameters. The calculations were performed for the (111) reflection in f.c.c. cobalt in the exact Bragg condition. The systematic reflections $(\bar{3}\bar{3}\bar{3}) \cdots (555)$ were taken into account in the calculations and the accelerating voltage was taken to be 150 kV. It can be seen from Fig. 2 that the

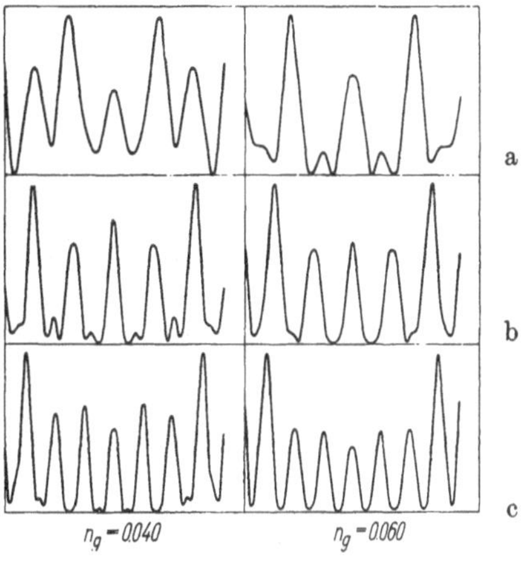

Fig. 2. Multi-beam bright-field profiles for an $\alpha = -2\pi/3$ stacking fault obtained for sets of absorption parameters corresponding to $n_g = 0.040$ and $n_g = 0.060$ where g is the (111) reflection. a), b), and c) correspond to crystal thicknesses of 3.5, 5.5, and 7.5 extinction distances, respectively

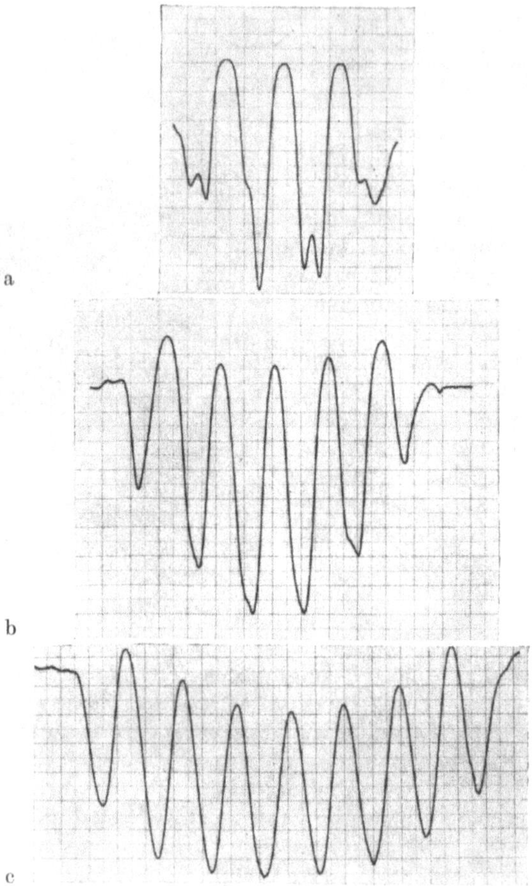

Fig. 3. Experimental bright-field profiles for the same values of crystal thickness as were used in obtaining the theoretical profiles in Fig. 2

disappearance of structure in the fringes, which consists of a transition from doublets to singlets, is dependent on the set of absorption parameters used. Fig. 3a, b, c shows a set of experimental profiles for the same values of crystal thickness as were used in obtaining Fig. 2. By a comparison of experiment profiles such as those in Fig. 3 with theoretical profiles such as those in Fig. 2, it is possible to obtain a set of anomalous absorption parameters which gives good agreement between theory and experiment. The value obtained in this way was $n_{111} = 0.060 \pm 0.005$, where n_{111} is defined in the expression for the complex part of the lattice potential $V_g' = n_g V_g$. Values of n_g for higher order Fourier coefficients were obtained by assuming that $n_g \propto g$ [12, 13]. The value of n_{111} obtained by the method just outlined is somewhat higher than the theoretical value of 0.050 given by Humphries and Hirsch [14]. This theoretical value is approximate and was obtained by interpolation from Fig. 2a of their paper noting the dependence of absorption parameters of different materials on atomic number.

It should be noted that an experimental value of n_0, corresponding to the normal absorption coefficient, is not required for purposes of the present investi-

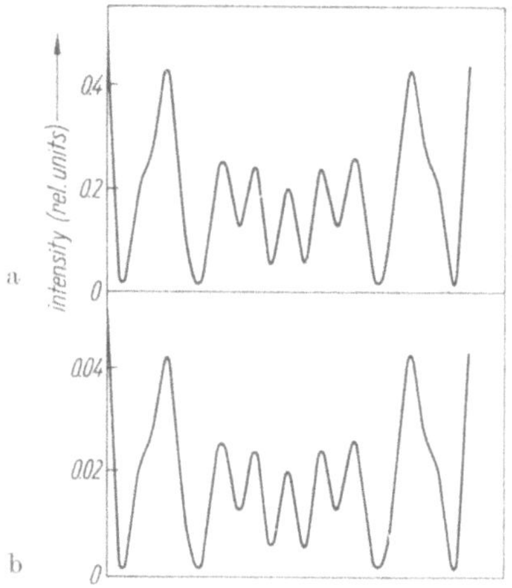

Fig. 4. Multi-beam bright-field profiles for an $\alpha = -2\pi/3$ stacking fault and a crystal thickness of five extinction distances. The anomalous absorption parameter in each case was $n_g = 0.060$ while the normal absorption parameters were (a) $n_0 = 0.01$ and (b) $n_0 = 0.06$

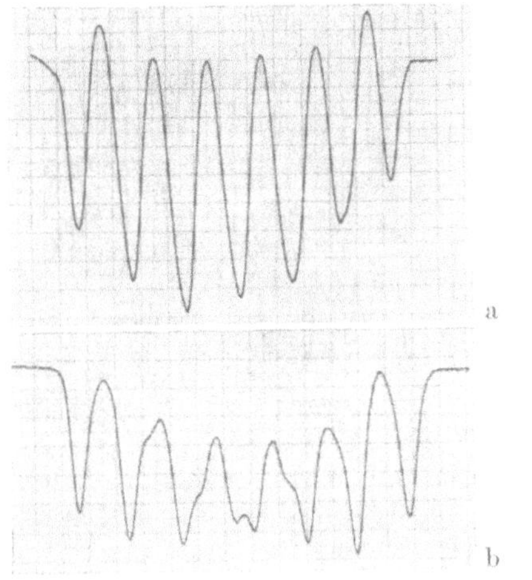

Fig. 5. Experimental bright-field profiles for an $\alpha = -2\pi/3$ stacking fault in crystals of the following thicknesses: a) 6.5 extinction distances, b) 7.0 extinction distances. The (111) set of systematic reflections only was excited

gation since the image contrast obtained is independent of the value of this parameter. This is illustrated in Fig. 4a, b which show theoretical multi-beam profiles for an $\alpha = -2\pi/3$ stacking fault at a crystal thickness of five extinction distances. The anomalous absorption parameter in each case was $n_{111} = 0.060$ while the normal absorption parameters in Fig. 4a, b were $n_0 = 0.01$ and 0.06 respectively. It is clear from these results that, although the overall intensity differs, stacking fault contrast is in fact independent of the normal absorption parameter used.

3.2 Fringe contrast in moderately thick crystals

The manner in which fringe contrast changes with crystal thickness in moderately thick crystals in the approximate thickness range of from five to nine extinction distances is illustrated in Fig. 5a, b. These figures show experimental bright-field profiles for an $\alpha = -2\pi/3$ stacking fault obtained at an accelerating voltage of 150 kV and at values of crystal thickness corresponding to 6.5 and 7 extinction distances. The crystal was oriented with the (111) reflection in the exact Bragg condition and, as nearly as was possible, with only the (111) set of systematic reflections in f.c.c. cobalt being excited. It can be seen from Fig. 5b that when the crystal is n extinction distances thick the fringes at the centre of the fault are doublets and exhibit relatively weak contrast. Fig. 5a, on the other hand, shows that when the crystal thickness is $(n + 1/2)$ extinction distances thick the fringes at the centre of the fault are singlets and exhibit relatively high contrast. In both cases fringes at the edge of the fault exhibit good contrast. This latter feature was found to be a characteristic of stacking fault contrast which was independent of crystal thickness so that in the comparisons of image contrast which follow, attention will mainly be given to fringes near the centre of the fault. Fig. 6 shows the results of both two- and multi-beam

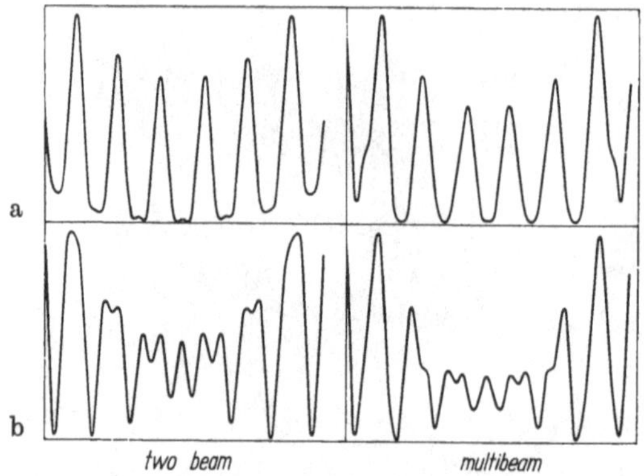

Fig. 6. Two-beam and multi-beam bright-field profiles for an $\alpha = -2\pi/3$ stacking fault in crystals of the following thicknesses: a) 6.5 extinction distances, b) 7.0 extinction distances. The (111) set of systematic reflections only was excited

calculations of stacking fault contrast at the same crystal thicknesses as were used in obtaining the experimental profiles in Fig. 5a, b. The first feature of interest to note in comparing Fig. 5 and 6 is that good agreement is obtained between experiment and the two-beam theory. That this is so can be seen by comparing the detailed shapes of both experimental and theoretical profiles as well as noting that the marked difference in contrast obtained experimentally between stacking faults in crystals of thicknesses n and $(n + 1/2)$ extinction distances is also predicted by the two-beam theory. A second point of interest to note from the theoretical profiles shown in Fig. 6 is that there are no significant differences in the results obtained from the two- and multi-beam calculations thus indicating that in the range of crystal thicknesses so far considered, systematic reflections have no significant effect on stacking fault contrast.

3.3 Fringe contrast in thick crystals

The next question to arise is whether or not the results presented in the previous paragraph for moderately thick crystals are also characteristic of thick crystals (i.e. thicknesses greater than about nine extinction distances). The use of extinction contours to obtain stacking fault profiles at a given value of crystal thickness was not possible for crystals in this range of thickness. The reason for this was that in crystals thicker than about nine extinction distances, the extinction contours were too faint to permit accurate determination of crystal thickness. This difficulty was overcome by determining from the extinction contours in relatively thin parts of the crystal, the distance perpendicular to the contours required to give a certain thickness increment. This distance would give the same thickness increment in thicker parts of the crystal, provided that a specimen with a uniform wedge was obtained. The uniformity of the wedge could of course be determined from the overall shape of the stacking fault image (such as the one shown in Fig. 1) since the fault runs from the top to the bottom surface of the crystal. Using this method to determine changes in crystal thickness with distance from the edge of the specimen, densitometer traces were taken at thickness increments of about 25 Å (approximately 1/10 of an extinction distance) in the approximate range of crystal thickness from 11.5 to 14.5 extinction distances. The experimental profiles, illustrated in Fig. 7a by the results obtained at a thickness of 12.5 extinction distances, were all similar in that the fringes at the centre of the fault exhibit weak contrast. The interest-

Fig. 7. Bright-field profiles obtained for an $\alpha = -2\pi/3$ stacking fault at a crystal thickness of 12.5 extinction distances. a) experimental profiles, b) two-beam profile, c) multi-beam profile taking systematic reflections into account. The (111) set of systematic reflections only was taken to be excited in each case

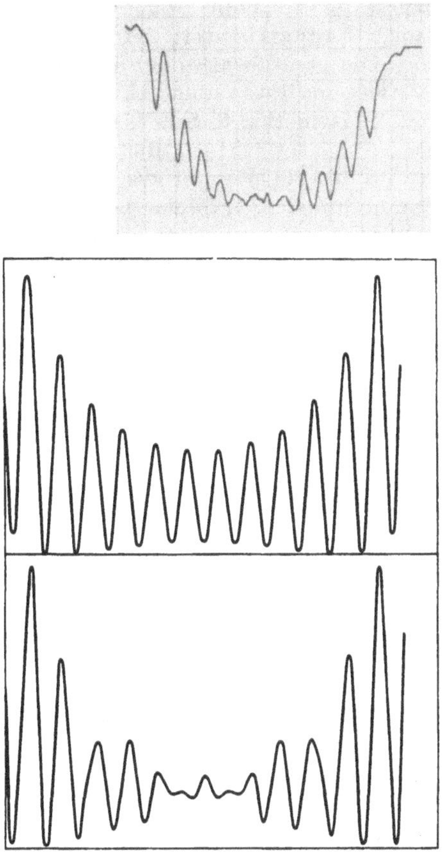

ing point to note from these results therefore is that the difference in contrast of stacking fault fringes in moderately thick crystals, discussed in Section 3.2, does not seem to persist in thick crystals.

In an attempt to understand these results calculations of stacking fault contrast in the range of crystal thickness from 11.5 to 14.5 extinction distances were performed. The results are illustrated in Fig. 7b, c which shows the profiles obtained at 12.5 extinction distances for both the two- and multi-beam cases. The two-beam calculations in Fig. 7b are in poor agreement with experiment showing that fringes at the centre of the fault should exhibit good contrast. The results of the calculations taking systematic reflections into account in Fig. 7c, on the other hand, are in good agreement with experiment showing that these fringes exhibit weak contrast. The results of the present investigation clearly show therefore that, for the (111) set of systematic reflections in f.c.c. cobalt, the two-beam theory adequately predicts the contrast obtained in moderately thick crystals. In thick crystals, however, agreement between theory and experiment could only be obtained by employing the multi-beam theory.

4. Discussion

In previous Section 3.3 it was shown that the (111) set of systematic reflections in f.c.c. cobalt have a deleterious effect on the contrast of fringes at the centre of a fault. This fact suggests that if a systematic set can be chosen for which effects of multiple reflections are minimal, good contrast should be obtained in much thicker crystals than would otherwise be possible. In order to test this hypothesis two- and multi-beam calculations of stacking fault contrast were performed assuming that only the (220) set of systematic reflections was excited and that the (220) reflection was in the exact Bragg condition. The two- and multi-beam profiles obtained from these calculations were very similar to those shown in Fig. 6 and are therefore not reproduced. The results showed that fringes at the centre of a fault will exhibit relatively high contrast in crystals of thickness up to about 10.5 extinction distances. Since the (220) extinction distance is approximately 500 Å this indicates that good contrast should be obtained in crystals of about 5000 Å in thickness. For the (111) set of systematic reflections it was found that good contrast in crystals of thickness equal to $(n + 1/2)$ extinction distances did not persist beyond about 8.5 extinction distances. Since the (111) extinction distance is about 250 Å, this result indicates that with the (111) set of systematic reflections excited, good contrast would not be obtained in crystals greater than around 2000 Å. These results are of particular interest in view of the work of Sheinin and Botros [15, 16] who showed that increased contrast in thick crystals results in better effective penetration. The above comparison of the contrast obtained when different systematic sets are excited therefore indicates that in choosing the conditions for obtaining optimum penetration, particular attention should be paid to the systematic row excited.

An additional point of interest can be seen by comparing the fringes at the edges of the fault in Fig. 7c with those at the centre of the fault. As already noted in the comparison of Fig. 7a, b, c, the fringes at the centre of the fault are multi-beam in character. The fringes at the edges of the fault, on the other hand, are two-beam in character. That this is so can be seen from the similarity of the fringes at the edges of the fault in Fig. 7b and c which shows the results given by the two- and multi-beam theories respectively. It can also be seen from the fact that the spacing of the fringes at the edges of the fault in the multi-beam profile in Fig. 7c is given by $1/(\gamma^1 - \gamma^2)$ indicating that, as in the two-beam theory, contrast is determined by interaction between Bloch waves 1 and 2. Clearly therefore the two-beam theory gives an adequate description of the contrast exhibited by stacking faults near the top and bottom surfaces. For stacking faults near the centre of the foil, however, the effects of systematic reflections must be taken into account.

5. Conclusions

The results of the present investigation have shown that stacking faults in moderately thick crystals of cobalt (thickness range from five to nine extinction distances) exhibit contrast of a two-beam character when the (111) set of systematic reflections is excited. This contrast is characterized by the fact that fringes at the centre of faults in crystals n and $(n + 1/2)$ extinction distances thick exhibit a marked contrast difference. In thicker crystals the effects of systematic reflections become important and these fringes exhibit weak contrast regardless of crystal thickness. The fringes at the edges of the fault, on the other hand, remain two-beam in character for all values of crystal thickness investigated.

These results illustrate the following points: first that the (111) set of systematic reflections have a deleterious effect on the contrast of fringes in bright-field images of stacking faults in f.c.c. cobalt. This is in contrast to results reported elsewhere [5] on the effects of systematic reflections on stacking fault contrast in gold. The second point to note from these results is that crystal thickness and defect depth are important factors in determining whether or not effects of the systematic reflections play an important role in determining image contrast. These factors are in addition to others such as accelerating voltage and crystal orientation which have been considered in the past.

Finally it has been shown that the deleterious effects of systematic reflections on fringe contrast can be avoided if the (220) systematic set only is excited thus giving rise to strong contrast in much thicker crystals than would otherwise be possible.

Acknowledgements

The authors wish to express their thanks to C. D. Cann, J. Andrew, and A. E. B. Monk for helpful discussions. They are also grateful to the National Research Council of Canada for financial support.

References

[1] M. J. Whelan and P. B. Hirsch, Phil. Mag. **2**, 1121 (1957).
[2] H. Hashimoto, A. Howie, and M. J. Whelan, Phil. Mag. **5**, 967 (1960).
[3] H. Hashimoto, A. Howie, and M. J. Whelan, Proc. Roy. Soc. **A269**, 80 (1962).
[4] R. Gevers, A. Art, and S. Amelinckx, phys. stat. sol. **3**, 1563 (1963).
[5] C. J. Humphries, A. Howie, and G. R. Booker, Phil. Mag. **15**, 507 (1967).
[6] C. D. Cann and S. S. Sheinin, phys. stat. sol. **30**, 791 (1968).
[7] S. S. Sheinin and K. Z. Botros, phys. stat. sol. (a) **3**, 271 (1970).
[8] S. S. Sheinin, phys. stat. sol. **37**, 683 (1970).
[9] S. S. Sheinin and K. Z. Botros, to be published.
[10] A. Howie and M. J. Whelan, Proc. Roy. Soc. **A263**, 217 (1961).
[11] G. H. Smith and R. E. Burge, Acta cryst. **15**, 182 (1962).
[12] A. J. F. Metherell and R. M. Fisher, phys. stat. sol. **32**, 551 (1969).
[13] S. S. Sheinin, K. Z. Botros, and C. D. Cann, phys. stat. sol. (a) **3**, 537 (1970).
[14] C. J. Humphries and P. B. Hirsch, Phil. Mag. **18**, 115 (1968).
[15] S. S. Sheinin, phys. stat. sol. **38**, 675 (1970).
[16] S. S. Sheinin and K. Z. Botros, J. appl. Phys. **42**, 1231 (1971).

(Received September 7, 1971)

phys. stat. sol. (a) 8, 459 (1971)
Subject classification: 14.3.2 and 14.4.1; 2; 22.6

Laboratoire d'Electronique et de Technologie de l'Informatique – Centre d'Etudes Nucléaires de Grenoble

Mechanism of Negative Resistance and Filamentary Conduction in Thermal Switching Devices

By
J. DUCHENE, G. ADAM, and D. AUGIER[1])

A particular coplanar thin film structure is analysed theoretically in order to determine the post-breakdown region parameters of a switching semiconductor device. Two types of conductivity versus temperature curves are used; firstly an exponential and secondly a transition-type one related to the VO_2 case. Uniform and filamentary current–voltage characteristics and temperature profiles are deduced from homogeneous and inhomogeneous computed solutions of the thermal equilibrium equation. Both are compared to experimental results obtained on Ovonic and VO_2 devices.

Une structure coplanaire particulière, en couche mince, est analysée théoriquement dans le but de déterminer les paramètres après avalanche d'un dispositif semiconducteur à commutation. Deux types de courbes conductivité–température sont étudiées; d'abord l'une exponentielle et ensuite une courbe présentant une transition, applicable au VO_2. Nous en déduisons des caractéristiques I–U uniformes et filamentaires et des profils de température déduits des solutions homogènes et inhomogènes calculées à partir de l'équation d'équilibre thermique. Les deux types sont comparés aux résultats expérimentaux obtenus avec des dispositifs Ovonic et des dispositifs à VO_2.

1. Introduction

Recently, many interpretations have been given for the threshold switching phenomena [1]. Electronic [2, 3] and thermal [4, 5] theories have been developed, although a thermal interpretation is sufficient to describe most of the features of the switching phenomenon [6, 7].

In accordance with that, we are trying to explain the post-breakdown region using a simple thermal model. This model assumes an increase in temperature produced by Joule heating in a semiconductor. In such a way a thermal avalanche can be produced which initiates the switching phenomenon.

Two different cases will be analysed:

Firstly, a model for an amorphous semiconductor material (A-type) with an electrical conductivity σ varying exponentially with temperature T, i.e.

$$\sigma = \sigma_0 \exp\left(\frac{-\Delta E}{kT}\right), \tag{1}$$

where σ_0 is the electrical conductivity for $1/T \to 0$, ΔE the activation energy for conduction ($\Delta E = 0.6$ eV), and k the Boltzmann constant.

Secondly, a model for a semiconductor to metal transition at critical temperature T_c (B-type). Vanadium dioxide VO_2 is considered here, with $T_c = 68$ °C. The conductivity versus temperature curve is the same as described elsewhere [8].

2. Theory

The calculation method is described hereafter and permits the comparison with experimental results obtained on coplanar type threshold switching devices.

---

[1]) Present address: Cofelec—Choisy le Roi (France).

The static thermal equilibrium equation is written (Fig. 1):

$$\frac{d^2T}{dx^2} + E^2 \frac{\sigma(T)}{K} = a^2 T , \tag{2}$$

where E is the electric field in the semiconductor, K its thermal conductivity, a^2 a constant characteristic of the heat flow in the substrate and $T(x)$ the temperature of the film in a direction perpendicular to the current flow. The assumptions contained in this equation are the following [9]:

The temperature and the electric field are constant through the thickness of the semiconducting film.

The heat flow through the electrical contacts (along the y-axis) is negligible.

The thermal conductivity of the semiconductor is independent of the temperature.

The trivial solution of equation (2) is related to an uniform conduction process through the sample $\left(T(x) = \text{constant}\right)$:

$$\sigma(T) = \frac{a^2 K}{E^2} T . \tag{3}$$

This equation is classical and gives the uniform current flow solution of the thermal avalanche effect (thermistor effect).

The inhomogeneous temperature profile is obtained from equation (2) by considering the functional dependency $T(x)$. It is related to the "filamentary" conduction process. We obtain from equation (2)

$$\left[\left(\frac{dT}{dx}\right)^2\right]_{-l_2/2}^{x} = 2 \int\limits_{T_s}^{T(x)} \left[a^2 T' - \frac{E^2}{K} \sigma(T')\right] dT' , \tag{4}$$

where l_2 is the electrode width and T_s the temperature at $x = -l_2/2$.

From this latter equation, we can determine, for each applied voltage, a relation between $T(x)$ and dT/dx. In order to obtain the temperature profile $T(x)$ it is then necessary to make a precise numerical integration, because of the weak dependence of dT/dx on the x coordinate in the filament region. We have done this calculation with a computer.

For both cases of interest (A and B), computed temperature profiles are shown in Fig. 2, for two values of the total current flowing in the sample. Those values were determined in the uniform and filamentary conduction cases using the following expression:

$$I = E\,S \int\limits_{-l_2/2}^{+l_2/2} \sigma(x)\, dx , \tag{5}$$

where S is the thickness of the semiconducting film. The corresponding I–U characteristics are plotted in Fig. 3 and Fig. 4. Another interesting theoretical result, in the B case, is the possibility to relate the width w of the semiconductor region supporting a temperature above 68 °C and corresponding values of the

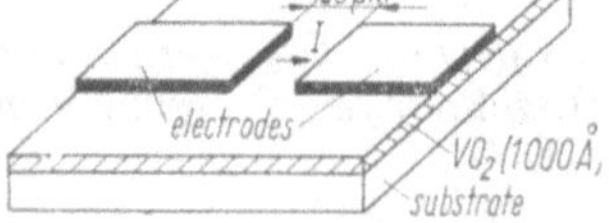

Fig. 1. Illustration of the coplanar structure to be analysed

current and of the voltage (Fig. 5). w is measured on a $T(x)$ profile corresponding to a given applied voltage.[2])

3. Discussion

It is possible to investigate the validity of these calculations by comparing these results with the experimental observations. Experimental temperature profiles are difficult to obtain, but approximate curves have been given by Stocker [10] for an amorphous material (A case) and they are in good qualitative

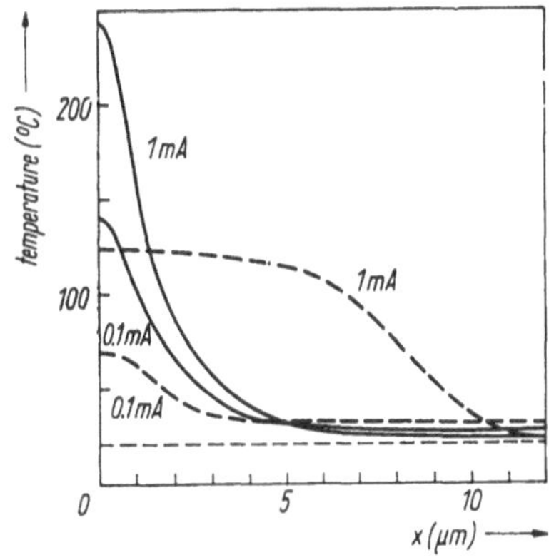

Fig. 2

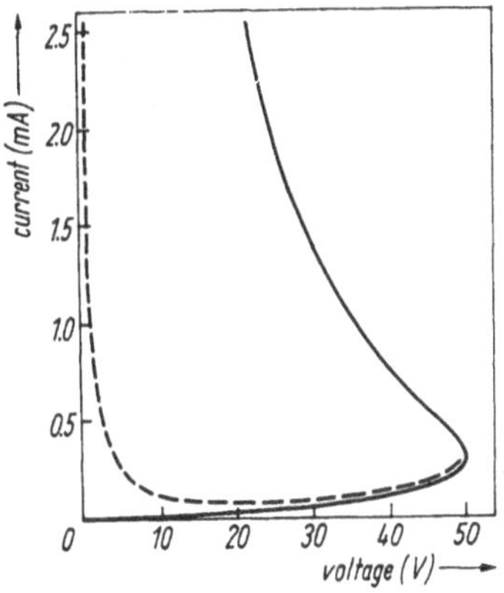

Fig. 3

Fig. 2. Temperature profiles calculated for two values of the total current flow. Full lines: semiconductor with A-type conductivity, $\Delta E = 0.6$ eV and σ (20 °C) $= 2.5 \times 10^{-2}\ \Omega^{-1}\ \text{cm}^{-1}$. Dashed lines: B-type transition material (The $\sigma(T)$ curve is shown in the insert of Fig. 4). T temperature increase above 20 °C, $K = 10^{-2}$ W/cm °K, $a^2 = 1.5 \times 10^7\ \text{cm}^{-2}$. The device dimensions are $l_1 = 50$ μm (length along current direction), $l_2 = 400$ μm (electrode width) and $s_2 = 1000$ Å (film thickness)

Fig. 3. Theoretical I–U curves for uniform (full line) and filamentary (dashed line) conduction (A-type)

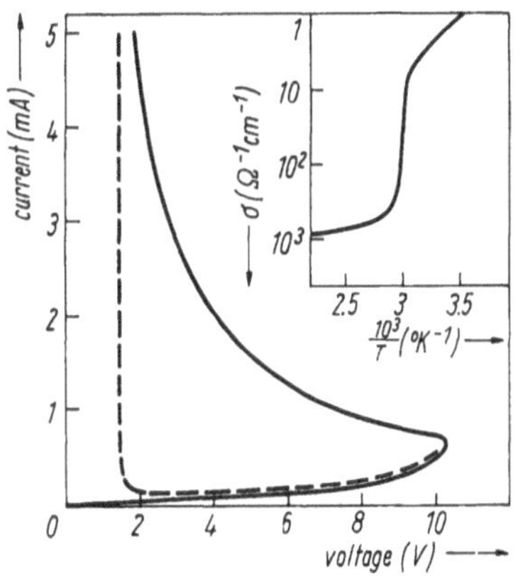

Fig. 4

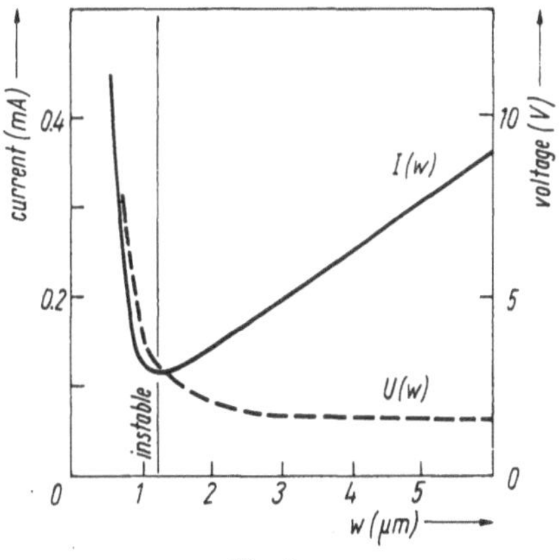

Fig. 5

Fig. 4. Theoretical I–U curves for uniform and filamentary conduction (full and dashed lines) (B-type). The inserted curve is the conductivity–temperature dependence

Fig. 5. Current and voltage versus filament width; curves deduced from the isothermal $T_c = 68$ °C in the temperature profiles of a transition device (B-type)

[2]) We note that for large values of the filament width, the voltage is nearly constant and the current increases linearly with the width. The region of small width is instable, as can be seen on the I–U characteristics.

agreement with our calculations. The case of VO_2 is very interesting because it presents a change in optical reflectance at the transition. We have so been able [11] to compare the width w of the switched region with the experimental corresponding values of voltage and current.[3]) On the other hand, I–U filamentary characteristics have been observed with amorphous semiconductor [1, 2] and VO_2 devices [12, 13]. Our calculations have shown that the discontinuous I–U characteristics can well be explained by the filamentary theory, with the too large values of current which were discussed above. The only difference with the theory is the positive dynamical resistance observed in the "on" region of some Ovonic devices [1]. We think this is due to the geometrical limitations of the device. In fact we have observed this feature in our coplanar VO_2 devices. By directly observing the filament, we have seen that the slope begins to be positive when the filament reaches the total inter-electrode volume. It cannot then develop laterally and that produces an increase of the filament temperature. This corresponds to a positive slope. In the Ovonic devices too, where the same process of lateral expansion of the filament exists [14], the positive slope can be due to these geometrical limitations.

4. Conclusion

The comparison between our computed results and the experimental observations seems to show that a thermal model can explain the switching and post-switching behaviour of a threshold amorphous semiconductor devices. However the problem of the existence of a filament remains unresolved. The thermal negative resistance devices correspond in fact to a uniform conduction since a filamentary one is related in all cases to a discontinuous switching. Then Ridley's [15] conclusions about the instability of a uniform type conduction do not seem sufficient and are to be discussed in some detail.

References

[1] S. R. Ovshinsky, Phys. Rev. Letters **21**, 1450 (1968).
[2] H. K. Henisch, Sci. American **221**, 30 (1969).
[3] D. C. Mattis, Phys. Rev. Letters **22**, 936 (1969).
[4] K. W. Böer and G. Döhler, phys. stat. sol. **36**, 679 (1969).
[5] H. Fritzche and S. R. Ovshinsky, J. non-crystall. Solids **4**, 464 (1970).
[6] K. W. Böer and S. R. Ovshinsky, J. appl. Phys. **41**, 2675 (1970).
[7] G. Adam and J. Duchene, Solid State Commun. **9**, 785 (1971).
[8] E. N. Fuls, D. H. Hensler, and A. R. Ross, Appl. Phys. Letters **10**, 199 (1967).
[9] C. N. Berglund and R. H. Walden, IEEE Trans. Electron Devices **17**, 137 (1970).
[10] H. J. Stocker, J. non-crystall. Solids **2**, 371 (1970).
[11] J. Duchene, M. Terraillon, P. Pailly, and G. Adam, Appl. Phys. Letters **19**, 115 (1971).
[12] M. Guntersdorfer, Solid State Electronics **13**, 369 (1970).
[13] R. H. Walden, IEEE Trans. Electron Devices **17**, 603 (1970).
[14] D. Schuocker, Appl. Phys. Letters **19**, 7 (1971).
[15] B. K. Ridley, Proc. Phys. Soc. **82**, 954 (1963).

(Received October 6, 1971)

[3]) The features of the experimental curves present a close similarity with the computed ones. However, the experimental values of current are higher. In fact, for the computation, we made the simplifying assumption that the electrodes were of infinite length. In practice, some more current can flow laterally between the electrode boundaries and also in the volume of the VO_2 film. We will found the same discrepancies on I–U characteristics, the measured current being somewhat larger than the computed one.

phys. stat. sol. (a) 8, 463 (1971)

Subject classification: 10; 22

Institut für Angewandte Physik, Technische Hochschule Wien

Dislocation Loops in Antimony Telluride

By

P. SKALICKY and E. WOLFGANG

The investigation of the diffraction contrast of large hexagonal dislocation loops in Sb_2Te_3 suggests that nearly all of them are multiple loops. A model which explains satisfactorily the contrast is described. At elevated temperatures the loops do not shrink but take a circular equilibrium shape. Dislocation loops with basal plane Burgers vectors which are probably due to a pinching-off of gliding dislocations are also observed.

Die Untersuchung des Beugungskontrastes von großen sechseckigen Versetzungsringen in Sb_2Te_3 zeigt, daß es sich nahezu immer um Mehrfachringe handelt. Es wird ein Modell beschrieben, das den experimentell beobachteten Kontrast befriedigend erklärt. Bei höheren Temperaturen verringern die Versetzungsringe nicht ihre Fläche, sondern nehmen eine kreisrunde Gleichgewichtsgestalt an. Versetzungsringe mit Burgersvektoren in der Basisebene wurden ebenfalls beobachtet. Sie sind wahrscheinlich auf Abschnürung von gleitenden Versetzungen zurückzuführen.

1. Introduction

Crystals of antimony telluride which are slowly cooled from the melt contain large hexagonal dislocation loops that show no stacking fault contrast. As suggested by Delavignette and Amelinckx [1] their origin is due to a deviation of Sb_2Te_3 crystals from stoichiometry. They contain an excess of antimony [2]. At higher temperatures more antimony atoms are found on tellurium sites than vice versa which causes an excess of vacancies in the Te^I–Te^I layers when the crystal is cooled from the melt. Due to the easier diffusion in the tellurium layers this leads to a precipitation of these vacancies in the form of loops. They are not quench loops because they are not caused by a thermally generated frozen-in supersaturation of point defects but by a deviation from stoichiometry. This explanation is also supported by the investigations of Bierly [3], who finds no dislocation loops in Bi_2Te_3 which is isomorphous to Sb_2Te_3 but shows a smaller deviation from stoichiometry. In the pseudo-binary alloy Bi_2Te_3–Sb_2Te_3 loops are therefore only observed up to a composition limit of 60% Bi_2Te_3. Since the assumption of single layer loops as suggested by Delavignette and Amelinckx is not sufficient to explain satisfactorily the observed diffraction contrast, further investigations were performed.

2. Electron Microscopy

2.1 Specimen preparation

Although thin foils of Sb_2Te_3 may be produced by cleavage, we used preferentially electropolishing. Such specimens are nearly dislocation-free and allow therefore a better investigation of isolated loops. The crystals were prepared by melting together 99.999% pure antimony and tellurium in a quartz tube under vacuum and continuous agitation. They were slowly cooled down to room temperature during 6 h. Discs of approximately 3 mm in diameter were cut from large crystal grains and electrolytically thinned until a small hole appeared. The electropolishing solution used was the following [4]:

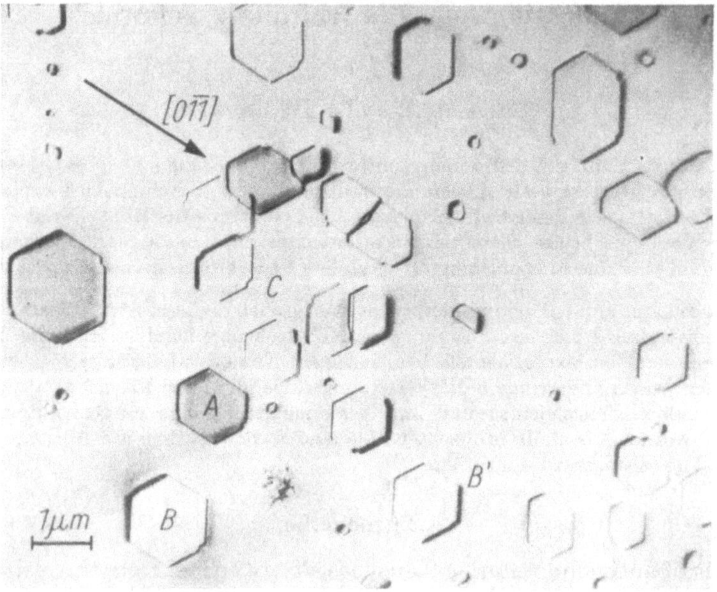

Fig. 1. Electropolished antimony telluride crystal slowly cooled from the melt containing large hexagonal loops

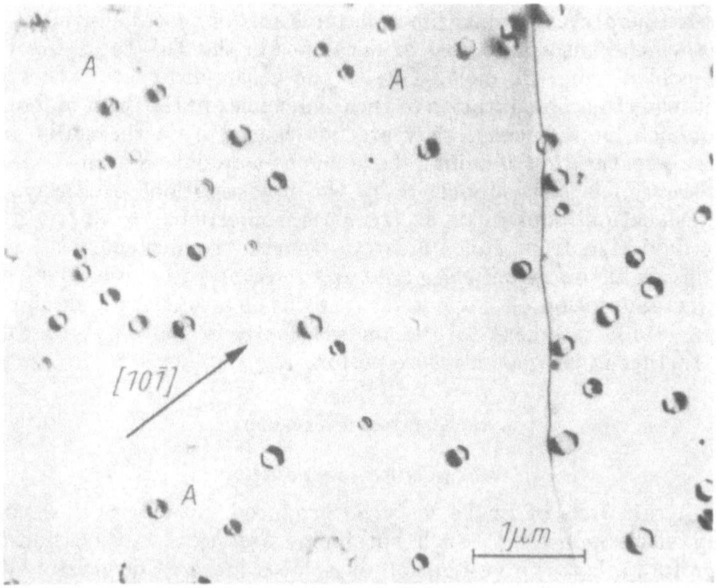

Fig. 2. Small dislocation loops in a crystal quenched from a temperature immediately below the melting point

415 ml NaOH (10%), 85 ml tartaric acid (40%) and 300 ml glycerine. All specimens were investigated in a Siemens Elmiskop I equipped with a goniometer stage at 100 kV.

Large dislocation loops with diameters up to 10 μm form when the crystals are kept at 50 to 70 °C below the melting point for some time. When the crystals are cooled down to room temperature, the loops will assume circular or oval shape. At very slow cooling rates hexagonal loops are formed (Fig. 1). Quenching of the crystals immediately after solidification produces small circular or hexagonal loops (Fig. 2).

2.2 Geometry and contrast of the dislocation loops

The structure of antimony telluride is a close packed layer structure ($R\bar{3}m$), the layer sequence is

$$\cdots -Te^{I}-Sb-Te^{II}-Sb-Te^{I}-Te^{I}-Sb-Te^{II}-Sb-Te^{I}-Te-^{I}\cdots .$$

According to Drabble and Goodman [5] weak van der Waal bonds exist between the Te^{I}–Te^{I} layers which explains the easy cleavage.

The compound shows a disorder of the antistructure type [6], i.e. antimony occupies tellurium sites and vice versa. This, and the fact that antimony and tellurium have nearly the same atomic scattering factor allows the diffraction patterns to be indexed using a small pseudo-cubic cell with $\alpha = 84.3°$ according to Francombe [7]. In the subsequent discussion we shall always refer to this reduced rhombohedral cell.

The condensation of excess vacancies in the form of loops is to be expected in the cleavage planes, i.e. the Te^{I}–Te^{I} layers. Since the stacking fault energy is rather high (no dissociation of dislocations is observed) the stacking fault inside the loops can be eliminated according to the reaction

$$\frac{a}{3}\langle 111\rangle + \frac{a}{3}\langle 2\bar{1}\bar{1}\rangle = a\langle 100\rangle . \quad (1)$$

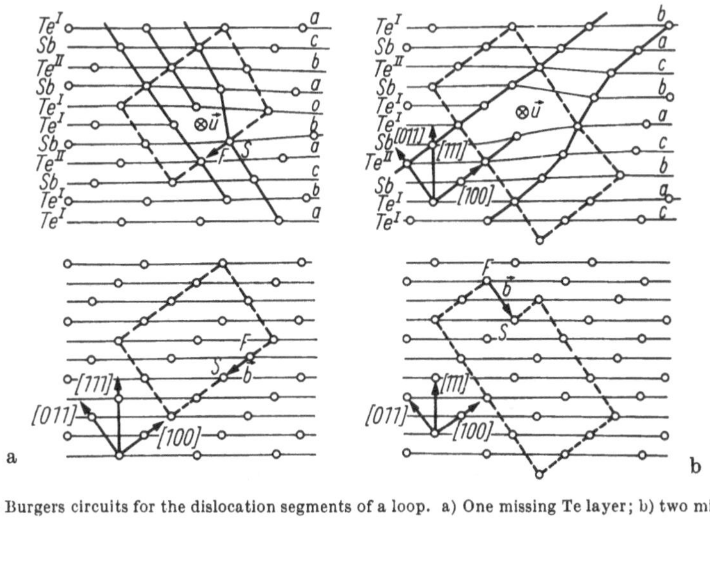

Fig. 3. The Burgers circuits for the dislocation segments of a loop. a) One missing Te layer; b) two missing layers

The remaining "chemical" stacking fault produces no contrast because of the small difference in the atomic scattering factors. Although it is energetically less favourable, the possibility of a double layer loop, i.e. of two missing Te layers, cannot be excluded. The geometrical stacking fault could also be eliminated by the reaction

$$\frac{2a}{3}\langle 111\rangle + \frac{a}{3}\langle \bar{2}11\rangle = a\langle 011\rangle\,. \qquad (2)$$

The corresponding Burgers circuits are shown in Fig. 3.

The segments of large loops can be regarded as isolated dislocations which run parallel to $\{110\}$ directions. For each loop two opposite segments have the character of pure edge dislocations and are therefore invisible under $\langle\bar{1}10\rangle$ diffraction conditions. The segments not parallel to the diffraction vector $\boldsymbol{g}$ show broad and widely different strong contrast. While the opposite segments of loop A in Fig. 1 show nearly the same contrast, it is strongly different for loop B. The contrast of the loop B′ is the mirror image of loop B. This is due

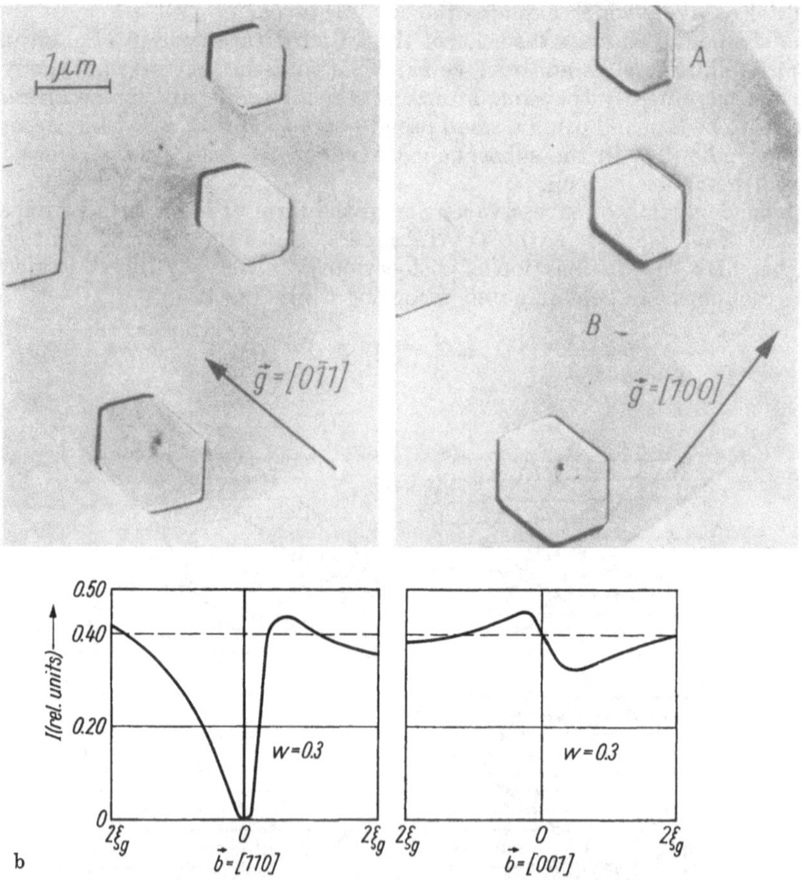

a

b

Fig. 4. Comparison of observed with calculated contrast of loop segments with either $\boldsymbol{b} = [110]$ or $\boldsymbol{b} = [001]$ Burgers vector. a) Micrographs; b) bright-field intensity profiles for $\boldsymbol{g} = [100]$ and an incident intensity of $I_0 = 1$. The width of the profiles is in units of the extinction distance ξ_g; $\boldsymbol{u} = [1\bar{1}0]$

to a diffraction effect. When the direction of the electron beam is reversed (e.g. by putting the foil upside down in the specimen holder) the contrast of the loops B and B′ will also be reversed. Photometer traces of such loop segments show frequently a double contrast which is not resolved in the prints. The double contrast of the loop C is due to the fact that it lies near the foil surface which was proved by taking stereo micrographs.

Quite generally, no good agreement could be found between the observed contrast and the dislocation contrast calculated for single dislocations and two-beam conditions. No big difference in the contrast of dislocations with ⟨100⟩ and ⟨110⟩ Burgers vectors can be expected for diffraction vectors which belong to the {111} zone, like ⟨$\bar{2}$11⟩ and ⟨0$\bar{1}$1⟩. A notable difference in the contrast should occur for a [100] diffraction vector but was not observed. Fig. 4 shows the calculated contrast profiles and the corresponding micrographs of dislocation loops for two-beam conditions with $\boldsymbol{g} = [100]$. A foil thickness of 3.5 ξ_g and a depth of the dislocation of 1.5 ξ_g below the surface of the foil was assumed (ξ_g is the extinction distance). Those segments of the loops which can be compared to the profiles show a stronger contrast than the calculations for $\boldsymbol{b} = [001]$ but do neither agree reasonably with the profile for $\boldsymbol{b} = [110]$.

It is furthermore notable that nearly all loops contain additional dislocation segments in the inside which are also invisible for $\boldsymbol{g} = \langle 1\bar{1}0\rangle$ diffraction vectors. An example for this is the loop A in Fig. 2. These observations and the contrast calculations suggest that most of the loops are multiple loops.

3. Model for a Multiple Loop

The most striking evidence for the occurrence of multiple loops are the additional "inside segments" which are nearly always found. As an example, we consider the contrast of the large isolated loop in Fig. 5. We can clearly

Table 1

Values of $\boldsymbol{g} \cdot \boldsymbol{b}$ and $\boldsymbol{g} \cdot \boldsymbol{b}_k$ (with the edge component $\boldsymbol{b}_k$ of the Burgers vector) for the two single dislocation loops of Fig. 5 and Fig. 7

u	$\boldsymbol{g} \cdot \boldsymbol{b}_1$	$\boldsymbol{g} \cdot \boldsymbol{b}_{k1}$	$\boldsymbol{g} \cdot \boldsymbol{b}_2$	$\boldsymbol{g} \cdot \boldsymbol{b}_{k2}$
		$\boldsymbol{g} = [0\bar{1}1]$		
[$\bar{1}$10]	0	−1/2	−1	−1/2
[0$\bar{1}$1]	0	0	−1	0
[$\bar{1}$01]	0	1/2	−1	−1
		$\boldsymbol{g} = [\bar{1}01]$		
[$\bar{1}$10]	−1	−1/2	0	−1/2
[0$\bar{1}$1]	−1	−1	0	1/2
[$\bar{1}$01]	−1	0	0	0
		$\boldsymbol{g} = [\bar{1}10]$		
[$\bar{1}$10]	1	0	−1	0
[0$\bar{1}$1]	1	1	−1	−1/2
[$\bar{1}$01]	1	1/2	−1	−1

distinguish between two separate loops from which alternatively one is invisible for the $\boldsymbol{g} = \langle\bar{1}01\rangle$ and $\boldsymbol{g} = \langle 0\bar{1}1\rangle$ diffraction vectors. In the corresponding drawings the contrast of the two loops is shown both isolated and superimposed. The two Burgers vectors are assumed to be $\boldsymbol{b}_1 = \langle 100\rangle$ and $\boldsymbol{b}_2 = \langle 010\rangle$. This agrees reasonably with the observed contrast and with Table 1, which gives the values for $\boldsymbol{g}\cdot\boldsymbol{b}$ and $\boldsymbol{g}\cdot\boldsymbol{b}_k$ (with the edge segment $\boldsymbol{b}_k$ of the Burgers vector) for both single loops. For example, the relatively weak contrast of the $\boldsymbol{u} = \langle\bar{1}10\rangle$ segment for $\boldsymbol{g} = \langle\bar{1}10\rangle$ has only terms which correspond to those of a pure screw segment. Four terms appear for the $\boldsymbol{u} = \langle\bar{1}01\rangle$ and $\boldsymbol{u} = \langle 0\bar{1}1\rangle$ segments under $\boldsymbol{g} = \langle\bar{1}10\rangle$ diffraction conditions which also show the strongest contrast in the brigth-field micrograph.

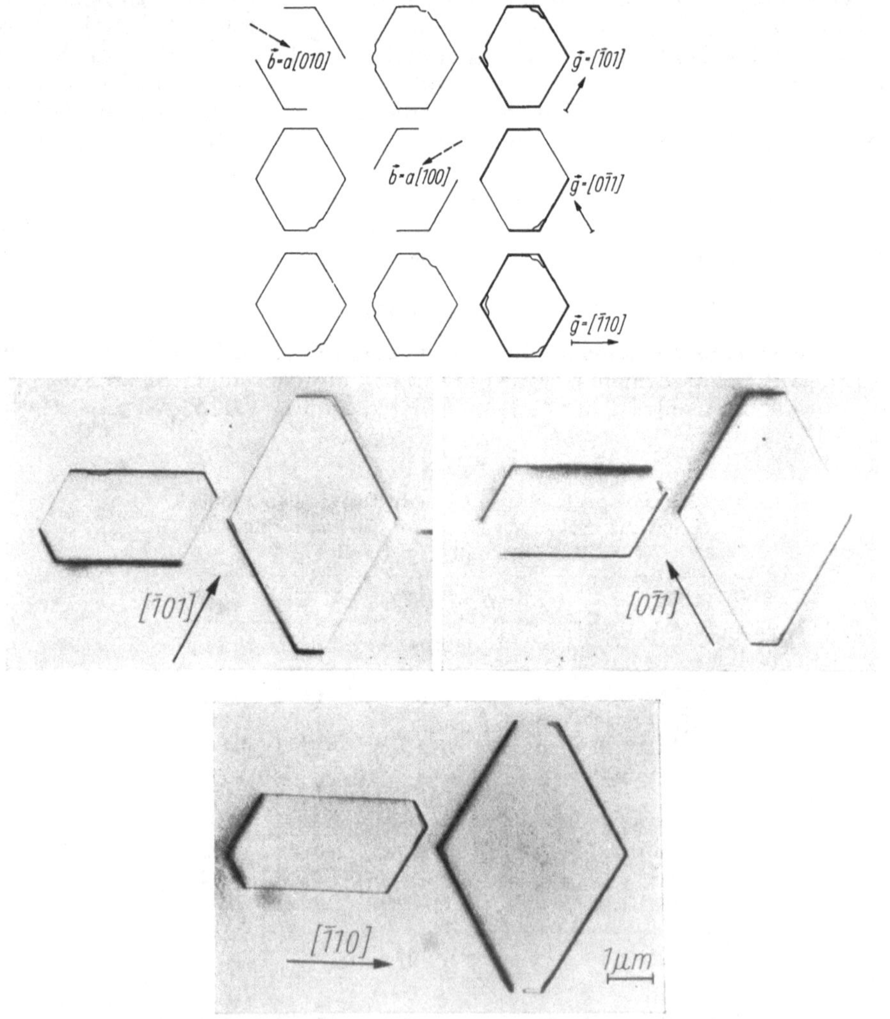

Fig. 5. The contrast of two large isolated dislocation loops under three different diffraction conditions. The drawings explain the contrast of the larger loop

If we assume however, that no reaction between the single loops has taken place, we cannot explain the fact that the $\langle\bar{1}10\rangle$ segment is completely invisible in the $\boldsymbol{g} = \langle\bar{1}10\rangle$ reflection. Only if a reaction according to

$$a\langle 100\rangle + a\langle 010\rangle = a\langle 110\rangle \tag{3}$$

takes place, then the resulting pure edge dislocation will be invisible in the $\boldsymbol{g} = \langle\bar{1}10\rangle$ reflection because in this case both conditions $\boldsymbol{g}\cdot\boldsymbol{b} = 0$ and $\boldsymbol{g}\cdot\boldsymbol{b}_{\mathrm{k}} = 0$ are fullfilled.

To investigate the possibility for such a reaction, we consider the expression for the interaction energy between two parallel straight dislocations according to Nabarro [8]:

$$W = -\frac{G}{2\pi}\left[\boldsymbol{s}_1\cdot\boldsymbol{s}_2\ln r + \frac{1}{1-\nu}\left(\boldsymbol{b}_{k1}\cdot\boldsymbol{b}_{k2}\ln r - \boldsymbol{b}_{k1}\cdot\boldsymbol{r}\cdot\boldsymbol{b}_{k2}\cdot\boldsymbol{r}\,\frac{1}{r^2}\right)\right]. \tag{4}$$

The screw- and edge components of the Burgers vectors of the two dislocations are $\boldsymbol{s}_1, \boldsymbol{s}_2$ and $\boldsymbol{b}_{\mathrm{k}1}$, $\boldsymbol{b}_{\mathrm{k}2}$, respectively. The vector $\boldsymbol{r}$ is the shortest distance between the dislocation lines. The force between the dislocations is given by $\boldsymbol{F} = -\operatorname{grad} W$. Since equation (4) is derived for isolated dislocations, it can only be applied to large loops where the segments can be treated as independent.

We consider first such segments of a double loop which can be completely invisible, like the short segments in Fig. 5. For $\boldsymbol{r} = r\langle 001\rangle$ the third term in equation (4) vanishes and we obtain an attractive force only (Fig. 6a). Such segments of the two single loops could therefore approach each other by combined gliding and climbing. Such a configuration is not possible for those segments, where only one of the single loops is invisible. For this case the screw component of one of the Burgers vectors vanishes and the two edge components are perpendicular to one another (Fig. 6b). Only the third term, which depends on the angle between the two glide planes, remains in equation (4). An attraction does not occur between such segments. We think that this may be an explanation for our observation that obviously a reaction takes place only in segments for which complete invisibility is found. The other segments of a composite loop consist of parallel single dislocations. A model of the dislocation loop in Fig. 5, drawn according to these considerations, is shown in Fig. 7. A con-

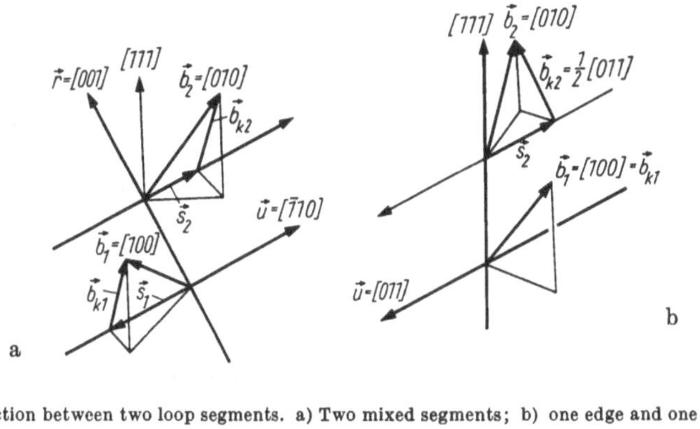

Fig. 6. Interaction between two loop segments. a) Two mixed segments; b) one edge and one mixed segment

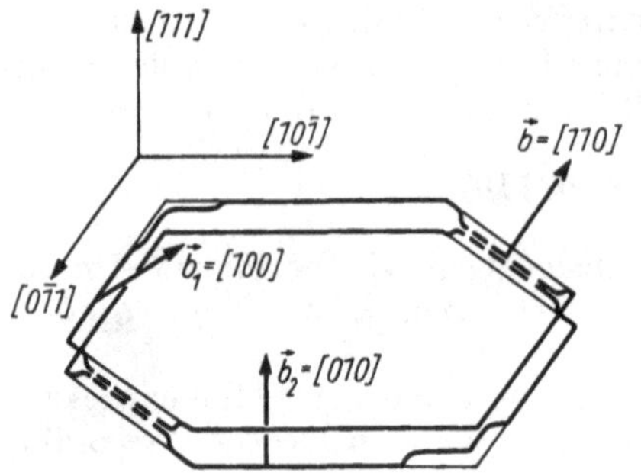

Fig. 7. Model of a double layer loop. The drawing refers to the large loop in Fig. 5

figuration of two sinlge loops with parallel Burgers vectors is unfavourable since they would repel each other.

Two single loops which form a multiple loop could in principle seperate by gliding on their respective glide prisms. If a reaction takes place in one of the segments the resulting dislocation has a new glide plane and will therefore block the gliding of the whole loop. We have in fact experimentally found only complete immobility of the loops.

3.1 Generation of the dislocation loops

Delavignette and Amelinckx [1] found that the centre of the prismatic loops is a preferred site for the nucleation of a second concentric loop. According to our observations we must assume that only multiple loops occur. Reactions between their segments happen frequently when they are still small and round. Some of the single layer loops which form multiple loops do not reach the same size. We think that this mechanism is responsible for the shape of the loops marked A in Fig. 2.

A similar explanation is proposed for the "inside segments" which are frequently observed in large loops (Fig. 5). They are very probably due to a smaller growth rate at the corners of one of the single loops. This could be caused by a too small concentration of vacancies or too slow diffusion. This explanation is also supported by the observation that these inside segments disappear when the specimens are heated.

4. Shape of Dislocation Loops

As already mentioned, the shape of the loops at room temperature depends on how the crystals are cooled from the melt. Crystals which are quenched from immediately below the melting point contain no loops because there is no time for vacancy diffusion. At 50 to 70 °C below the melting point the loops are circular or oval. If the crystal is quenched from this temperature, these shapes are preserved (Fig. 2). Hexagonal loops appear when the crystal is cooled slowly to room temperature. Many of the loops contain superjogs at their corners where the vacancy diffusion is not sufficient to smooth them out (Fig. 1). When the crystals are kept for several hours at a temperature of about 70 °C below the melting point, many concentric loops, which are multiple loops themselves, appear (Fig. 8). We have observed up to six concentric loops with sizes up to 10 μm.

Fig. 8. Concentric loops in a crystal that was kept several hours at a temperature of about 70 °C below the melting point

Fig. 9. A circular loop hindered from reaching its equilibrium shape

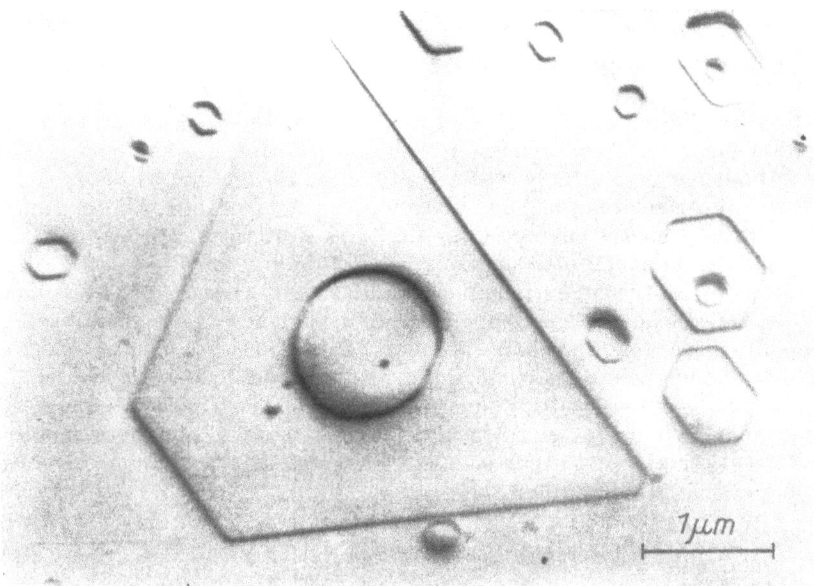

The concentric loops in Fig. 9 are an interesting configuration. The fast growth of the large, nearly triangular loop has obviously greatly reduced the vacancy supersaturation inside it. The smaller circular loop could therefore not reach the equilibrium shape. It remained in a shape which is characteristic for temperatures immediately below the melting point although all other loops in this foil have assumed a hexagonal shape.

The temperature dependence of the shape of the loops was also investigated by heating specimens, which already contained hexagonal loops, during the examination in the microscope. Fig. 10 shows the same loops before and after the specimen was heated to a temperature near the melting point. The foil was heated until the thinnest parts started to melt and kept at that temperature for 30 min. The loops did not shrink but rounded off independently of their

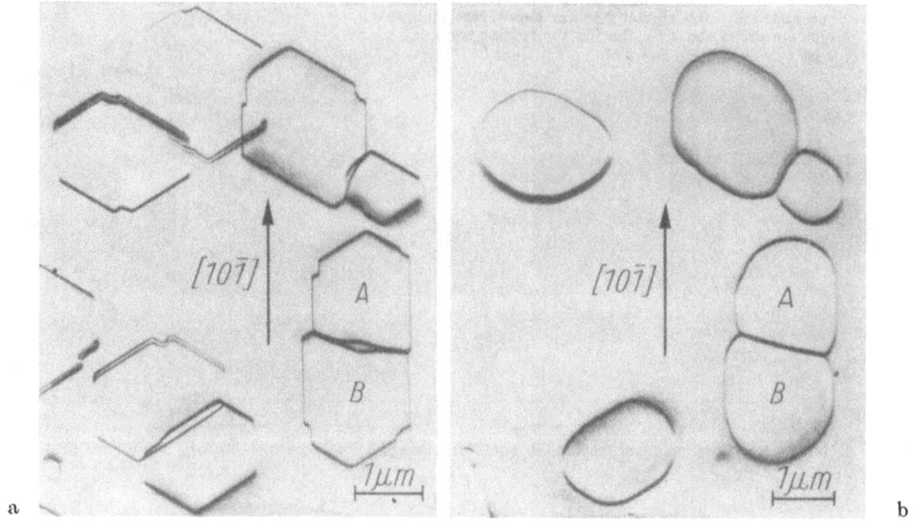

Fig. 10. Large loops. a) Before, and b) after a heat treatment near the melting point in the microscope

size. This can be regarded as further evidence for the assumption that these dislocation loops are due to the non-stoichiometry and are not quench loops. Some of the loops have disappeared after the heat treatment. They lay so close to the crystal surface that they could leave the foil. In Fig. 10a they show the double contrast which is characteristic for loops close to the surface.

The equilibrium shape of a single dislocation loop is an elongated hexagon. Since the energy of the edge segments is higher than that of the mixed segments, the loops tend to shorten their edge segments [1]. The edge segments will also presumably climb somewhat more rapidly than the other segments. The loops are therefore elongated parallel to the projection of the Burgers vector on to the basal plane. For a multiple loop consisting of two single loops, the edge segments of which make an angle of 120°, the shortest segments should be those where a reaction has occurred. These are the edge segments of the resulting loop and this is actually observed.

5. Interaction between Dislocation Loops

Dislocation loops which lie on top of each other or close together in the same plane can repel or attract each other depending on their Burgers vectors. The two loops A and B in Fig. 10 repel each other and were therefore handicapped during their growth. Since they cannot glide in the basal (111) plane, they cannot "make way" for another. As a consequence, they are both flattened on one side after the heat treatment which caused the rounding-off. The repelling segments lie parallel to each other at an equilibrium distance.

An attraction between loop segments can also be a handicap during loop growth. The micrograph with $\boldsymbol{g} = \langle 10\bar{1} \rangle$ contrast in Fig. 11 shows that one of the presumably two dislocations which form the $\langle 0\bar{1}1 \rangle$ segment of loop A bulges out due to the attraction of loop B. This is obviously also a multiple loop. It seems however, that all of its dislocation segments are attracted by loop A and were therefore hampered during their climbing.

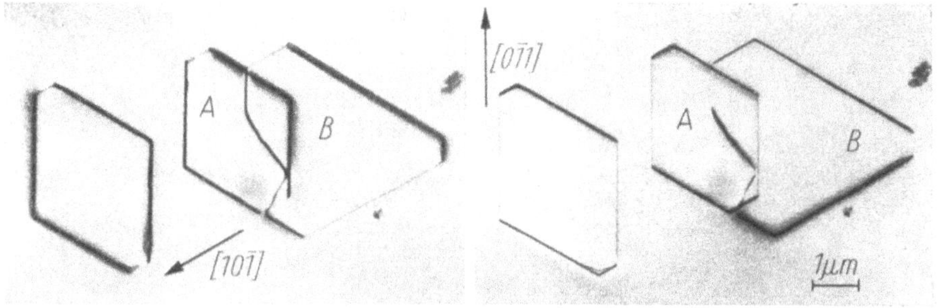

a b

Fig. 11. Interaction between two dislocation loops photographed under two different diffraction conditions to make all segments visible. a) $\boldsymbol{g} = \langle 101 \rangle$; b) $\boldsymbol{g} = \langle 011 \rangle$

Because of the large number of dislocation segments which are involved in such a process it is very difficult to obtain a quantitative analysis of the forces between them. Since loops which are influenced by each other must lie close together, it was also not possible to determine their distance by stereo measurements.

6. Non-Prismatic Dislocation Loops

Hexagonal dislocation loops are sometimes surrounded by one or more circular loops which lie in the same plane. Such dislocation rings (Fig. 12) are invisible under $\boldsymbol{g} = \langle \bar{2}11 \rangle$ diffraction conditions. They show the same contrast as dislocations with basal plane Burgers vectors and are never observed isolated. Such dislocation rings repel each other since they have parallel Burgers vectors. We find that the distance between them is independent of the temperature.

The most probable generation mechanism for such dislocation rings is the pinching-off of dislocations which glide in the (111) planes. The sessile hexagonale loops are obstacles for gliding basal plane dislocations. We assume that the dislocations forming the circular loops are generated by the same dis-

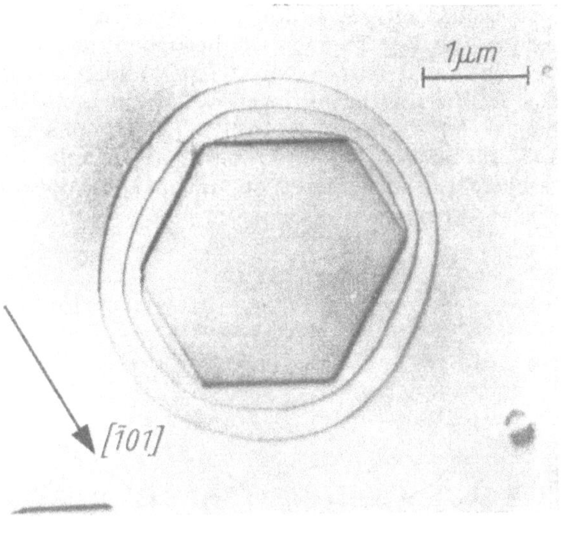

Fig. 12. Circular dislocation rings with basal plane Burgers vectors surrounding a hexagonal loop

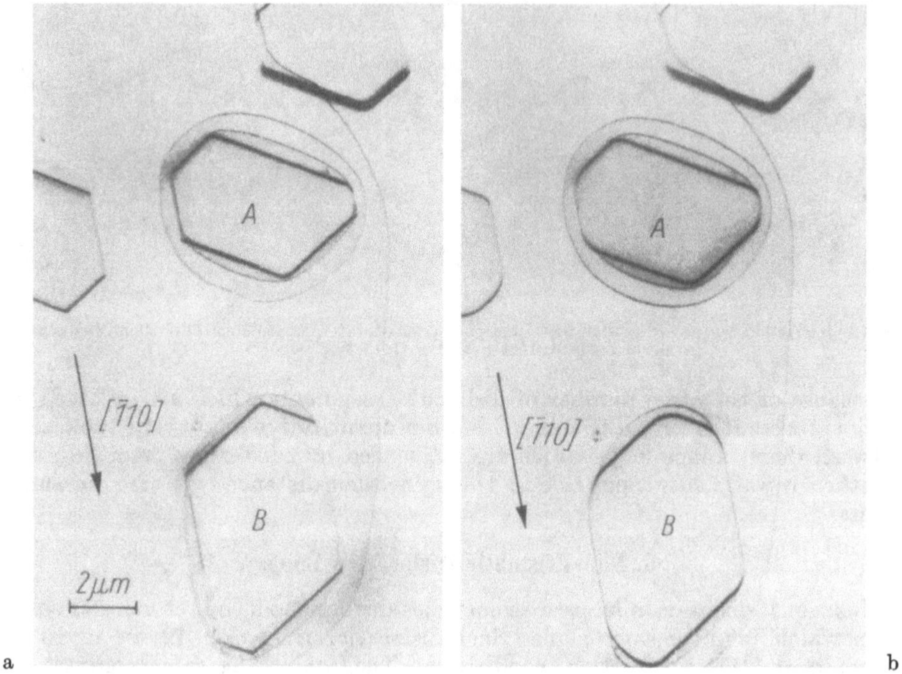

Fig. 13. Hexagonal loops and circular rings. a) Before, and b) after a heat treatment in the microscope

location source and are hampered by the prismatic loop in their glide plane. Other loops which are visible in the same micrograph lie obviously in different height and do therefore not interfere with the gliding dislocations. We could not observe a dislocation source.

The two hexagonal loops A and B in Fig. 13 lie in nearly the same plane and are therefore an obstacle for the long gliding dislocation which has not yet been pinched off. The foil was heated in the microscope in the same way as described above. Thereby the hexagonal loops rounded off, while the circular rings did not alter their shape. This is due to the fact that at room temperature an equal distribution of jogs around the loop is not energetically favourable. The dislocation rings with basal plane Burgers vectors do not contain jogs but kinks and remain therefore round independent of the temperature. The higher energy of the jogs causes an unequal jog distribution around the prismatic loops and therefore the hexagonal shape with straight segments.

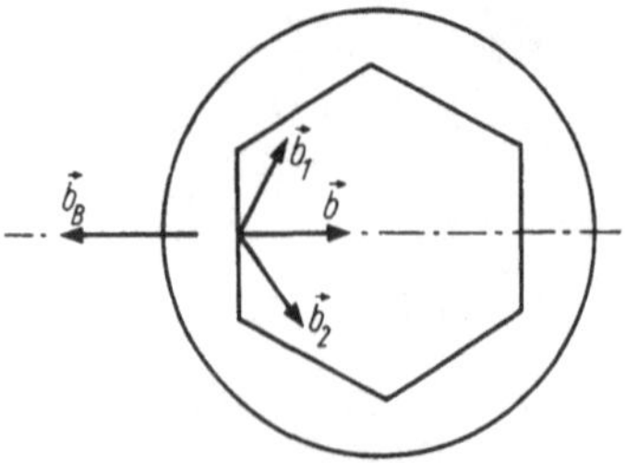

Fig. 14. The Burgers vectors of the loops in Fig. 13

The innermost circular rings are attracted by the hexagonal loop, as can be seen in Fig. 13b, which was taken after the heat treatment. The Burgers vectors of this loop configuration as determined from contrast experiments are shown in Fig. 14.

7. Conclusions

From the observation of the contrast of large prismatic dislocation loops in Sb_2Te_3 we conclude that nearly all loops are multiple loops. They consist of at least two single loops which show reactions in some segments. A model for a multiple loop such as the one in Fig. 7 can be successfully employed to explain the observed contrast.

The equilibrium shape of the dislocation loops but not their size depends on the temperature. This can be regarded as further evidence for the assumption that these loops are not quench loops but accommodate for the non-stoichiometry of the crystals.

The dislocation rings with basal plane Burgers vectors, which were also observed, have their origin very probably in the pinching-off of gliding dislocations. It was already noted by Delavignette and Amelinckx [1] that such a picture may account for the large influence of small changes in stoichiometric composition on the mechanical properties of such a compound.

Acknowledgements

The authors wish to thank Prof. Amelinckx for stimulating discussions and Prof. Lihl for his interest and encouragement. Financial support by the Oesterreichischer Forschungsrat and the Oesterreichische Nationalbank is also gratefully acknowledged.

References

[1] P. Delavignette and S. Amelinckx, Phil. Mag. **6**, 601 (1961).
[2] G. Offergeld and J. van Cakenberghe, J. Phys. Chem. Solids **11**, 310 (1959).
[3] J. N. Bierly, V. Internat. Congr. Electron Microscopy, Philadelphia 1962, J-14.
[4] D. M. Moon and R. Stickler, Prakt. Metallographie **4**, 137 (1967).
[5] J. R. Drabble and C. H. L. Goodman, J. Phys. Chem. Solids **5**, 142 (1958).
[6] F. A. Kröger and H. J. Vink, Solid State Phys. **3**, 310 (1956).
[7] M. H. Francombe, Phil. Mag. **10**, 989 (1964).
[8] F. R. N. Nabarro, Theory of Crystal Dislocations, Clarendon Press, Oxford 1967.

(Received August 6, 1971)

phys. stat. sol. (a) 8, 477 (1971)

Subject classification: 20.2; 22.2.1; 22.2.4

Zentralinstitut für Optik und Spektroskopie der Deutschen Akademie der Wissenschaften zu Berlin (a),
and P. N. Lebedev Physical Institute, Academy of Sciences of the USSR, Moscow (b)

On the Energy Dependence of the Internal Losses in Injection Lasers

By

H. BACHERT (a), P. G. ELISEEV (b), A. KEIPER (a), and S. RAAB (a)

The method of external optical coupling was used for the estimation of the energy dependence of the internal optical losses in injection lasers of different types. It is shown that in GaAs lasers with homojunctions and with single heterostructures the internal optical losses increase when the photon energy increases due to an influence of the band-to-band absorption within the passive layers. But no significant influence of the dispersion of the internal losses was observed in lasers with double heterostructures. The analysis of the waveguiding effect leads to the conclusion that either the internal losses are energy-independent or that they decrease in some cases when the photon energy increases in such a laser.

Для оценки спектральной зависимости внутренних оптических потерь в инжекционных лазерах разных типов использован метод вариации внешней оптической обратной связи. Показано, что в GaAs-лазерах на гомопереходах и на односторонних гетероструктурах вследствие влияния междузонного поглощения в пассивных слоях оптические потери увеличиваются с ростом энергии фотона, тогда как в лазерах с двойной гетероструктурой не наблюдается сколько-нибудь существенное влияние дисперсии внутренних потерь. Анализ волноводного эффекта приводит к выводу, что в таких лазерах внутренние потери либо не зависят от энергии фотона, либо несколько уменьшаются с её ростом.

1. Introduction

In recent years many experiments have been performed with injection laser devices using external selective elements to improve the spectral output of lasers and laser arrays [1 to 5]. The introduction of these elements leads to a change of the threshold condition. It has been shown in a previous paper [6], that for external optical coupling of the GaAs diffusion diode lasers the change of the threshold current density and the shift of the stimulated emission energy cannot be explained in the simple model with constant coefficient of internal optical losses α_i. Obviously, the energy dependence of α_i must be taken into account.

In this paper we present the results of such investigations with all types of the GaAs injection lasers in order to compare the behaviour of α_i in the various optical structures. We discuss also the nature of the energy dependence of α_i.

2. The Optical Coupling Experiments with Lasers of Different Types

In the case of the absence of $\boldsymbol{k}$-selection rule, assuming an exponential state density of electrons in the conduction band and a δ-function for the state density of holes in the valence band the optical gain can be expressed by the following equation:

$$g = C \exp\left(-\frac{E_-}{E_0}\right) \tanh\left(\frac{E_-}{2\,kT}\right), \tag{1}$$

where E_0 is the parameter of the state density of electrons, $E_- = E_g + E_{fc} + E_{fv} - \hbar\,\omega$, E_g is the bandgap, $\hbar\,\omega$ is the emitted laser energy, E_{fc} and E_{fv} are the energies of the quasi-Fermi level in the conduction and valence bands, respectively, C is an experimental constant. E_- can be written for the maximum of the optical gain

$$E_{-\,\max.} = 2\,kT \operatorname{artanh}\left[\frac{1}{E_0}\left(\sqrt{(kT)^2 + E_0^2} - kT\right)\right]. \tag{2}$$

The internal optical losses α_i are given by

$$\alpha_i = \alpha_{c_0} + \alpha_{bb}\,, \tag{3}$$

where α_{c_0} represents an energy-independent and α_{bb} the energy-dependent part of losses. α_{bb} can be expressed by [6]

$$\alpha_{bb} = \alpha_{bb_0} \exp\left(\frac{\Delta E}{E_0}\right). \tag{4}$$

Because of this α_i is a monotonically increasing function of the energy. Therefore, the maximum of the stimulated emission cannot coincide with the maximum of the optical gain. This connection between the internal optical losses and the optical gain (Fig. 1) shows that the energy of the stimulated emission is lower than the energy of the gain maximum.

In the case of definite change of transmission losses this model can be applied in order to find an agreement between the experimental and theoretical results.

By the extrapolation of the experimental values for $1/l \to 0$ (l length of the Fabry-Perot resonator) it is possible to estimate $j_{th._0} = \alpha_{i_0}/\beta_0$, where $\alpha_{i_0} = \alpha_{c_0} + \alpha_{bb_0}$. Using the equation [7]

$$\Delta E = [(kT)^2 + E_0^2]^{1/2} \ln \frac{j_{th.}}{j_{th._0}} \tag{5}$$

a hypothetical emission energy E^0 can be calculated for laser diodes without transmission losses (see also Fig. 2). With the values $j_{th._0}$, α_{i_0}, and E^0 the optical

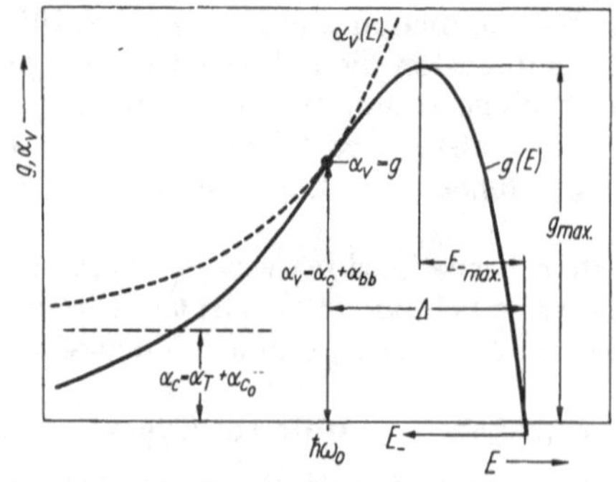

Fig. 1. The dependence of the optical gain and losses at the laser threshold on the emission energy (schematically)

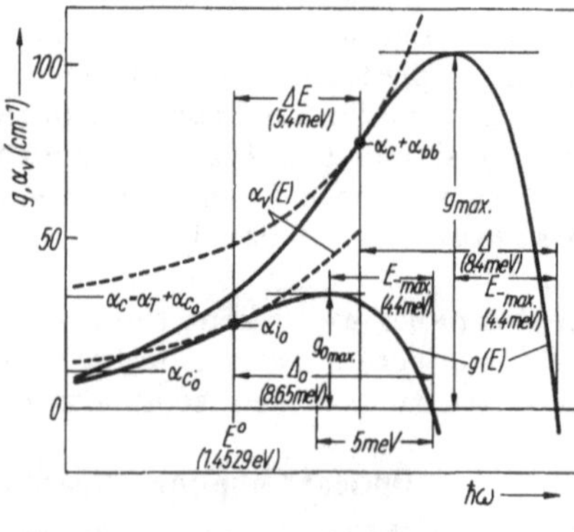

Fig. 2. Calculated curves for optical gain g_0 and g vs. the emission energy. $1/l = 0$: $\alpha_{c_0} = 10.4\ \text{cm}^{-1}$, $\alpha_{i_0} = 24.6\ \text{cm}^{-1}$, $g_{0\,\max.} = 33.6\ \text{cm}^{-1}$; $1/l = 20\ \text{cm}^{-1}$: $\alpha_c = 33.4\ \text{cm}^{-1}$, $\alpha_c + \alpha_{bb} = 77.0\ \text{cm}^{-1}$, $g_{\max.} = 104\ \text{cm}^{-1}$

gain in connection with energy-dependent internal losses $\alpha_i(E)$ can be determined by (1). These calculations are the base for a mathematical connection between optical gain and energy dependent internal losses. Fig. 2 shows calculated curves for g_0 and g, where Δ can be estimated by

$$\sinh\left(\frac{\Delta}{kT}\right) = \frac{\alpha_v E_0}{\alpha_c kT} \tag{6}$$

and $\alpha_v = \alpha_T + \alpha_i$, α_T are the transmission losses. Using the equation [7]

$$\ln \frac{g_{\max.}}{g_{0\,\max.}} = \left[\frac{(kT)^2 + E_0^2}{E_0}\right]^{1/2} \ln \frac{j_{\text{th.}}}{j_{\text{th.}_0}} \tag{7}$$

it is possible to estimate $j_{\text{th.}}$ for given transmission losses.

Now the connection between the optical gain and the internal optical losses in the case of the external optical coupling can be considered in detail. In Fig. 3 (the values for optical coupling are designated by *) it can be seen why a small change of the transmission losses causes an essential lowering of laser threshold and therefore an adequate shift of emission energy to lower energies. The experimental values used in Fig. 2 and 3 are from injection lasers prepared by the conventional diffusion technique.

$g_{\max.}$ can be determined by equation (1) if Δ and α_v are known. Further investigations show that injection lasers with higher internal losses have also a higher portion of energy-dependent losses. Because of this the external optical coupling effects a stronger decrease of the threshold current density (Fig. 4a). In Fig. 4a we used diffused injection lasers with different internal losses.

The investigations were extended to epitaxial injection lasers. We used epitaxial injection lasers with $E_0 = 5.1$ meV and internal losses $\alpha_i = 21\ \text{cm}^{-1}$. These values are not very different from the values of diffused laser diodes (Fig. 4b).

Further investigations of the external optical coupling have been carried out with heterojunction injection lasers. The used single-heterostructure (SHS) laser diodes are of the type nGaAs–pGaAs–p$Ga_{1-x}Al_xAs$ with a pGaAs layer 3 to 6 μm thick and internal losses $\alpha_i = 38\ \text{cm}^{-1}$. We also observed a distinct energy dependence of the internal losses which can be concluded from the strong threshold current lowering. Because of the relatively large thickness of the pGaAs layer the effect of band–band losses was essential.

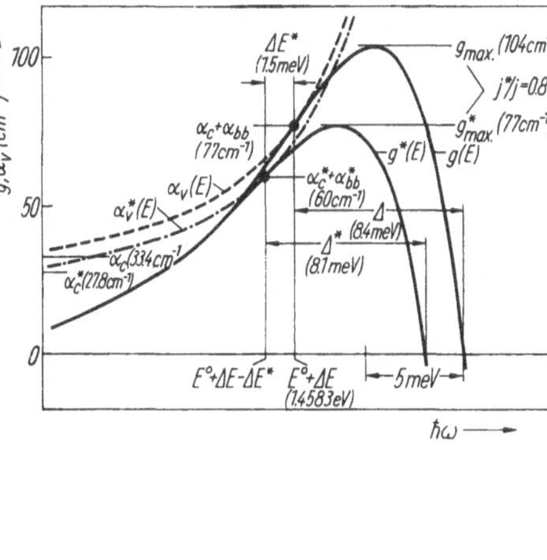

Fig. 3. Optical gain and losses with and without optical coupling as a function of the emission energy. $1/l = 20\ \text{cm}^{-1}$: $\alpha_T = 22.8\ \text{cm}^{-1}$, $\alpha_T^* = 17.4\ \text{cm}^{-1}$, $\alpha_{bb} = 43.8\ \text{cm}^{-1}$, $\alpha_{bb}^* = 32.0\ \text{cm}^{-1}$; $E_{-\max.} = 4.4$ meV

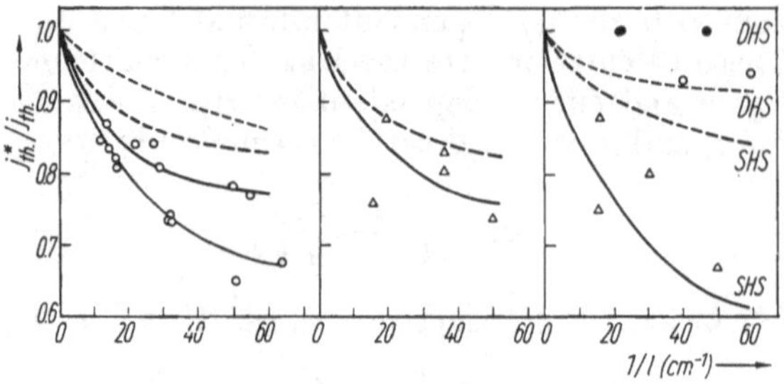

a b c

Fig. 4. Threshold current density lowering vs. reciprocal diode length at 80 °K. a) Diffused laser diodes. Theoretical curves without (– – –) and with (———) energy dependence of α_i for $\alpha_{i0} = 53\ cm^{-1}$, $E_0 = 4.1$ meV, ○ experimental values. Theoretical curves without (— — —) and with (———) energy dependence of α_i for $\alpha_{i0} = 24.6\ cm^{-1}$, $E_0 = 4.8$ meV, ○ experimental values. b) Epitaxial laser diodes. Theoretical curves without (— — —) and with (———) energy dependence of α_i, △ experimental values. c) Heterostructure laser diodes. Theoretical curves without (— — —) and with (———) energy dependence of α_i for SHS diodes, △ experimental values. Theoretical curves without energy dependence of α_i for DHS diodes (– – –), $d = 1.08\ \mu m$, ○ experimental values, ● experimental values for $d = 0.8\ \mu m$

The value of E_0 amounted to 7 meV. This higher value allows a stronger shift of the stimulated emission energy to lower energy which explained the strong threshold current lowering (Fig. 4c). The influence of the higher value of E_0 in the exponent of energy-dependent losses (4) was not so strong.

Similar experiments have been made with double-heterostructure (DHS) injection lasers of the type $nGaAs–nGa_{1-x}Al_xAs–pGaAs–pGa_{1-x}Al_xAs$. These laser diodes were produced with the well-known stripe geometry. The width of these stripes was about 50 μm. The thickness d of the optical active region amounted for one kind of samples to $d = 1.05\ \mu m$ and for another to $d = 0.8\ \mu m$. At such low values of d the divergence of the emitted radiation increased considerably [8]. For this reason R_2 in $\alpha_T = 1/\ln (R_1 R_2)^{-1/2}$ has an estimated value of 0.4 for $d = 1.05\ \mu m$ in the case of the optical coupling by an external mirror. On these conditions the change of the threshold current densities in dependence on the reciprocal resonator length was calculated without the energy-dependent internal losses. Because of the low internal losses of $\alpha_i = 10\ cm^{-1}$ the energy dependence of these losses could be pointed out experimentally. The experimental results (Fig. 4c) were only due to the change of the transmission losses caused by external optical coupling.

For injection laser with double heterostructure and optical active region $d = 0.8\ \mu m$ the divergence of the stimulated emission increased so strong that external optical coupling cannot influence the threshold current density under the applied experimental conditions.

3. Considerations of the Nature of the Energy Dependence of the Internal Optical Losses

The experimental results of the previous section may be resumed in the following way. Diffusion and epitaxial lasers (homojunction lasers) as well as SHS lasers demonstrated the significant contribution of the energy-dependent part of internal optical losses. Obviously, α_i increases when photon energy rises. On the contrary, DHS lasers are found to be in an agreement with the assumption of constant α_i. The origin of the spectral dependence of α_i is the band-to-band absorption of laser radiation in the two passive regions of the homojunction lasers and in the passive n-region of the SHS lasers. As to DHS lasers, the absorption is excluded evidently due to a) the shift of the absorption edge to higher energy in the passive $Ga_{1-x}Al_xAs$ regions, b) the more close optical confinement, preventing the deep penetration of the radiation in a passive region. It is necessary to consider other reasons for the energy dependence of α_i. This quantity can be expressed by a combination of the absorption coefficients in the passive regions, α_n and α_p (for n- and p-regions, respectively) and of the coefficient for the active layer α_d relating to dissipative losses (only to the practically irreversible absorption):

$$\alpha_i = \alpha_d \Gamma + \frac{\alpha_n + \alpha_p}{2} (1 - \Gamma) , \tag{8}$$

where Γ is the fraction of the radiation flux, which is confined to the active layer. The band-to-band transitions can appear in the second term of the expression (8) and lead to the positive value of $\partial\alpha_i/\partial(\hbar\,\omega)$. α_d can have a negative slope of the energy dependence, if the free-carrier absorption is the predominating process. Usually this dependence is rather weak. At last, the confinement parameter Γ also can have an energy dependence due to different dispersion of the dielectric constants ε of the active and passive regions. In the homojunction lasers this effect may be neglected, but in heterojunction lasers it must be considered especially. Thus, if we use the linear interpolation between ε for GaAs and AlAs [9 to 13] we obtain the waveguide effect in a double heterostructure which strongly depends on the photon energy. It disappears when the energy is small enough, as it is shown in Fig. 5. We performed an approximate estimation of the α_i for a homojunction laser and DHS lasers using a three-layer optical model with symmetrical distribution of ε. The result is presented in Fig. 6. It is seen that for DHS lasers the energy dependence of α_i either is

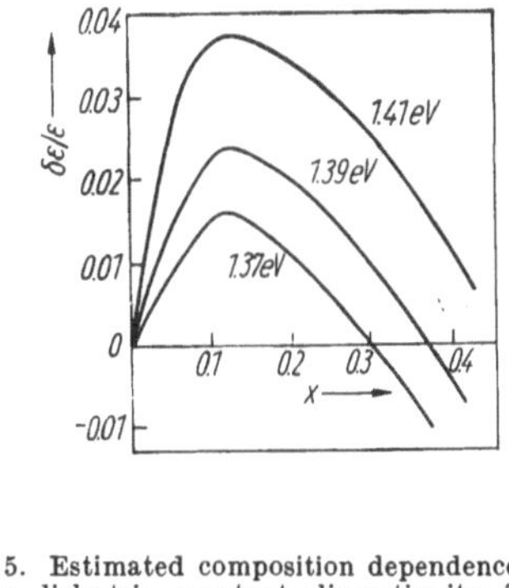

Fig. 5. Estimated composition dependence of the relative dielectric constant discontinuity $\delta\varepsilon/\varepsilon$ in the heterojunction $GaAs-Ga_{1-x}Al_xAs$ using refractive index data for GaAs and AlAs from [9 to 11]

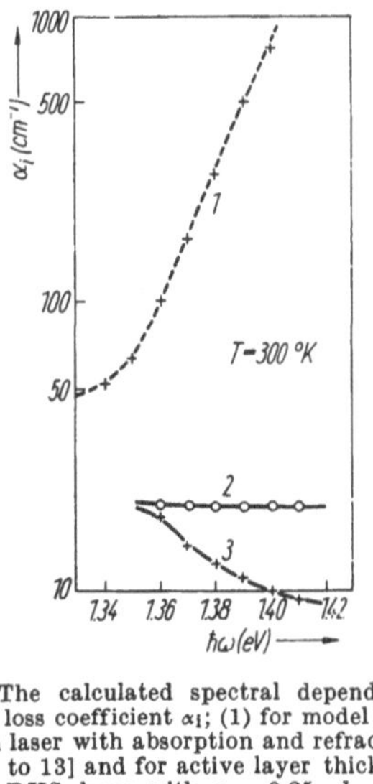

Fig. 6. The calculated spectral dependences of the internal loss coefficient α_i; (1) for model of the homojunction laser with absorption and refraction data for GaAs [9 to 13] and for active layer thickness 1.5 μm; (2) for DHS laser with $x = 0.25$, $d = 0.5$ μm, and $(\alpha_n + \alpha_p)/2 = \alpha_d = 20\ cm^{-1}$; (3) for DHS laser with the same parameters except α_d, which is equal to $5\ cm^{-1}$

absent (when $\alpha_d = (\alpha_p + \alpha_n)/2$) or has an opposite slope in comparison to the same dependence in the homojunction lasers. It is necessary to notice that the dispersion of Γ affects the net-gain curve, and the low-energy part of the curve can be suppressed.

4. Conclusion

We have found a similar behaviour of homojunction and single-heterojunction lasers in external optical coupling experiments. This is in agreement with the model of energy-dependent losses α_i. This approximation was found to be acceptable for the interpretation of experimental data. It is necessary to accent this conclusion because most of the investigators have used a more simplified model of lasers with energy-independent internal losses.

We have observed a different behaviour in case of double-heterostructure lasers, namely, α_i seems to be energy-independent. This is supported by the analysis of the waveguide effect in the laser. It has been estimated theoretically that either α_i has no energy dependence, or α_i decreases slightly when $\hbar\omega$ increases. The latter prediction is closely connected with the assumption of the strong dispersion of the dielectric constant discontinuity at the heterojunction.

References

[1] P. G. ELISEEV, I. ISMAILOV, M. A. MANKO, and W. P. STRAKHOV, Zh. eksper. teor. Fiz., Pisma **9**, 594 (1969).

[2] D. ACKERMANN, P. G. ELISEEV, A. KEIPER, M. A. MANKO, and S. RAAB, Kvantowaya Elektronika **1**, 85 (1971).

[3] D. ACKERMANN, H. BACHERT, P. G. ELISEEV, A. KEIPER, M. A. MANKO, and S. RAAB, Laser und ihre Anwendungen, Internat. Konf., Dresden 1970, VEB Deutscher Verlag der Wissenschaften, Berlin 1970 (p. 705).

[4] H. BACHERT, P. G. ELISEEV, M. A. MANKO, and S. RAAB, Zh. prikl. Spektr. **13**, 232 (1970).

[5] H. D. EDMONDS and A. W. SMITH, IEEE J. Quantum Electronics **6**, 356 (1970).

[6] H. BACHERT and S. RAAB, phys. stat. sol. **29**, K175 (1968).

[7] G. H. B. THOMPSON, Symposium on GaAs, Reading 1968 (p. 201).

[8] L. M. DOLGINOV, P. G. ELISEEV, L. D. LIBOV, I. S. PINSKER, E. L. PORTNOI, G. G. KHARISOV, and E. G. SHEVCHENKO, Kratkie Soobshcheniya po Fizike **12**, 63 (1970).

[9] D. MARPLE, J. appl. Phys. **35**, 1241 (1964).

[10] F. STERN, Phys. Rev. **133**, A1653 (1964).

[11] T. H. MINDEN, Appl. Phys. Letters **17**, 358 (1970).

[12] W. J. TURNER and W. E. REESE, J. appl. Phys. **35**, 350 (1964).

[13] I. KUDMAN and T. SEIDEL, J. appl. Phys. **33**, 771 (1962).

(Received September 17, 1971)

phys. stat. sol. (a) 8, 483 (1971)

Subject classification: 20.1; 22.9

Department of Physics, State University College, Plattsburgh, New York

Optical Properties of a Ladder Polymer, Polymethylpolyarsine, from Reflectance Studies

By

J. E. Lewis

The reflectance spectrum of a ladder polymer, polymethylpolyarsine $(CH_3As)_x$, has been measured in the spectral range 13 to 0.17 μm (0.095 to 7.3 eV), and its optical parameters, including the absorption coefficient, have been obtained by using Kramers-Kronig inversion. Analysis of the optical parameters indicates that the compound is possibly semiconducting, and that the arsenic atoms do not lie on exactly equivalent sites, so confirming the structure proposed by Daly and Sanz.

On a measuré le spectrum de reflection d'un polymère en chaînes, polyméthylpolyarsine $(CH_3As)_x$, dans le domaine de 13 à 0,17 μm (0,095 à 7,3 eV), et obtenu ses paramètres optiques, y compris le coefficient d'absorption, par le moyen de l'inversion Kramers-Kronig. L'analyse des paramètres optiques montre que le composé est probablement un semiconducteur, et que les atomes d'arsenic ne se trouvent pas sur des sites cristallographiques exactement equivalents. Ceci confirme la structure proposée par Daly and Sanz.

1. Introduction

The synthesis of a purple, insoluble form of arsenomethane, with a melting point of 204 °C, has been described in two recent papers [1, 2]. The compound forms microcrystallites of a typical size $0.15 \times 0.15 \times 0.05\ mm^3$, and appears a wine red color when a thin section, such as the edge of a crystallite, is viewed between crossed Nichols. A detailed X-ray analysis of the compound, using powder, oscillation, and Weissenberg photographs, has been carried out by Daly and Sanz [3]. Their results indicate that purple arsenomethane crystallizes in the orthorhombic system, the structure being polymeric, and that the arsenic atoms lie on an infinite coplanar ladder. For this reason we prefer to refer to the purple compound as polymethylpolyarsine, to distinguish it from the yellow form of the compound, X-ray analysis of which has shown to consist of discrete (CH_3As_5) molecules containing non-planar five membered arsenic rings [4].

The ladder structure, after Daly and Sanz, is shown in Fig. 1. The arsenic atoms form a ladder parallel to the a-axis of the orthorhombic unit cell, and these ladders lie in planes perpendicular to the c-axis, being bonded together by long-range van der Waals forces. Within one ladder the arsenic–arsenic rung distance (2.4 Å) is close to the bond length in elemental arsenic and yellow arsenomethane, whereas the inter-rung distance of 2.9 Å is shorter than the next-nearest neighbor contacts in grey arsenic or the 1–3 cross ring contact in arsenomethane. Each arsenic atom is also bonded to the carbon atom of a methyl group, the carbon–arsenic bond length being similar to that in arsenomethane (1.97 Å). It would seem that we have trivalent arsenic, making one bond to a carbon atom of the methyl group, one bond to an arsenic atom on

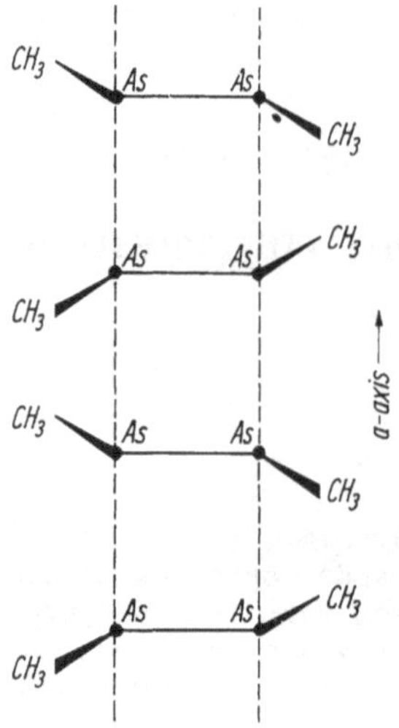

Fig. 1. Schematic representation of one polymer ladder chain in polymethylpolyarsine, after Daly and Sanz. The methyl groups alternately project above and below the plane of the As ladder

the same rung, and two half bonds to the two arsenic atoms at 2.9 Å immediately above and below on the same upright. The long arsenic bond would then correspond to a one electron bond with a bond order of about 0.5.

We undertook an examination of the optical properties of polymethylpolyarsine to elicit more information on this unique polymer ladder compound, especially to see if the arsenic atoms lie in equivalent sites or not, a point that had been raised by Daly and Sanz. We also conjectured that the compound might show semiconducting properties, due to the long half bonds between the arsenic atoms in the ladder being possibly favorable to the production of free carriers.

2. Experimental

The samples of polymethylpolyarsine used in this study were prepared by Rheingold, who will be shortly publishing details on their synthesis [5]. The polycrystalline samples were of high purity and of a size such that a polished face of dimensions 8×3 mm² could be obtained by gentle hand polishing with diamond paste, the final grade being less than 0.25 μm.

The compound is too brittle to permit sections thin enough for transmission work to be prepared, so the necessary spectrum was obtained in the reflectance mode, in the wavelength range 13 to 0.17 μm (energy range 0.095 to 7.3 eV). The infrared from 13 to 2.5 μm was covered using a Perkin-Elmer 221G spectrophotometer, the near infrared, visible, and ultraviolet (2.8 to 0.17 μm) being measured with a Carey 15 instrument.

The sample was held in the beam using a Barnes micro specular reflectance holder, model 126, the radiation hitting the sample at an angle of incidence of about 11°. As the reference beam was merely attenuated, allowance had to be made for the loss of energy in the sample beam due to the multiple reflections undergone in the Barnes holder, being six reflections at aluminum surfaced mirrors and the reflection at the sample. A dummy run, with a further aluminum mirror replacing the sample, was made under identical conditions to the run with the sample and the absolute reflectance of the sample was calculated from the following expression, given by simple considerations of the geometry of the holder:

$$R(\lambda) = \frac{R_s(\lambda)}{[R_m(\lambda)]^{6/7}}.$$

$R(\lambda)$ is the absolute reflectance of the sample at wavelength λ, R_m is the reflectance of the holder in the dummy run (all mirrors), and R_s is the reflectance of the holder with the sample in place.

3. Kramers-Kronig Analysis and Extrapolation Procedures

The use of the Kramers-Kronig relation in the analysis of reflectance data is now fairly well established. Starting with the experimental curve of reflectance $R = |r|^2$ vs. energy E, where $E = h\nu$ is the photon energy, and the Fresnel equation for normal incidence reflection

$$r(E) = |r|\, e^{i\theta} = \frac{n - i k - 1}{n - i k + 1} \tag{1}$$

(n and k are the real and imaginary parts of the complex refractive index $n - i k = c/u$, c being the velocity of light in vacuum, u its velocity in the sample at energy E), the phase shift θ can be computed from the Kramers-Kronig relation in the integral form

$$\theta(E_0) = \frac{1}{\pi} \int_0^\infty \frac{d \ln |r|}{dE} \ln \left| \frac{E + E_0}{E - E_0} \right| dE\,. \tag{2}$$

The complex reflectance r can be related to the optical constants n and k by separating the real and imaginary parts of (1) to give

$$n = \frac{1 - |r|^2}{1 + |r|^2 - 2 r \cos\theta}\,,$$

$$k = \frac{(n + 1)\, |r| \sin\theta}{|r| \cos\theta - 1}\,.$$

We used the computer program as given by Kreiger et al. [6], modifying it slightly to fit the requirements of our computer. The program essentially computed the phase angle θ at the desired wavelength λ from the experimental data by using (2), and then printed out values of the optical constants n and k, ε_1 and ε_2 (the real and imaginary part of the complex dielectric constant ε), the energy loss function $-\operatorname{Im}\varepsilon^{-1}$, and the absorption coefficient α, at the wavelength λ. These other parameters are related to n and k by

$$\varepsilon = \varepsilon_1 + i\varepsilon_2\,, \quad \varepsilon_1 = n^2 - k^2\,, \quad \varepsilon_2 = 2 n k\,, \quad -\operatorname{Im}\varepsilon^{-1} = \frac{\varepsilon_2}{\varepsilon_1^2 + \varepsilon_2^2}\,, \quad \alpha = \frac{4\pi k}{\lambda}\,.$$

The Kramers-Kronig relation is an integral over all photon energies, and as experimental data are limited in spectral range, to use the integral it is necessary to extrapolate the reflectance curve to all regions of the spectrum in order to evaluate the optical constants accurately in the energy range under study. Moreover, the extrapolation used must be such that the phase angle θ lies in either the third or fourth quadrant, as values outside this range will result in negative absorption coefficients α, an unphysical possibility.

At very high energies the behavior of the reflectance can be deduced from the asymptotic expression for the dielectric constant to be

$$R(E) = \frac{\text{const}}{E^4}\,. \tag{3}$$

At very low energies and in the region between the highest energy datum point and where equation (3) is valid, no expression is generally available and careful attention must be given to the form of the extrapolation so that meaningful optical parameters are obtained. In our study equation (3) was used at energies greater than 25 eV. For the other regions empirical extrapolations were employed so as to give zero phase angle (hence zero absorption coefficient) at 0.1 and 6.5 eV, and positive values of absorption coefficient at all other values of the spectrum. It was found in practice that the values of the optical parameters so obtained did not depend too sensitively on the details of the extrapolation used, an encouraging sign.

4. Experimental Results

The experimental reflectance spectrum of polymethylpolyarsine is shown by the full line in Fig. 2. The accuracy is about 5%. The dashed line shows the final extrapolation used at high and low photon energies. The spectrum is rich in structure, with a series of peaks in the infrared, and a further series of peaks in the visible and ultraviolet. For the computer analysis the experimental curve was approximated to by a series of straight line segments. About 200 of these data pairs (R, E) were fed into the computer, which then calculated the values of the optical parameters at various chosen energies in the same spectral region. The results are shown in Fig. 3 and 4, where a smooth curve has been drawn through these points.

The spectral dependence of the absorption coefficient α is shown in Fig. 3. The absorption coefficient is essentially zero for energies below 0.097 eV, then suddenly rises to form a narrow absorption line centered at 0.099 eV. At slightly higher energies, starting at 0.104 eV, a typical band spectrum occurs, on which are superimposed two series of structure, one in the infrared, the other in the visible and ultraviolet. A further band edge occurs at higher energies giving rise to absorption coefficients greater than 10^6. The variation of the real and imaginary parts of the dielectric constant and the energy loss function in the same spectral region are also given and are displayed in Fig. 4. In general these follow the structure observable in the absorption coefficient, with ε_2 zero at 0.1 and 6.5 eV. The energy loss function has its maximum at 3.9 eV, in the ultraviolet, associated with a series of peaks in the visible and the ultraviolet.

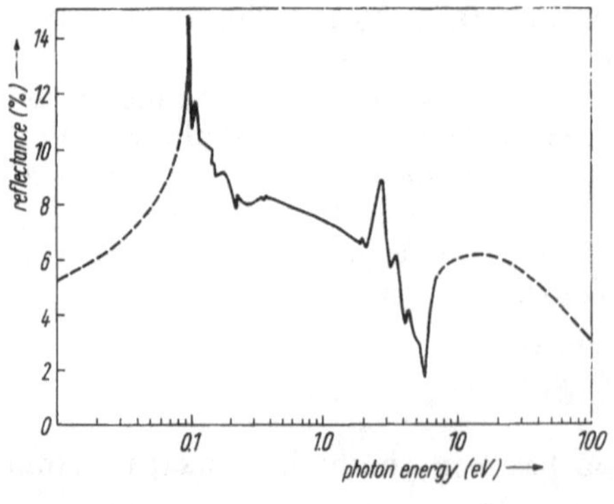

Fig. 2. The variation of the reflectance R with photon energy E for polymethylpolyarsine, as determined experimentally. The dashed portion of the curve is the extrapolation used in the Kramers-Kronig inversion

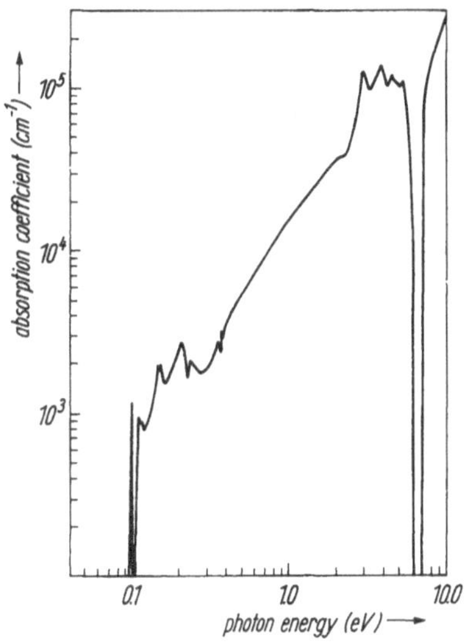

Fig. 3. The variation of the absorption coefficient with photon energy E, obtained from Fig. 2 using the Kramers-Kronig inversion technique

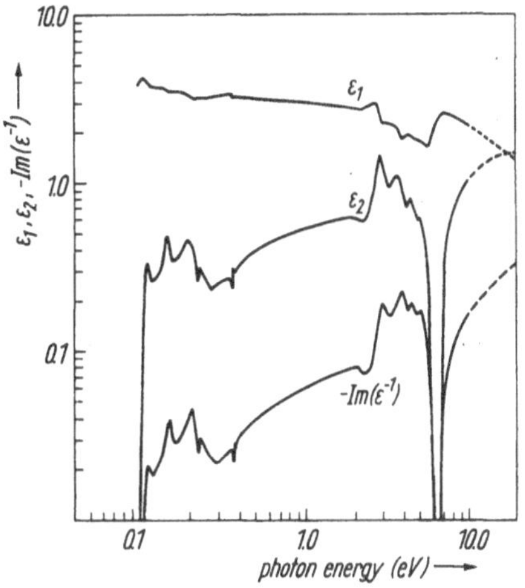

Fig. 4. The variation of the real (ε_1) and imaginary (ε_2) parts of the complex dielectric constant ε, and the variation of the energy loss function $-\text{Im}(\varepsilon^{-1})$ with photon energy E. These curves have been obtained from Fig. 2 using the Kramers-Kronig inversion technique

5. Interpretation

The general shape of the absorption coefficient versus energy curve (Fig. 3) can be interpreted in terms of two bands, the first with its edge at about 0.1 eV (12.4 μm), the second at the much higher energy of 6.8 eV (0.18 μm). The energy gap between the two bands is about 0.8 eV. The edges of each band are quite steep and can be fitted to the classical expression for direct transitions across a forbidden gap $\alpha = \text{const}\,(\hbar\,\omega - E_g)^{1/2}$ only over a short energy range. The assumption of a band edge at 0.1 eV would make polymethylpolyarsine a semiconductor, although, at such a low value of energy, attribution of this band edge to a bond failure would be hazardous; possibly it is due to the presence of defect clusters in the material. The edge at 6.8 eV is most probably associated with the C–H or the C–As bonds.

The structure seen on the absorption coefficient falls into three categories.

The absorption line centered at $E = 0.099$ eV is anomalously sharp due to the extrapolation used in the Kramers-Kronig analysis, and the lack of experimental data in the immediate vicinity of the line at lower energies. The line is most probably due to some phonon mode in the crystal, but the characteristic C–As stretch mode is too low in energy to be identified with it, being nearer to 0.06 eV.

The series of peaks in the infrared are all associated with various vibration modes of the crystal complex. The well resolved doublet at 0.360 and 0.366 eV is due to the symmetric and antisymmetric modes of stretching of the C–H bond, whereas the broad peak at about 0.2 eV would be associated with the first harmonic of the phonon line at 0.099 eV. The resolved doublet at 0.1475 and 0.1512 eV is due to the deformation of the methyl groups, the splitting being caused by the groups being situated in non-equivalent sites in the crystal structure. This would be in agreement with the conclusions of the X-ray analysis of the compound by Daly and Sanz [3]. The other doublet at 0.110 and 0.114 eV is as yet unassigned, but could be due to the symmetric and antisymmetric modes of methyl wag, a similar doublet at this energy being observed in methyl arsine [7].

The series of peaks in the visible and ultraviolet are the strongest in the crystal and occurring where they do, in the center of a band, can only be assigned to Van Hove type singularities, that is, to kinks in the density of states where the energy bands in momentum space touch the Brillouin zone boundary, the so called critical points. The disposition of these absorption peaks completely explains the color of the compound. In the red part of the visible spectrum the absorption coefficient is about 10^4 cm^{-1} whereas a sudden rise occurs in the blue to $\alpha = 1.3 \times 10^5$ cm^{-1}. In other words, much more of the red is transmitted by the compound than the blue, with the result that thin sections of the material appear a deep shade of red. In reflected light the opposite occurs, blue being reflected better than red, and the compound then appears purple.

The energy loss function $-\mathrm{Im}\,(\varepsilon^{-1})$ closely follows the behavior of the absorption coefficient, having its main peaks in the blue-violet part of the visible and in the ultraviolet. These are due to the Van Hove type singularities occurring at these energies in the band structure. The imaginary part of the complex dielectric constant, ε_2, has a similar variation to the energy loss function. They are both zero at 0.1 and 6.5 eV, as is to be expected in the forbidden gap between two bands. The real part of the dielectric constant, ε_1, behaves normally, dropping successively in value as it passes through the dispersion regions of the infrared, the blue, and the ultraviolet. A slight rise occurs in the forbidden gap region around 6.0 eV, but thereafter ε_1 continues to decrease asymptotically to unity at very high photon frequencies.

6. Conclusions

From analysis of the reflection spectrum, by using the Kramers-Kronig inversion technique, we have obtained the relevant optical constants in the regions of interest. From the behavior of these constants we have been able to show that the organic compound polymethylpolyarsine is probably a semiconductor, and that the arsenic sites are not quite equivalent in the crystal structure. Some information on the band structure of the material has been obtained, if our interpretation of the major peaks in the absorption spectrum as being due to Van Hove singularities is correct. However, further work needs to be done before such a conclusion can be considered absolute.

Acknowledgements

The author would like to thank Dr. A. Rheingold for the use of his samples of polymethylpolyarsine and for the many useful discussions with him concerning this work. He would also like to acknowledge the help extended to him by Dr. G. Slack and Dr. H. R. Philipp of the General Electric Research Laboratories, Schenectady, with the Kramers-Kronig computer program. Finally, part of this research was carried out with the aid of a State University of New York Faculty Research Fellowship, for which the author is extremely grateful.

References

[1] A. L. RHEINGOLD and J. M. BELLAMA, Chem. Commun. 1058 (1969).
[2] J. C. SUMMMERS and H. H. SISLER, Inorg. Chem. **9**, 862 (1970).
[3] J. J. DALY and F. SANZ, Helv. chim. Acta **53**, 1879 (1970).
[4] J. H. BURNS and J. WASER, J. Amer. Chem. Soc. **79**, 859 (1957).
[5] A. L. RHEINGOLD, to be published.
[6] E. L. KREIGER, D. J. OLECHNA, and D. S. STORY, General Electric Technical Note No. 63-RL-3458 G (1963).
H. R. PHILIPP and E. A. TAFT, Phys. Rev. **136**, A1445 (1964).
[7] A. B. HARVEY and M. K. WILSON, J. chem. Phys. **44**, 3535 (1966).

(Received September 27, 1971)

phys. stat. sol. (a) 8, 491 (1971)

Subject classification: 20.3; 16; 22.4; 22.4.1

Laboratoires de Recombinaisons Radiatives dans les Solides, Faculté des Sciences de l'Université de Reims et U.E.R. 52, Université de Paris VI

Sur l'étude des processus quadratiques en fluorescence et en photoconductivité

Par

E. Grillot, M. Egée et G. Mary

L'étude de la cinétique des processus quadratiques des recombinaisons des porteurs de charge s'est trouvée jusqu' ici limitée aux conditions d'excitation par impulsions rectangulaires et ses résultats restaient difficilement interprétables en termes de probabilité de transitions. En effet, sous excitation modulée sinusoïdalement $I = 1/2\ (I_m\ (1 - \cos \omega t))$, le phénomène est régi par une équation de Mathieu. Nous avons calculé à l'aide d'un ordinateur la solution approximative, expérimentalement très convenable, de cette équation. La probabilité de recombinaisons y apparait caractérisée par un paramètre p proportionnel à $\sqrt{\gamma I_m}$ (γ étant spécifique des centres impliqués) que les courbes ici présentées permettent de déterminer par la très commode méthode de phase. La méthode d'amplitude est également examinée.

Researches concerning the kinetics of quadratic recombinations remain up to now limited to rectangular pulse excitation conditions and their results are difficultly interpretable in terms of transition probabilities. Under sinusoïdal excitation, as known, this phenomena obeys a Mathieu equation. Using a computer, the approximate, but experimentally very convenient solution of this equation was calculated. The recombination probability appears to be characterised by a parameter p proportional to $\sqrt{\gamma I_m}$ (γ being specific for the implied centres). The curves here presented allow to determine it, using the very convenient phase method. The amplitude method is also discussed.

1. Introduction

Les centres luminogènes responsables de la fluorescence des cristaux II–VI ont fait l'objet de nombreuses hypothèses successives. Certaines d'entre elles, parmi les plus récentes, proposent des schémas susceptibles d'expliquer la grande largeur des bandes spectrales d'émission, qui subsiste à très basses températures, en mettant en jeu, directement ou indirectement, les probabilités des recombinaisons des porteurs de charge impliquées dans la fluorescence. Il semble donc qu'il suffise de déterminer ces probabilités pour confronter les théories proposées aux données expérimentales.

Cependant, cette comparaison n'apparait relativement aisée que dans le cas où les mécanismes de recombinaisons ont indubitablement un caractère spontané. Alors l'étude de leur cinétique conduit à la définition d'un temps de vie, inverse de la probabilité du processus de disparition de l'état métastable, traduite statistiquement par une exponentielle. Or, en photoconductivité comme en fluorescence, il en va très fréquemment autrement: les mécanismes n'apparaissent pas comme des fonctions exponentielles du temps. On envisage alors, dans un second stade, des processus quadratiques prédominants. La probabilité des recombinaisons dépend ainsi, d'une part des caractères propres des deux porteurs de charge (nature, mobilité, etc.) et d'autre part de leurs concentrations respectives en présence. La cinétique des processus, après établissement ou arrêt brusque de l'excitation, conduit alors à une hyperbole en chaque point de laquelle on ne peut définir qu'une „durée de vie instantannée" dont le terme caractérisant la constitution des partenaires dans la probabilité de recombinaison ne peut être convenablement déduit.

La méthode impulsionnelle d'étude de la cinétique de conductivité ou de fluorescence sous excitation lumineuse ou cathodique, qui consiste à soumettre périodiquement le cristal à des impulsions rectangulaires et à recueillir sur l'écran d'un oscillographe ou à l'aide d'un analyseur multicanaux les informations sur la montée ou le déclin du phénomène étudié, était la seule théoriquement utilisable pour des processus complexes, et notamment quadratiques. Mais elle est pratiquement d'un emploi difficile pour des phénomènes, soit très rapides (durée de vie inférieure à 10^{-7} s), soit de faible intensité, comme on l'observe par exemple en fluorescence dans les régions marginales des bandes spectrales d'émission, pourtant d'un intérêt souvent considérable [1, 2]: le bruit de fond, important par rapport au signal, ne permet plus une analyse précise des courbes.

Les méthodes de phase ou d'amplitude ne présentent pas ces inconvénients, mais ne pouvaient jusqu'ici être significatives que pour des processus monomoléculaires. Le présent travail a pour but de montrer que leurs résultats sont également interprétables [3] pour des mécanismes, soit purement quadratiques, soit intermédiaires.

2. Méthode de phase

2.1 Cas de processus spontanés

En fluorescence, Duschinsky [4] a montré que la durée de vie τ des processus d'émission dus à des transitions spontanées, qui caractérise la montée comme le déclin exponentiels, peut être mesurée par la différence de phase φ qui s'établit en régime permanent entre une excitation sinusoïdale de pulsation ω et le signal luminescent qui en résulte. On a simplement:

$$\tau = \frac{1}{\omega} \operatorname{tg} \varphi \,. \tag{1}$$

2.2 Cas de processus quadratiques

Cette méthode de phase, très avantageuse pour l'étude des processus très rapides [5] cesse malheureusement d'être applicable dès qu'il ne s'agit plus d'un processus purement monomoléculaire, c'est à dire lorsque τ n'est pas indépendant de ω. La notion même de probabilité de la transition ne conserve dès lors plus de signification intrinsèque, caractéristique des seuls centres luminogènes.

Dans un processus purement quadratique, accompli sous l'effet d'une excitation de nature déterminée et d'intensité I à l'instant t, la variation dn en un temps dt du nombre n des paires de porteurs de charge opposées résulte de la comparaison de deux termes: création de ces paires, dépendant du coefficient k d'absorption de l'excitation et du coefficient β traduisant l'efficacité statistique de chaque quantum absorbé et, d'autre part, recombinaison proportionnelle à n^2. En fluorescence, le coefficient γ de ce dernier terme est la somme de γ' et de γ'' se rapportant respectivement aux recombinaisons, soit radiatives, soit non-radiatives. γ', seul facteur spécifique des centres luminogènes et le rapport

γ'/γ'' peuvent être considérés déterminés dans des conditions précises données d'excitation. Soit la relation classique des processus bimoléculaires:

$$\frac{dn}{dt} + \gamma\, n^2 = \beta\, k\, I \tag{2}$$

d'intégration triviale lorsque I est constant à partir d'un temps pris comme origine, en choisissant comme variable auxiliaire

$$y = \exp \gamma \int_0^t n\, dt$$

on a

$$n = \frac{1}{\gamma\, y}\frac{dy}{dt} = \frac{p}{\gamma}\,\frac{e^{pt} - A\, e^{-pt}}{e^{pt} + A\, e^{-pt}} \tag{3}$$

en posant $p = \sqrt{\beta\, \gamma\, k\, I}$ et A étant une constante.

Mais il n'y a plus de solution rigoureuse lorsqu'on s'écarte de ces strictes conditions. En particulier, sous excitation sinusoïdale

$$I = \frac{1}{2}\, I_m\,(1 - \cos \omega\, t) \tag{4}$$

(2) se ramène à une équation de Mathieu.

Pour contourner cette difficulté principielle, jusqu'ici obstacle à une étude précise des processus quadratiques, nous avons recherché une solution approximative, expérimentalement valable, par recours à un ordinateur, en généralisant les calculs de Ryvkin [6]. Si à l'instant $t = 0$, l'intensité d'excitation passe brusquement de zéro à I_1: dans (3): $A = 1$. Si ensuite, à l'instant t_1, nouvelle origine des temps t', où le nombre des paires de porteurs a atteint la valeur $n = n_1$, on substitue brusquement à I_1 une nouvelle intensité d'excitation I_2, l'évolution de n avec le temps devient:

$$n = \frac{p_2}{\gamma}\,\frac{(p_2 + \gamma\, n_1)\, e^{p_2 t'} - (p_2 - \gamma\, n_1)\, e^{-p_2 t'}}{(p_2 + \gamma\, n_1)\, e^{p_2 t'} + (p_2 - \gamma\, n_1)\, e^{-p_2 t'}} = \frac{p_2}{\gamma}\ \mathrm{th}\,(p_2\, t' + \alpha)$$

avec

$$p_2 = \sqrt{\beta\, \gamma\, k\, I_2} \qquad \text{et} \qquad \coth \alpha = \frac{1}{n_1}\,\frac{p_2}{\gamma}. \tag{5}$$

L'approximation réalisée a consisté à substituer à la sinusoïde d'excitation (4) une succession de 100 créneaux dans chaque période et à calculer de proche en proche, successivement, à l'aide d'un ordinateur IBM de type 330, à un facteur de proportionnalité près, l'intensité du signal fluorescent qui doit en résulter. A la $i^{\text{ème}}$ étape, de durée $\Delta t = \pi/50\, \omega$, le nombre de paires est porté à n_i, condition initiale pour l'étape suivante. L'intensité J_i de l'émission fluorescente, traduisant la recombinaison de n_i paires dans ce processus purement quadratique s'exprime donc par $\gamma'\, n_i^2$, proportionnel à $\gamma\, n_i^2$:

$$J_i \sim \frac{p_i^2}{\gamma}\left[\frac{1 - \dfrac{p_i - \gamma\, n_i}{p_i + \gamma\, n_{i-1}}\, e^{-2p_i \Delta t}}{\dfrac{p_i - \gamma\, n_i}{p_i + \gamma\, n_{i-1}}\, e^{-2p_i \Delta t}}\right]^2. \tag{6}$$

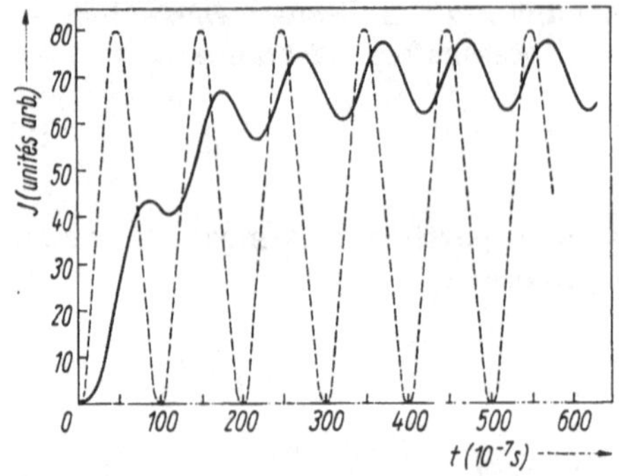

Fig. 1. Intensité du signal fluorescent (courbe en trait plein) à une frequence d'excitation de 100 kHz (courbe en pointillés)

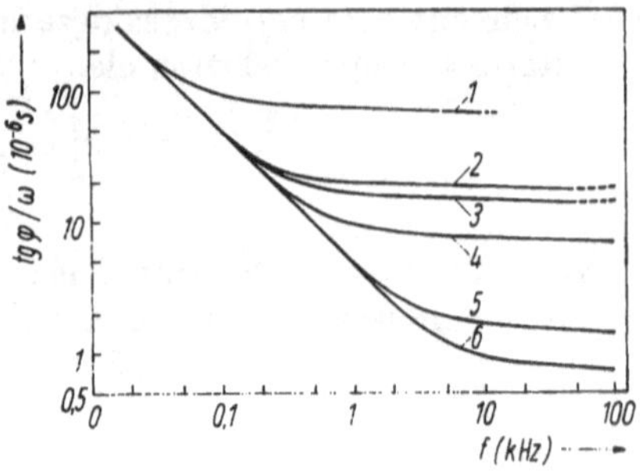

Fig. 2. $\tau = (1/\omega)$ tg φ en fonction de la fréquence de l'excitation. Les diverses courbes sont établies pour les valeurs de $1/p$ suivantes: (1) 1×10^{-4}, (2) $2{,}5 \times 10^{-5}$, (3) 2×10^{-5} (4) 1×10^{-5}, (5) 2×10^{-6}, (6) 1×10^{-6} s

La solution comporte d'abord un terme d'établissement avant de tendre vers un régime permanent. C'est ce que montre la Fig. 1 sur l'exemple d'une fréquence d'excitation de 100 kHz: la courbe en pointillés représente cette excitation tandis que celle en trait plein reproduit, sur une échelle d'ordonnées arbitraire, l'intensité du signal fluorescent. On voit que les calculs doivent être poussés pendant un assez grand nombre de périodes, dépendant essentiellement du paramètre $p = \sqrt{\beta \gamma k I_m}$ qui représente réellement le terme de la probabilité de recombinaison dans ces processus quadratiques.

Lorsque le régime permanent peut être considéré comme atteint, l'intensité de cette émission fluorescente apparait bien comme une fonction périodique de même pulsation ω que l'excitation et présentant par rapport à celle-ci un retard de phase φ bien déterminé pour chaque valeur du paramètre p défini ci-dessus. Une certaine distorsion est manifeste par rapport à une sinusoïde. Cependant, les phasemètres utilisés expérimentalement peuvent sélectionner la composante fondamentale de ce signal en rejetant bruit et harmoniques ce qui conserve un sens précis aux mesures.

Dans des conditions bien définies d'excitation d'un luminophore donné, on peut donc encore déduire par la méthode de phase, comme pour les processus spontanés, un paramètre ayant la dimension d'un temps $\tau = (1/\omega)$ tg φ. Mais sa valeur dépend de la fréquence comme de l'intensité I_m d'excitation. C'est pourquoi il nous a fallu reprogrammer l'ordinateur pour toute une série de valeurs, non seulement de la pulsation ω mais aussi du paramètre de probabilité $p = \sqrt{\beta \gamma k I_m}$.

La Fig. 2 met en évidence la rapide monotone décroissance de τ, porté en ordonnées en échelle logarithmique, en fonction de la fréquence en kHz de l'excitation. Dans un tel diagramme, nous obtiendrions, pour un mécanisme monomoléculaire, une droite d'ordonnée constante pour chaque valeur de $1/p$: la confrontation avec les courbes expérimentales de τ en fonction de cette fréquence permet donc de décider du caractère réel du processus: spontané, quadratique, intermédiaire ou complexe. Les résultats indiquent en outre qu'un déphasage supérieur à $\pi/2$ est parfaitement plausible à des fréquences d'excitation relativement grandes pour de faibles intensités locales d'excitation (par exemple optique) si le processus est quadratique, ce qui est évidemment impossible pour les mécanismes monomoléculaires.

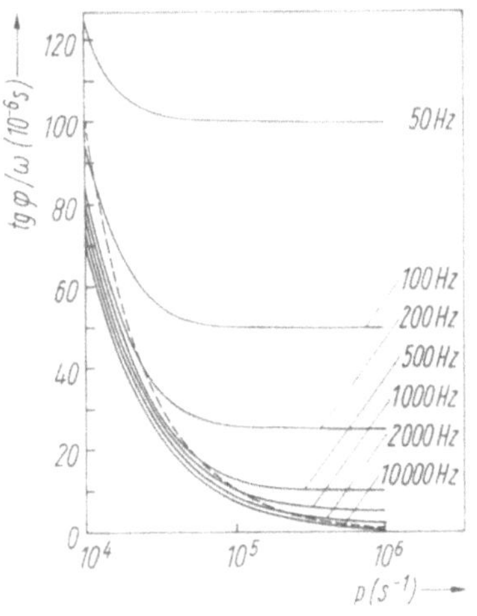

Fig. 3. $\tau = (1/\omega)\ \mathrm{tg}\,\varphi$ **en fonction du paramètre p dans toute la gamme de fréquences de l'excitation. La courbe en pointillés correspond au mécanisme spontané**

La Fig. 3 constitue une abaque directement utilisable pour déduire la valeur du paramètre p de la mesure de τ dans toute une gamme de fréquences de l'excitation. La courbe en pointillés y est l'hyperbole qui donnerait τ (indépendant de la fréquence d'excitation) en fonction de p pour un mécanisme spontané. Qualitativement, τ varie bien de la même manière dans les processus quadratiques.

3. Méthode d'amplitude

La méthode d'amplitude, basée sur la comparaison des composantes alternatives et continues des signaux de fluorescence ou de conductivité obtenus sous excitations modulées à différentes fréquences a été jusqu'ici beaucoup moins employée. Pourtant, un fluoromètre utilisant ce principe a permis de mesurer des durées de vie situées entre 10^{-7} et 10^{-9} s [7]. La cathodoconductivité de cristaux de sulfure de cadmium a été également étudiée de cette manière [8].

3.1 Cas de processus spontanés

Sous excitations modulées sinusoïdalement, d'intensités I_m constantes et de fréquences progressivement constantes, l'amplitude du signal luminescent subit un effet d'aplatissement [4]. Ce phénomène s'explique simplement pour un processus spontané de recombinaisons, l'équation régissant l'amplitude J de l'émission étant

$$\frac{dJ}{dt} + \frac{J}{\tau} = \beta\, k\, \frac{I_m}{2}\,(1 - \cos\omega\, t)\,, \tag{7}$$

où τ est la durée de vie radiative. La solution est

$$J(t) = J_0\left[1 - \frac{1}{\sqrt{1 + \omega^2\,\tau^2}}\cos\,(\omega\, t - \varphi)\right] \tag{8}$$

avec $J_0 = \frac{1}{2}\beta\, k\, I_m$ et $\mathrm{tg}\,\varphi = \omega\,\tau$.

Ainsi, pour une valeur I_m de l'intensité d'excitation, le rapport des amplitudes des signaux luminescents résultant de processus spontanés obtenus, soit en excitation modulée de pulsation ω, soit en excitation continue est

$$\frac{J_\omega}{J_0} = \frac{1}{\sqrt{1 + \omega^2\,\tau^2}}\,. \tag{9}$$

Cette loi de variation est représentée sur la Fig. 4 par les courbes en pointillés 1, 3, 5 et 8, tracées pour différentes valeurs de τ.

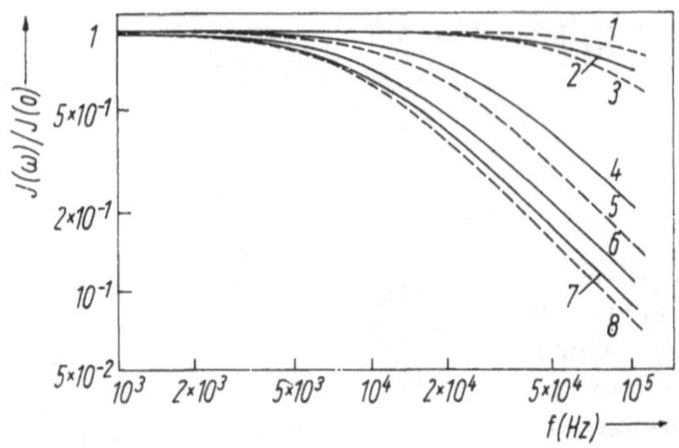

Fig. 4. Rapport des amplitudes des signaux luminescents en excitation modulée de pulsation ω et en excitation continue, en fonction de ω. Les courbes en pointillés sont tracées pour différentes valeurs de τ: (1) 10^{-6}, (3) 2×10^{-6}, (5) 10^{-5}, (8) 2×10^{-5} s; les courbes en trait plein pour différentes valeurs de $1/p$: (2) 2×10^{-6}, (4) 10^{-5}, (6) 2×10^{-5}, (7) $2{,}5 \times 10^{-5}$ s

3.2 Las de processus quadratiques

Dans le cas des processus quadratiques, on retrouve les mêmes difficultés de résolution des équations que pour la méthode de phase. Mais les calculs que nous avons effectués pour celle-ci et les courbes telles que celles de la Fig. 1 prolongées jusqu'au stade du régime permanent nous permettent de répondre également à cette question. Pour chaque valeur du paramètre $p = \sqrt{\beta \gamma k I_m}$ et pour des fréquences de modulation successives, nous calculons le rapport de l'amplitude de la composante alternative du signal fluorescent à celle que l'on obtiendrait sous excitation continue de même intensité I_m. C'est ainsi qu'ont été construites les courbes en trait plein 2, 4, 6 et 7 de la Fig. 4 correspondant à diverses valeurs du paramètre p. On remarquera que, pour des valeurs identiques de τ en processus spontanés et de $1/p$ pour des mécanismes quadratiques, les courbes (couples 2 et 3, 4 et 5, 6 et 8) ont la même allure et ne s'écartent pas beaucoup l'une de l'autre.

4. Conclusions

On voit que la méthode d'amplitude présente un certain nombre d'inconvénients: outre qu'elle est peu commode et peu précise, elle ne permet pas de caractériser correctement les processus de recombinaisons. Cependant, il apparait ici que le recours à cette méthode reste qualitativement justifiée dans certains cas, comme l'ont fait pour l'étude du sulfure de cadmium Stössel et Zimmermann [8], en dépit de ce que leurs propres résultats sur l'influence de l'intensité d'excitation, aussi bien que les études de cathodoconductivité que nous avons nous-mêmes également entreprises sur CdS à 77 °K en utilisant la méthode de phase [9] dénotent un caractère bimoléculaire certain du processus. Le principe même de la méthode d'amplitude conduit à des conclusions importantes. Lorsqu'un luminophore est susceptible, sous une excitation donnée, de fournir diverses émissions correspondant à des valeurs différentes du terme γ' typique des centres, celles correspondant aux processus les plus rapides (c'est à dire de moindre valeur de $1/p = \sqrt{1/\beta \gamma k I_m}$) subiront moins que les autres dans le spectre l'effet d'aplatissement lors d'un accroissement de la fréquence de modulation de l'excitation. C'est ce raisonnement qui a conduit récemment à montrer [10] l'intérêt d'emploi d'excitations diversement modulées permettant de mettre en évidence des bandes d'émission jusque là non soupçonnées.

Mais c'est surtout la méthode de phase, dont l'emploi et la signification se trouvaient jusqu'ici cantonnés aux processus spontanés, qui apparait pouvoir être désormais étendue aux mécanismes quadratiques, intermédiaires ou complexes.

Dans la probabilité de recombinaison, on a mis ci-dessus en évidence l'importance du paramètre $p = \sqrt{\beta \gamma k I_m}$ qui tient compte, par β et k de la nature de l'excitation, par I_m de son intensité et par γ de la constitution des centres de recombinaisons radiatives (γ') et non radiatives (γ''). Ainsi se trouve confirmée l'explication [11] de notre observation d'une variation de 1 à 100 des valeurs de τ de mêmes échantillons de CdS(Ag) à la même température, soumis à des excitations de mêmes modulations mais, soit cathodiques, soit photoniques. L'influence de β, k et I_m y fait varier l'intensité locale d'excitation, augmentant considérablement le paramètre p en cathodofluorescence.

Le présent travail justifie également la méthode que nous avons antérieurement proposée [11, 2, 3] pour étudier les larges bandes d'émission fluorescente dans les cristaux activés des composés II–VI. La répartition énergétique spectrale de fluorescence ne constitue pas une information suffisante pour décider de la constitution des centres luminogènes. La distribution spectrale de la probabilité de recombinaisons radiatives, à l'intérieur de chaque bande d'émission, nous parait un élément complémentaire indispensable. En dépit du fait que les processus de recombinaisons sont indiscutablement autres que spontanés, nous avons admis que, pour une excitation donnée, sinusoïdale de pulsation ω et d'intensité I_m déterminées, le temps τ déduit du déphasage φ selon (1) reste une fonction monotone décroissante du terme γ' qui, dans la probabilité de transition, caractérise les centres luminogènes. La Fig. 3 démontre que cette proposition est incontestable. Grâce à cette abaque, nous pouvons déduire de τ la valeur du coefficient p qui caractérise la probabilité de recombinaison pour chaque intervalle étroit découpé dans la bande d'émission. Ainsi, les résultats des calculs ci-dessus, complétant nos mesures de cinétique effectuées antérieurement à l'intérieur des bandes spectrales d'émission des luminophores ZnS et CdS remettent en question tous les modèles de centres luminogènes jusqu'ici proposés pour eux.

Bibliographie

[1] E. Grillot, M. Egée et G. Mary, C.R. Acad. Sci. (France) **272 B**, 147 (1971).

[2] E. Grillot, M. Bancie-Grillot et E. Egée, C.R. Acad. Sci. (France) **270 B**, 1423 (1970).

[3] M. Bancie-Grillot, Y. Charreire et J. C. de Vos, C.R. Acad. Sci. (France) **271 B**, 457 (1970).

[4] F. Duschinsky, Z. Phys. **81**, 23 (1933).

[5] Y. Charreire et E. Egée, Ann. Univ. et A.R.E.R.S., Reims **7**, 68 (1969).

[6] S. M. Ryvkin, Photoelectric Effects in Semiconductors, Consultants Bureau, New York 1964 (p. 13).

[7] J. B. Birks et D. J. Dyson, J. sci. Instrum. **38**, 282 (1961).

[8] W. Stössel et W. Zimmermann, phys. stat. sol. **30**, 311 (1968).

[9] Non encore publié.

[10] M. Bancie-Grillot, M. Egée et G. Mary, phys. stat. sol. (a) **4**, 81 (1971).

[11] M. Bancie-Grillot, M. Egée, Y. Charreire et E. Grillot, C.R. Acad. Sci. (France) **268 B**, 1385 (1969).

(Received October 1, 1971)

phys. stat. sol. (a) 8, 499 (1971)

Subject classification: 18.2.1; 21.1; 21.1.1

Institute of Physics, Technical University, Warsaw

Electron Microscopic Investigation of Stripe Domain Structure in Ni–Fe Films at High Temperatures[1)]

By

I. B. PUCHALSKA

The Lorentz method was used to study 83% Ni–17% Fe films deposited at 45° angle and 1500 Å thick. At about 300 °C it was found that the weak stripe domain structure becomes a strong one. Subsequently, at elevated temperatures of 490 and 400 °C the stripe domains disappear and upon cooling, the normal domain structure (no stripe) appears. These phenomena are believed to be related to the disappearance of the anisotropy component normal to the plane of the film.

A l'aide de la microscopie électronique, dans les conditions de Lorentz on a observé les lames minces de 83% Ni–17% Fe d'epaisseur 1500 Å formées par évaporation à l'angle d'incidence de 45°. En 300 °C les "faibles" domaines de ruban se sont transformées en domaines de ruban "fortes". En chauffant les lames jus qu' au 490 et 400 °C toutes les domaines se sont dissipées, mais ils réapparurent quand la temperature est tombée jusqu' au 250 °C. Et alors ce n'étaient plus les domaines en ruban. On explique ce phénomène par l'extinction de la composante d'anisotropie perpendiculaire à la surface de la lame.

1. Introduction

As is known, [1, 2] permalloy films of 2000 to 4000 Å thickness display a "weak" stripe domain structure if specimens are deposited at angles of incidence of 0° to 40° and a "strong" one at 50° to 70°. Both weak and strong structures as well as 360° walls occur in intermediate films formed at an angle of incidence of 45°.

The present investigation is concerned with thinner films about 1500 Å thick which have also a weak and strong stripe domain structure depending on the angle of deposition. It is noteworthy, however, that whereas the weak structure was as regular as in thicker films observed previously [1 to 3], the strong stripe domains evidenced some deviations from regularity (Fig. 1, 2).

The perpendicular anisotropy component due to the columnar structure in the thicker (> 1000 Å) films is responsible for the existence of weak stripe domains in non-magnetostrictive films. A columnar structure may also occur in thinner films depending on the evaporation conditions of the samples. Fujiwara and Sugita [4] have shown that anisotropy perpendicular to the plane of the film can occur even at the critical thickness of 1000 Å in films deposited onto a substrate at 300 °C, depending on the vacuum at which the film forms. Tsukahara [5] has observed stripe domains in film less than 1000 Å thick.

An experimental model of the strong stripe domain structure has been given in [2], whereas calculations for this model have been presented by Sukiennicki [6]. Sukiennicki states that the condition for a strong stripe domain

1) The greatest part of the work presented in this paper was carried out at the Laboratoire d'Electrostatique et de Physique du Métal C.N.R.S., and Laboratoire de Physique du Solide C.E.N., Grenoble, France.

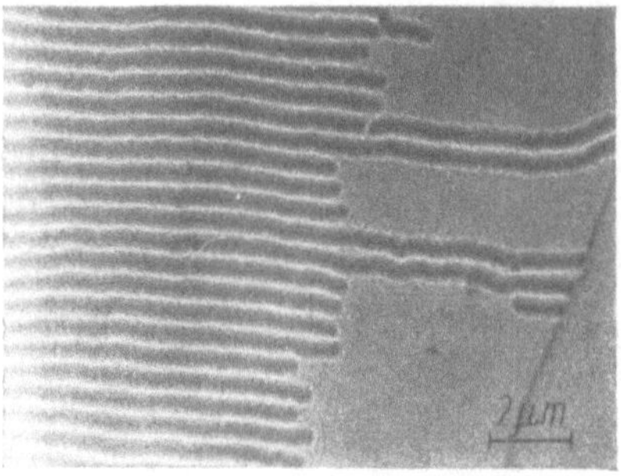

Fig. 1. Strong stripe domain structure in the 83% Ni–17% Fe film. Angle of incidence $\theta = 45°$, thickness T = 3000 Å

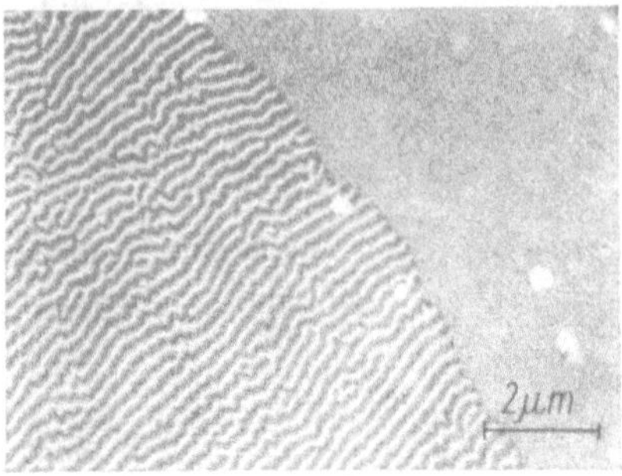

Fig. 2. Strong stripe domain structure in the 83% Ni–17% Fe film. θ = 45°, T = 1500 Å

structure to appear at a certain film thickness is a considerably high anisotropy constant $K_\perp$. The exact direction of the easy axis is not known in thick films in which the columns are inclined by a considerable angle, i.e. greater than 45°, in other words, in films where *only* strong stripe domains occur. This problem calls for, first and foremost, accurate measurements of K_α (α is the angle between the easy axis and the normal to the film plane), which have hitherto not been made by anyone.

It may be assumed that whereas the easy axis in the films deposited at an angle of 0° is perpendicular to the plane, in the case of films evaporated at angles exceeding 40° the easy axis juts out beyond the plane of the film and perhaps coincides with the direction of the columns or is close to the direction suggested by Tsukahara [5].

2. Experimental Data

1500 Å films of 83% Ni–17% Fe were evaporated from a Al_2O_3 crucible and deposited on a glass covered with a film of 1% collodium at an angle of 45°. During evaporation a vacuum of 10^{-5} to 10^{-4} torr was maintained. The substrate temperature during evaporation rose from room temperature to 100 °C. The distance between the crucible and the middle of the glass plate was 14 cm, and the evaporation rate was 1.5 Å/s. The thickness of the films was measured by means of the calibrated frequency of a quartz crystal placed in the evaporator. The sample was removed from the evaporator after cooling to room temperature.

The observations were carried out with a Philips EM 200 electron microscope operating at 100 kV with a temperature attachment. The films were studied in the range from room temperature to 500 °C.

3. Results

Both weak and strong stripe domain structures appeared in the films at room temperature (Fig. 3a and 4a). Upon heating the structure remained unchanged up to a temperature of 300 °C for the specimen shown in Fig. 3 (total heating time 1.5 h) and up to 280 °C for the specimen in Fig. 4 (total heating time 45 min). At these temperatures (cf. Fig. 3b and 4b, respectively) the weak stripe domains turned into strong ones. On further heating, at 370 °C (Fig. 3c) the contrast of the picture changed considerably and the structure was scarcely visible. It disappeared totally at a temperature of 490 °C (Fig. 3d), and at 400 °C in the case of the second specimen. The observations of the Lorentz

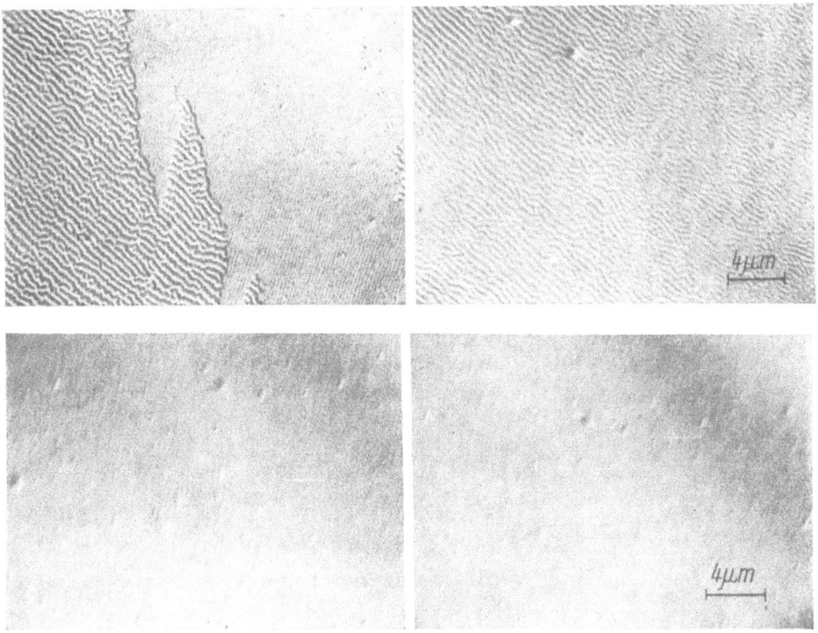

Fig. 3. Domain structure in 83% Ni–17% Fe at various temperatures, a) room temperature, b) +300 °C, c) +370 °C, d) +490 °C, e) after cooling +250 °C. Total heating time $1^1/_2$ h. T = 1500 Å, $\theta = 45°$

pictures were continued during cooling (about 60 min). At a temperature of about 250 °C there appeared ordinary (non-stripe) elongated domains lying in the same direction as the strong domains observed before (i.e. prior to heating), that is to say in the direction of the incident vapour (Fig. 3e and 4c).

4. Discussion

a) Many authors (e.g. [7 to 10]) have observed and described the rotation of the planar easy axis in non-magnetostrictive thin films hundreds of Å thick at a temperature of about 300 °C. Roth [10] emphasized that the main cause of this rotation of the planar easy axis is the change of the direction of ordered pairs of Fe atoms in the Ni matrix in the entire film owing to the displacement of the grain boundaries. In our case, the displacement of column boundaries as well as grain boundaries can take place in a film 1500 Å thick, and this is decisive for the transformation of the weak stripe domain structure into a strong one which in this case is energetically more advantageous.

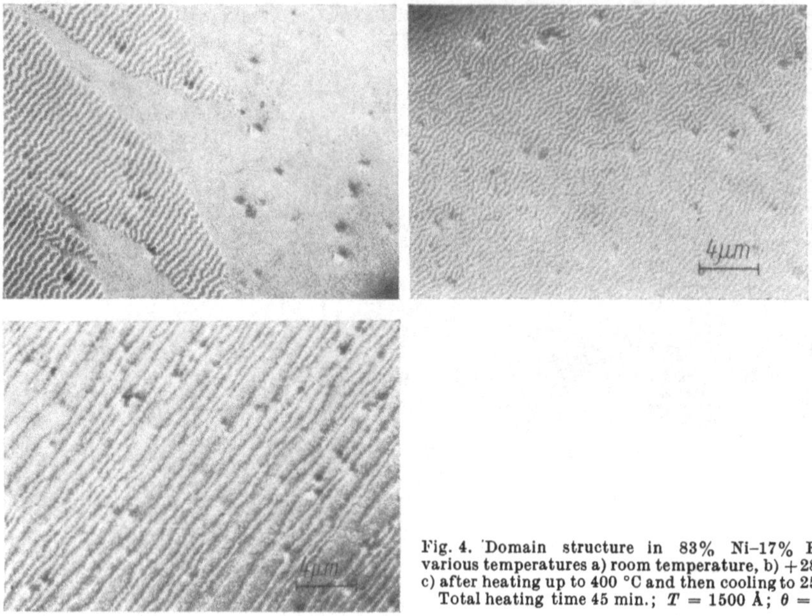

a b c

Fig. 4. Domain structure in 83% Ni–17% Fe at various temperatures a) room temperature, b) +280 °C, c) after heating up to 400 °C and then cooling to 250 °C. Total heating time 45 min.; $T = 1500$ Å; $\theta = 45°$

Hence, it is reasonable to suppose that the transformation of weak stripe domains into strong ones at the temperatures 300 and 280 °C, respectively, may perhaps be due to a change in the ratio of the diameter of the columns to the distance between columns [11]. This in turn may cause a change in the value of the constant K_α for the easy axis of anisotropy jutting out of the plane of the film.

b) The disappearance of the stripe domain structure, preceded by a loss of picture contrast in the case of the samples presented at temperatures of 490 and 400 °C, respectively, occurred below the Curie point which for bulk permalloy 83% Ni–17% Fe is about 530 °C [12]. No doubt, this phenomenon is related to the destruction of columns at these temperatures.

c) After cooling, the observed films were seen to develop long, narrow domains which no longer had the character of stripe domains. The columns, destroyed by heating turned into large crystallites and "thick" films now turned into "thin" films in the sense

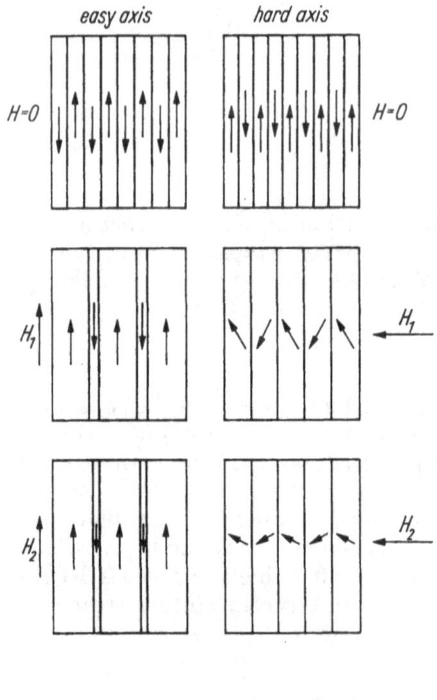

Fig. 5. Distribution of magnetization in 83% Ni–17% Fe films after heating and then being cooled. $T = 1500$ Å; $\theta = 45°$; $H_2 > H_1$

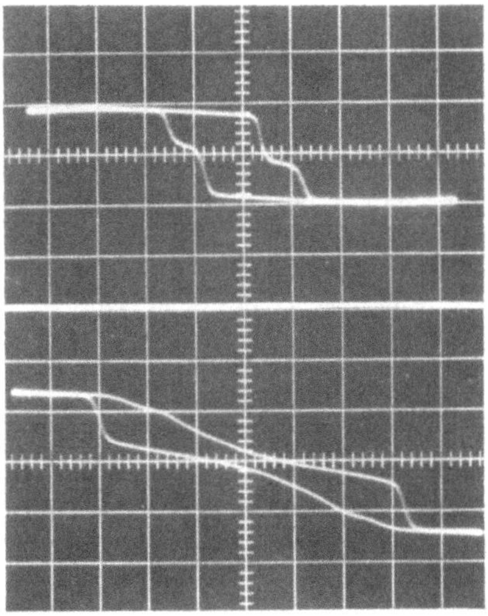

a

b

Fig. 6. Hysteresis loop of a 83% Ni–17% Fe film before heating. $T = 1500$ Å, $\theta = 45°$. The abscissa is 50 Oe/div. a) H perpendicular to the vapour stream, b) H parallel to the vapour stream

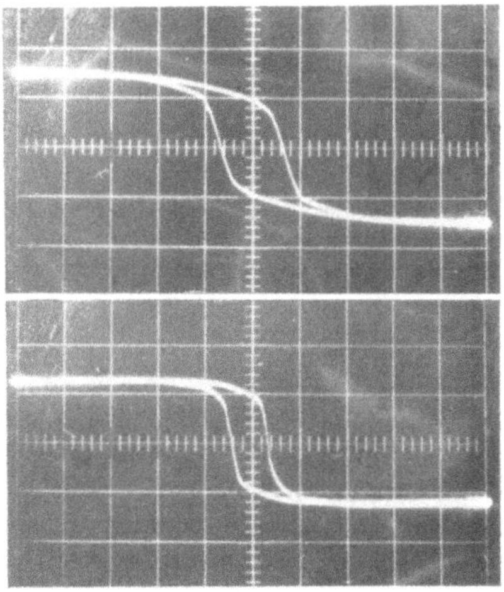

a

b

Fig. 7. Hysteresis loop of a 83%Ni–17%Fe film after heating up to 490 °C and then cooling back to room temperature. $T = 1500$ Å, $\theta = 45°$. The abscissa is 22.5 Oe/div. a) H perpendicular to the vapour stream, b) H parallel to the vapour stream. (The loops were made in the Dep. of Magnetics, Inst. Phys., Acad. Sci., Poland)

that the magnetization lay in the plane of the film. The distribution of magnetization in such films is shown in Fig. 5. The walls formed are believed to be of the Bloch type since for these film thickness they are more advantageous than Néel walls [13].

Fig. 6 and 7 show the magnetization curves for the observed film before and after heating, and then cooling to room temperature, respectively. As can be seen, the shape of the hysteresis loop changed. The step shape of the hysteresis loop before heating is related to the two stripe domain structures occurring in the film and will be the subject of a separate paper. After heating the magnetization curves of the film have no longer a step shape according to the fact that

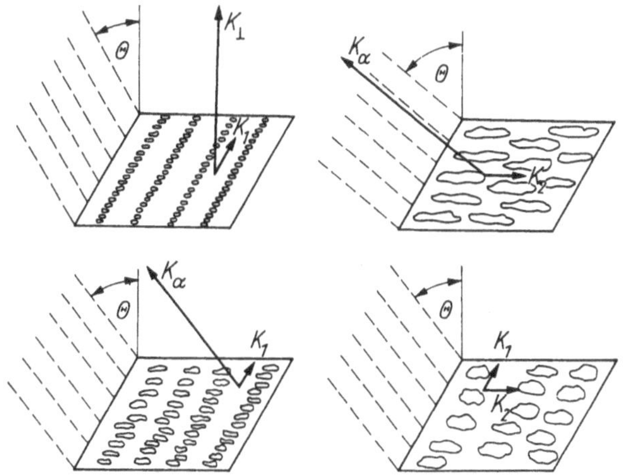

a

b

c

d

Fig. 8. Schematic representation of the axis directions for films deposited at various angles. $D > 1000$ Å. a) $\theta = 0$ to 40 °, b) $\theta > 50°$, c) $\theta = 45°$ no heating, d) $\theta = 45°$ after heating up to ≈ 400 °C and then cooling to room temperature

the stripe domain structure vanished. Comparing the hysteresis curves in Fig. 7a and 7b, we see that the loops are very similar to each other, i.e. K_1 differs little from K_2. Fig. 8 shows schematically the axis directions for films evaporated at various angles with particular attention to the 45°-film, before and after heating. The value of H_c in easy direction (in agreement with the evaporation direction) is 9 Oe, whereas in the hard direction it is higher, about 15 Oe, which is the opposite to the classical behaviour of easy- and hard-axis hysteresis loops.

Differences in the values of M_s appear in the case of remagnetization in the easy and hard direction, with a value higher in the hard direction. This would imply that unswitched domains exist at these values of the applied field H. This would explain the diagram in Fig. 5. When the film is switched into the easy direction, narrow domains antiparallel to the direction of the external field remain unswitched for these values of the applied field. When the film is magnetized in the hard direction, the angles of the walls change and the average value of the magnetization in the direction of the applied field is greater than M_s in the easy direction.

4. Conclusions

A transformation of the weak stripe domain structure into a strong one at about 300 °C may be due to a change of the value of the constant K_α for the easy axis of anisotropy jutting out of the film plane. The disappearance of the stripe domain structure at certain temperatures depending on the heating time (490 and 400 °C in our case) is related to the destruction of the columns. The formation of an elongated domain structure when the film is cooled to about 250 °C is interpreted as transformation of the thick film with columnar structure into a film composed of large crystallites.

Acknowledgements

The author is highly indebted to Prof. L. Néel, whose hospitality at the Laboratoire d'Electrostatique et de Physique du Métal C.N.R.S., and Laboratoire de Physique du Solide C.E.N. in Grenoble, France made this work possible. My thanks are due to R. H. Wade, A. Sukiennicki, and T. Tymosz for helpful discussions. I am also very grateful to Prof. S. Szczeniowski for his critical reading of the manuscript.

References

[1] I. B. Puchalska and R. P. Ferrier, Thin Solid Films (Netherlands) **1**, 437 (1967/68).
[2] R. P. Ferrier and I. B. Puchalska phys. stat. sol. **28**, 335 (1968).
I. B. Puchalska, Acta phys. Polon. **36**, 589 (1969).
[3] T. Koikeda, K. Suzuki, and S. Chikazumi, Appl. Phys. Letters **4**, 22 (1964).
[4] H. Fujiwara and Y. Sugita, IEEE Trans. Magnetics **4**, 22 (1968).
[5] S. Tsukahara, J. Phys. Soc. Japan **28**, 62 (1970).
[6] A. Sukiennicki, phys. stat. sol. **29**, 417 (1968).
[7] I. B. Puchalska and R. J. Spain, V. Internat. Congr., Electron Microscopy, Philadelphia 1962, Vol. II-6, Academic Press, 1962.
[8] P. Weber and W. Ruske, phys. stat. sol. **17**, 185 (1966).
[9] I. B. Puchalska and A. Sukiennicki, phys. stat. sol. **17**, K1 (1966).
[10] M. Roth, J. appl. Phys. **41**, 1286 (1970).
[11] T. Iwata, R. Prosen, and B. E. Gran, J. appl. Phys. **37**, 1285 (1966).
[12] R. S. Tebble and D. J. Craik, Magnetic Materials, Wiley-Interscience, 1969 (p. 97).
[13] M. Prutton, Thin Ferromagnetic Films, Butterworths, London 1964.

(Received August 9, 1971)

phys. stat. sol. (a) 8, 505 (1971)

Subject classification: 10; 22.9

Australian Defence Scientific Service, Defence Standards Laboratories, Melbourne (a), and Department of Physics, Monash University, Clayton Austrialia (b)

An X-Ray Topographic Study of Defect Structures in Cyclotrimethylene Trinitramine

By

I. T. McDermott (a) and P. P. Phakey (b)

Low-defect-density crystals of cyclotrimethylene trinitramine have been grown from solution by slow cooling and examined by X-ray transmission topography (Lang method). Well-defined pendellösung fringe systems, line defects radiating from the seed crystal, and "granular" structures have been observed. The extinction distance for a given reflection has been calculated from the fringe spacing and compared with the value derived from the structure factor. The line defects are shown to be grown-in dislocations whose character and Burgers vector have been determined in some cases. An explanation is suggested for the "granular" structures.

Kristalle von Zyklo-Trimethylen-Trinitramin mit niedriger Defektkonzentration wurden aus der Lösung durch langsames Abkühlen gezüchtet und mit Röntgenstrahltransmissionstopographie (Lang-Methode) untersucht. Wohldefinierte Pendellösung-Streifensysteme, Liniendefekte, die aus dem Kristallkeim ausstrahlen, und ,,Granular"-Strukturen wurden beobachtet. Die Extinktionsweite für eine vorgegebene Reflexion wurde aus dem Streifenabstand berechnet und mit dem Wert verglichen, der aus dem Strukturfaktor berechnet wurde. Es wird gezeigt, daß die Liniendefekte ursprüngliche Versetzungen sind, deren Charakter und Burgersvektoren in einigen Fällen bestimmt wurden. Eine Erklärung für die ,,Granular"-Struktur wird vorgeschlagen.

1. Introduction

Cyclotrimethylene trinitramine (commonly known as RDX) crystallizes in the orthorhombic system with $a = 13.22$ Å, $b = 11.61$ Å, and $c = 10.72$ Å [1] and yields well formed crystals from a number of solvents. Connick and May [2] reported on the habit of RDX crystals grown from acetone, cyclohexanone, and dimethyl formamide (DMF), and studied the geometry of etch pits produced on the major faces by these same organic solvents. They also demonstrated the correspondence of etch pits with dislocations in these crystals using the well established techniques of sequential etching, from the matching of etch pits on pairs of cleaved faces, from polygonization produced by heating, and from the increase in etch pit density following plastic deformation. The correspondence between etch pits on a crystal face and emerging dislocations in this material has been directly observed using X-ray transmission topography [3].

Gross [4] showed that inclusions of solvent in RDX often form net-like arrays with individual inclusions associated with dislocations. Silva and Gross [5] studied RDX crystals grown from cyclohexanone solution. These crystals had become milky from exposure to various radiations. Using shadow carbon replica techniques, electron microscopy revealed tubular features which stood proud of the cleaved and etched surface of these crystals. These features were about 1 μm in length with wall thickness of 20 nm and their surface density of

10^{12} m^{-2} was of the right order to cause the milky appearance by Tyndall scattering.

Defect structures have therefore been studied to some extent in the material RDX, and it is the purpose of this report to disclose further details observed by X-ray transmission topography [6].

2. Experimental

Suitable crystals with low defect density were prepared by the slow cooling of a saturated solution of RDX in DMF [3]. Prismatic crystals were formed from this solution usually with {111} faces dominant and {210} faces secondary in facial area. For topography, crystals 5 to 10 mm in linear dimension and up to 5 mm thick were used rather than thin sections. The three-dimensional dislocation line arrangements were then made more obvious when viewed stereoscopically with *hkl* and $\bar{h}\,\bar{k}\,\bar{l}$ topographs. MoK_α radiation was used for which the linear absorption coefficient $\mu = 0.186$ mm^{-1}. Since the thickness t of the crystals used was 5 mm or less, the product μt was always less than unity.

From the results of Connick and May [2], the geometry of etch pits on the faces of the crystals was used to determine crystallographic faces and directions for the purpose of orienting crystals for topography.

3. Observations and Discussion

3.1 Pendellösung effect

The sequence of crystals examined was labelled RDX (A), RDX (B), etc. Fig. 1 shows a portion of a topograph of RDX (E) using the 220 reflection which has the third-largest calculated structure factor $|F_c| = 136$ [1]. In Fig. 1 we see a system of fringes resulting from the pendellösung effect where wedge-shaped material is formed by the (210) and (010) faces of the crystal. Fourteen fringes can be seen, and from the angle between these faces (computed from the axial data) and the geometry of the experimental arrangement an extinction distance of about 160 μm can be deduced.

The extinction distance may be calculated from the theoretical expression

$$t_e = \frac{mc^2}{e^2} \frac{\pi V (\cos\theta_0 \cos\theta_g)^{1/2}}{\lambda F}, \tag{1}$$

where t_e is the extinction distance, V the unit cell volume, λ the X-ray wavelength, F the structure factor for the operating reflection, θ_0 and θ_g the angles between the entrance surface normal and the incident and diffracted beams [7]. Using the calculated structure factor of Harris [1] in the above expression gives a value of 120 μm for the 220 reflection. The differing values of extinction distance may be accounted for by the following errors. Firstly, the structure factors of Harris give a reliability factor of 13%, and an error of this magnitude is to be expected in the theoretical calculation from expression (1). Secondly, the crystal faces forming the wedge were not planar because of heavy etching (to remove the highly defective outer crystalline layers formed during removal from the growth bath), and consequently the angle between the faces was variable. This gives rise to perhaps a similar error in the direct calculation. Despite these errors, the agreement is good, and the results show that the image widths of dislocations in topographs, which are of similar dimension to the extinction

Fig. 1. Pendellösung fringesystem in a natural wedge of RDX (E) formed by the (210) and (010) faces. The lowermost arrow which is normal to the intersection of these faces indicates the direction used to determine the average fringe spacing. Heavy etching of the crystal has rounded the crystal edges thus causing a curving of the fringes. The diffraction vector is indicated by $\boldsymbol{g}$. 220 reflection, $MoK\alpha_1$ (50 ×)

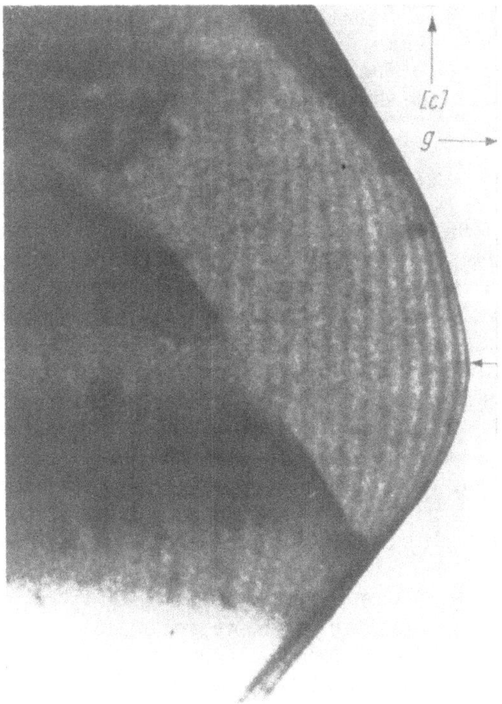

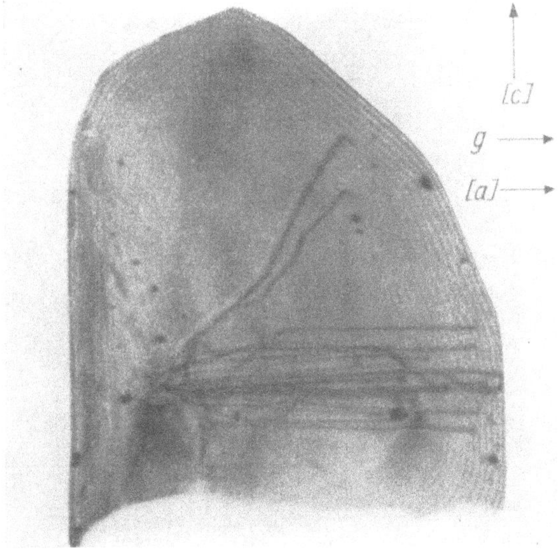

Fig. 2. Grown-in dislocations in RDX(E) radiating from the seed crystal together with peripheral pendellösung fringe system. 200 reflection, $MoK\alpha_1$ (10 ×)

distance [8] are extensive. Thus bulk crystals such as those under review must necessarily have low dislocation densities in order to be resolved. Since most other reflections are weaker than the 220 type, their structure factors are smaller and the corresponding extinction distance larger by equation (1). Hence image widths from these reflections will be generally larger.

Fig. 2 shows another topograph of RDX (E) using the 200 reflection for which $|F_c| = 100$. A system of pendellösung fringes is evident in the peripheral regions of the topograph. This demonstrates the success of the crystal growing technique since the pendellösung effect indicates a highly perfect crystal matrix from the X-ray point of view.

3.2 Line defects

The topograph of Fig. 2 also shows a group of line defects in high contrast which run from the seed crystal (on the left side) across to the right. Since these line defects have been found to vary in contrast for different reflections, and also since they are associated with etch pit formation at the surface, they have the characteristics of dislocations. That they radiate from the seed crystal implies that they are associated with crystal growth. These line defects are therefore interpreted as "grown-in" dislocations.

Stereoscopic inspection of 200 and $\bar{2}00$ topograph pairs of RDX (E) show that some of these dislocations run in the [100] direction whilst others appear to run normally to the {210} faces. This latter observation could not be checked with certainty for this crystal because of its shape and the relative smallness of the {210} faces. However, a topograph of another crystal with well developed {210} faces has shown bundles of dislocations emerging approximately normally to these faces. The pair of lines in the upper right position of Fig. 2, although not straight, tend to be approximately normal to the (111) face.

That the dislocations may align approximately to the normals to crystal faces is to be expected since, by image forces, dislocations should always arrive normally to a free surface if they are free to take their most stable configuration [9]. Accordingly since each segment of a dislocation line was at some stage of

Fig. 3. Additional dislocations in RDX(E) are observed running approximately in the [001] or *c*-direction. The upper left portion of the topograph is missing because of slit delimitation. 111 reflection, Mo$K\alpha_1$, (10×)

crystal growth at the surface (or growth horizon) it follows that the whole line should tend to be aligned to a face normal.

These lines are also double-imaged due either to parallel dislocation pairs, or to the effect of double-lobed misorientation surrounding single dislocations [10]. The latter explanation seems the correct one, since there is no observable double imaging of these lines in topographs from the stronger 220 reflection. In this case the image width has contracted (by virtue of the larger structure factor); whereas if the double image were due to a dislocation pair, the double lines would retain the same separation but with higher resolution.

In Fig. 2, there are no dislocations evident in the [001] or c-direction, and this also proves to be the case for other topographs taken about this direction as a zone (e.g. 210, 220, and 020). However, dislocations running approximately in the [001] direction are evident in topographs using reflecting planes inclined to this direction e.g. 111 (Fig. 3), 221, and 332. Hence the Burgers vector of these dislocations is directed along [001] and is most likely $\boldsymbol{b} = c$ [001] since the c-axial length is the shortest translational vector in this direction for the RDX structure. These dislocations are also of predominantly screw character because they are approximately aligned along their Burgers vector.

In the topographs using reflections 111, 221, and 332, the dislocation lines that run from left to right with high contrast in the 200 topograph of Fig. 2 become low in contrast. This suggests that they have a predominantly screw character with their Burgers vector in the direction normal to the (200) plane i.e. the a-axis. Accordingly they should vanish or become very low in contrast in topographs using reflections with the a-axis as zone. Unfortunately, RDX (E) became mosaic as a result of handling during reorientation and so grossly defective that individual defects could no longer be resolved. In order to determine the nature of dislocations running in the a-direction, other crystals were sequentially examined. Fig. 4 is a $\bar{2}00$ topograph of RDX (N) with only seven well

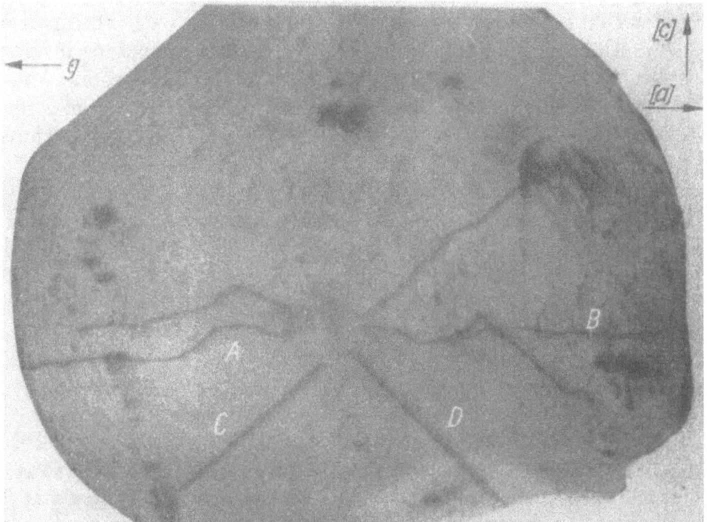

Fig. 4. Topograph of RDX(N) showing seven dislocation lines radiating from the seed crystal. $\bar{2}00$ reflection, $MoK\alpha_1$, (10×)

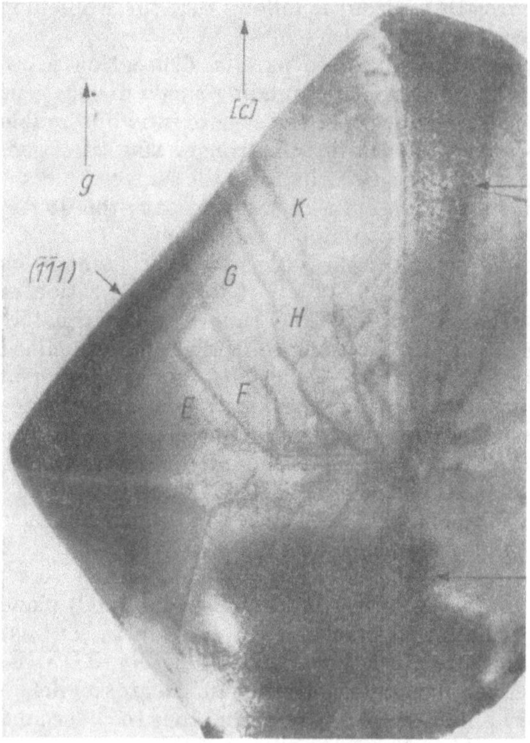

Fig. 5. Topograph of RDX(N) showing five disloations E, F, G, H, and K, not visible in Fig. 4. The *a*-axis, which is normal to **c**, does not lie in the plane of this figure, but is inclined approximately 30° to this plane, and is directed to the right. The unlabelled arrows indicate the granular structures. 002 reflection, $MoK\alpha_1$ (10×)

resolved and well spaced dislocations visible. Two of these dislocations, labelled A and B, appear to run mainly in the *a*-direction when viewed stereoscopically. These two dislocations are not visible in reflections with the *a*-direction as a zone i.e. 002 (Fig. 5), 040, 021, and 023. They are therefore predominantly screw in character with $\boldsymbol{b} = a$ [100]. The character of the remaining five dislocations of Fig. 4 has not been satisfactorily solved, owing mainly to poor quality topographs from other reflections. The two straight dislocations marked C and D appear to run in the $\bar{1}\bar{1}\bar{1}$ and $1\bar{1}\bar{1}$ directions respectively.

Fig. 5 is an 002 topograph also of RDX (N) in which five more dislocation lines E, F, G, H, and K are visible. These dislocations emerge on the $(\bar{1}\bar{1}1)$ face, but are out of contrast for the 210, 200, and 220 reflections. Hence they have their Burgers vector in the [001] direction. Since the direction of these dislocations is neither along nor at right angles to the Burgers vector, they must be of mixed character with $\boldsymbol{b} = c$ [001].

3.3 Granular structures

Also in Fig. 5 is a granular effect (arrowed) which is either invisible or very low in contrast for reflections from planes belonging to the *c*-axis zone. Examination of the stereoscopic pairs 002 and $00\bar{2}$ shows that the granularity arrowed near the bottom of the figure is close to the surface of the crystal where it had previously been mounted in plasticine. This effect is probably due to strains set up in the crystalline surface as a result of impurities absorbed from the plasticine. However, the granularity in the upper part radiates from the seed crystal and diverges in a fan-like manner through the crystalline volume. Unlike the former granularity, this is associated with crystal growth and it may be due to the tubular features described by Silva and Gross [5]. However, it seems more likely to be due to the sectorial distribution of chemical impurities [11]. By means of this mechanism a crystal may develop a face in the presence of an impurity in the solution. This face then becomes a growth horizon which selectively absorbs the impurity. In the present case a small (001) face has been formed, and entrapped solvent or impurity such as water (which the solvent DMF readily absorbs) may have been absorbed in this face during growth in the fast growing c-direction. Growth by cooling would then set up a strain because of different thermal properties between the impurity and the crystal matrix. Since the contrast produced by this strain is zero or very low in reflections from planes belonging to the c-axis zone, it follows that the strain distortion has its major component along the c-axis.

4. Conclusion

It has been shown that RDX crystals of high quality from the X-ray point of view may be prepared by slowly cooling a seeded solution. X-ray topography has yielded results some of which are similar to those published by Emara et al. [12] for the case of potash alum. The line defects observed in topographs are grown-in dislocations which may vary in direction or remain straight and align with crystallographic directions. Those that align approximately with the a- or c-direction are predominantly screw in character, whereas those that emerge at the $(\bar{1}\bar{1}1)$ face are mixed. The granular-like structures are probably the result of strains caused by absorbed impurities, with distortion parallel to [001]. Agreement between observed and theoretically calculated extinction distances is reasonable when probable errors are taken into account.

References

[1] P. M. Harris, Structure of Trinitroaromatic Crystals and Related Substances, AFOSR-TR-59-165, Ohio State University Research Foundation, Columbus, Ohio 1959.
[2] W. Connick and F. G. J. May, J. Crystal Growth **5**, 65 (1969).
[3] I. T. McDermott and P. P. Phakey, J. appl. Cryst., to be published.
[4] K. A. Gross, J. Crystal Growth **6**, 210 (1970).
[5] S. R. Silva and K. A. Gross, Abstracts, Australian Conf. Electron Microscopy, Australian Academy of Science, Canberra 1968, Paper 34.
[6] A. R. Lang, Acta cryst. **12**, 249 (1959).
[7] A. R. Lang, An Investigation of Crystal Imperfections by X-Ray Diffraction, Office of Aerospace Research — Final Report Contract AF 61 (052)-449, University of Bristol, 1966 (p. 102).
[8] A. R. Lang, Advances in X-Ray Analysis, Vol. 10, Eds. J. B. Newkirk and G. R. Mallett, Plenum Press, New York 1967 (p. 91).
[9] J. Friedel, Dislocations, Pergamon Press, 1964 (p. 45).
[10] A. C. McLaren, C. F. Osborne, and L. A. Saunders, phys. stat. sol. (a) **4**, 235 (1971).
[11] D. P. Grigorev, Ontogeny of Minerals, Oldbourne Press, London 1965 (p. 56).
[12] S. H. Emara, B. R. Lawn, and A. R. Lang, Phil. Mag. **19**, 7 (1969).

(Received October 1, 1971)

phys. stat. sol. (a) 8, 513 (1971)

Subject classification: 1.4; 2; 20.1; 22.9

TH Karl-Marx-Stadt, Sektion Physik/Elektronische Bauelemente

Investigations of the Structure of Thin Fluorocarbon Films by X-Ray Diffraction and Infrared Spectroscopy

By

H. GIEGENGACK and D. HINZE

The X-ray diffraction method was tested for the investigation of the structure of thin fluorocarbon films grown in a glow discharge in tetrafluoroethylene. It could be shown that the computed electron densities are strongly related to the structure of the well known linear molecule of polytetrafluoroethylene. The results are supported by infrared absorption spectra. Foreign atoms and chain branching being a reason for the amorphous structure of the films were ascertained.

An in einer Glimmentladung in Tetrafluoräthylen hergestellten dünnen Fluorkohlenstoffschichten wurde die Möglichkeit der Anwendung der Röntgenbeugung zur Strukturuntersuchung erprobt und gefunden, daß die erhaltenen Elektronendichteverteilungskurven sich in enge Beziehung zur Struktur des linearen Polytetrafluoräthylen-Moleküls bringen lassen. Infrarotabsorptionsspektren stützen dieses Ergebnis. Fremdatome und Kettenverzweigungen, die eine Ursache für die Amorphität der Schichten sind, können nachgewiesen werden.

1. Introduction

In order to check the hypotheses on the mechanism of growth of thin organic polymer films [1, 2] and to interpret the results of studies regarding the applications of these films in microelectronics [3], a profound knowledge of the structure is indispensable.

To acquire information on the atomic short-range order of thin fluorocarbon films we tried out the X-ray method since this method will permit a more exact quantitative evaluation than electron diffraction. The present studies are confined to the wide-angle range of the scattering diagram. Diffraction experiments on polytetrafluoroethylene (PTFE) were carried out additionally to reveal possible structural similarities of both materials.

To obtain the amorphous state of the polymer it was heated to a temperature above the melting point.

Infrared absorption spectra were taken in order to determine the kind of atoms and groups of atoms contributing to the film structure, the bonding and coupling conditions existing between them and, to a certain extent, the order of the molecules and to supplement the results of the X-ray diffraction.

2. Experiment

The 1 to 15 μm thick polymer films used in the scattering experiments were produced on glass in a gas discharge at a pressure of $p = 1$ Torr, a discharge frequency of 400 kHz and glow current densities of 4 mA/cm^2. The gas used was tetrafluoroethylene (C_2F_4).

X-ray scattering curves were obtained by a Philips vertical goniometer (molybdenum K_α radiation) with crystal monochromator and proportional counter tube.

The low scattering intensities made some difficulties. Therefore measurements were carried out by means of the step-by-step system.

The scattering sample consisted of a large number of crushed fluorocarbon films.

The PTFE measurements were made in a high-temperature chamber immediately above the temperature of the melting point where the sample changes its appearance from white to glassy-translucent and where the very strong (100) reflection of the scattering diagram which according to [4] remains almost unchanged up to the melting point breaks down.

The infrared absorption spectra were recorded with a dual-trace instrument (Beckman IR 12) in the 300 to 4000 cm^{-1} wave number range. To this end the polymer films were either produced on KBr discs or used in a self-supporting way. For comparison we used spectra given in literature (see for instance [13]) and spectra of common teflon sheet recorded by us, respectively.

3. Evaluation

The scattering curves obtained up to $s = 11$ Å^{-1} have been corrected regarding background, polarization, and absorption and normalized according to the high-angle region method. The simultaneously registered Compton scattering radiation was theoretically calculated according to [5] and subtracted. For the atomic scattering factors f_m the values valid for neutral atoms as given by Cromer and Waber [6] were used. The deviation of the f^2 values of free and covalently bonded C atoms does not exceed 1.5% [7], whereas the ionic bond proportion of the C–F bond, according to an empirical formula given by Hannay and Smith [8], is approximately 30%. This was not taken into consideration since theoretically obtained values for the Compton scattering of ions could not be found in literature.

When using molybdenum radiation a dispersion correction of the f-values is not required [9].

Evaluation was carried out with the help of the method developed by Warren et al. [10].

The following equation applies to the radial electron density distribution of multi-atomic systems:

$$4\pi r^2 \sum_m \bar{K}_m \varrho_m(r) = 4\pi r^2 \varrho_0 \sum_m \bar{K}_m + \frac{2r}{\pi} \int_0^\infty s\, i(s) \sin rs \, ds .$$

In this equation r is the spacing from an arbitrarily selected atom, $K_m = f_m/f_e$ is the actual number of electrons of the m-th atom, $f_e = \sum_m f_m / \sum_m Z_m$, $\varrho_m(r)$ is the effective electron density at a distance r from the atom m, ϱ_0 is the average electron density, $s = 4\pi \sin\theta/\lambda$ with 2θ being the scattering angle and λ the wavelength, $i(s) = (I_{ee}(s) - I_k)/f_e^2$ with $I_k = \sum_m x_m f_m^2$, I_{ee} is the normalized coherent scattering intensity distribution in electron units and x_m the proportion of the atom type m in the structure unit.

Fig. 1. Corrected X-ray scattering diagrams;
– – – PTFE at 300 °C,
——— C–F films

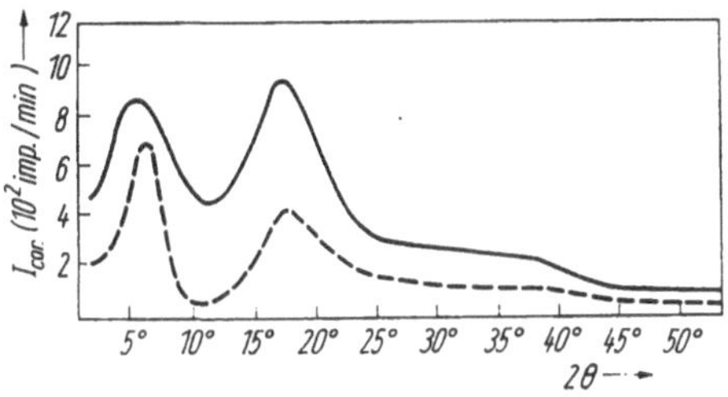

$\int_{r_1}^{r_2} \sum_m \overline{K}_m \, 4\,\pi\, r^2 \, \varrho_m \, dr$ is the squared total effective number of electrons in the volume of the spherical shell between r_1 and r_2.

The Fourier integral for 80 different r-values was computed by means of an electronic computer (type "Odra").

4. Results and Discussion

4.1 X-ray diffraction

The corrected diffractometer scattering diagrams of massive PTFE at 300 °C and flourocarbon films at room temperature are plotted in Fig. 1.

The positions of the maxima are given in Table 1. This table moreover, gives values which have been obtained during measurements with CuK_α

Table 1

Position of the maxima in the scattering diagrams

	s (Å^{-1})		
	1. max.	2. max.	3. max.
fluorocarbon films (MoK_α)	0.848	2.69	5.67
fluorocarbon films (CuK_α)	0.85	2.77	5.60
PTFE (300 °C)	1.00	2.70	5.75
fluorocarbon films (electron diffraction)	—	2.45	5.57

radiation and from transmission electron diffraction patterns of single films. The latter ones have been taken for checking purposes and to determine the exact position of the third maximum. The first maximum in the electron scattering diagram is covered by the primary beam.

The evaluation of the measurements made with MoK_α radiation resulted in the radial electron density distribution curve given in Fig. 2. Because of the very low intensities especially at large scattering angles and the resultant uncertainty regarding the scattering curve behaviour no analogue curve could be obtained for measurements made with copper radiation.

In Fig. 2 all interatomic spacings occurring within the chain molecule of the PTFE have been plotted in the range from $r = 0$ to $r = 9$ Å. They have been calculated on the basis of the structure data given in [11] and permit the comparison of the structures of film and massive materials. A sphere with a diameter of $2\,r = 18$ Å covers almost the entire identity period of a chain molecule. (Along a distance of 19.5 Å the chain is twisted by 180°. The 16th CF_2 group

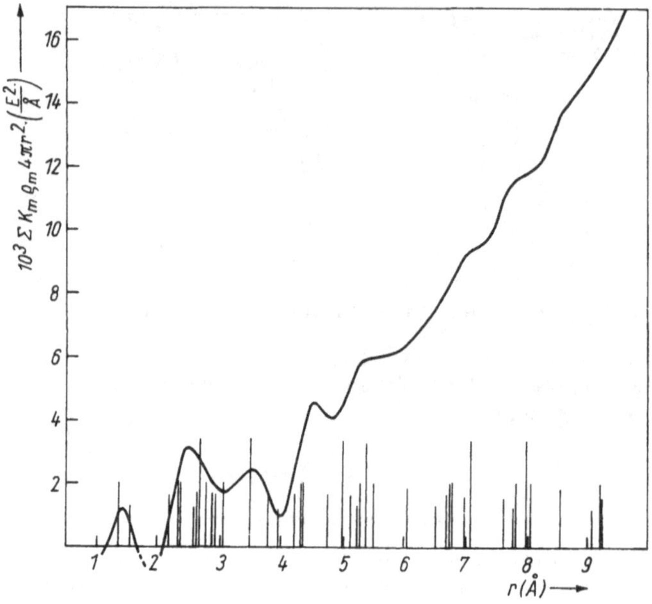

Fig. 2. Radial distribution of electron density calculated from the scattering diagram of the polymer films, and atom spacings calculated from the PTFE chain structure

lies exactly over the one that has been selected first.) The length of the spacing marks is a measure of the portion shared by the atoms in the respective spacings in the distribution of electron density.

It may be seen that the shape of the curve and the groups of lines in the examined r-range have common features. The first maximum of the distribution curve is at 1.4 Å. Its position coincides very well with the shortest spacings that occur in the molecule chain, that is to say the spacing between a fluorine and a carbon atom within a CF_2 basic module, which amounts to $r = 1.36$ Å, and the spacing between contiguous carbon atoms of the zig-zag framework with $r = 1.55$ Å. The area of the peak comprises the electrons of both spacings. With $K_C = 5.56$ and $K_F = 9.22$, 2 F atoms round one C atom make up an area of $A_1 = 206$ El² and 2 C atoms round one C atom on area of $A_2 = 124$ El². The overall theoretical area of $A = 330$ El² agrees well with the experimentally obtained value of 345 El². However, the negative values of the distribution curve between the first and the second maximum render this experimental value uncertain.

The following double maximum has its peaks at 2.45 Å and 3.50 Å. Whereas it includes the groups of spacings following $r = 2.20$ Å to $r = 4$ Å, neither the shape of the curve nor the area ($A = 3870$ El²) may be explained solely by the structural elements of the chain. The first peak should be at about $r = 2.8$ Å and the overall area at $A_{th.} = 2396$ El². In that range of spacings, however, atoms of the adjacent chains become apparent whose electrons increase the peak area and determine the profile of the maximum. The maximum at $r =$ $= 2.45$ Å may be attributed to the shortest spacing of the fluorine atom helixes of adjacent chains at a close packing with $r_{FF} = 2.4$ Å, and the second maximum to the distance from the F helix to the carbon atom helix of the respective adjacent chains with $r_{F-C} = 3.5$ Å. Likewise the third maximum at $r = 4.55$ Å may be interpreted as being formed by the atoms of the three spacings occurring in the chain at $r_{FF} = 4.23$, $r_{C-F} = 4.34$, and $r_{C-F} = 4.35$ Å, and by a contribution of carbon atoms from the adjacent chains, since the shortest spacing of the C helixes of adjacent molecules is $r = 4.75$ Å.

The following peaks at $r = 5.35$, 6.95, and 7.75 Å are formed by the accumulation of adjacent atoms within the chain in these distance ranges. With large distances an influence of atoms of the intermolecular environment is not to be expected.

All in all the distribution curve of electron density implies that the carbon and fluorine atoms of the films occur in form of CF_2 groups (first maximum of the distribution curve) which polymerize into chains of different length that on an average are longer than the identity period length of the PTFE chain.

The similar shape of the scattering curves of molten PTFE and fluorocarbon films suggests the affinity of the structures. While, however, the scattering intensity of the PTFE which is known to be a branchless linear polymer decreases to nearly zero between the prism reflection (100) and the pyramid reflections (*hko*) and this low minimum is maintained also in the amorphous state as regards chain orientation at high temperatures, in the case of fluorocarbon films the minimum is relatively flat which may be caused by the higher degree of disorder due to the existence of chain branching and foreign atoms. Chain branching may result from the forming of a secondary radical during a transfer reaction with subsequent addition of a monomer molecule. From the energetic point of view the latter step does not differ considerably from the addition reaction of a monomer to a chain end, but according to [12] the transfer reactions are energetically at a disadvantage as against the normal chain growth.

Chain branching may furthermore originate from bonds broken in the completed chain through impinging charge carriers.

The inclusion of foreign atoms will be discussed in detail later on.

4.2 Infrared spectroscopy

The infrared absorption spectrum of polymer films is shown in Fig. 3. The position of the bands, their relative intensities and the assignments are shown in Table 2. All bands are rather wide, therefore they are partly overlapping and the assignment is complicated. From the occurrence of wide bands, just as before from the occurrence of diffuse diffraction reflections in the X-ray and electron diffractions, it may be concluded that the polymer in the thin film is amorphous.

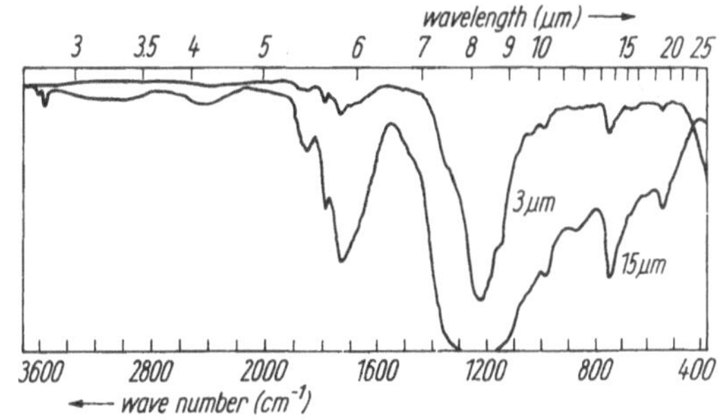

Fig. 3. Infrared absorption spectrum of thin fluorocarbon polymer films

Table 2
Table of the polymer film bands

ν (cm^{-1})	relative intensity	assignment
3555	w	OH stretch
1850	s	C=O stretch
1783	s	C=C stretch
1725	s	C=O stretch (acid group)
1450	sh	CF_2 asymmetrical stretch, A_2
1340	s	CF stretch
1221	vs	CF_2 asymmetrical stretch, E_1
1158	vs	CF_2 symmetrical stretch, E_1
1034	w	?
981	m	"CF_3"
740	s	"amorphous" PTFE
650	w	CF_2 wagging, A_2
532	m	CF_2 wagging or CF_2 rocking vibration

The strong band at 740 cm^{-1} may also be seen in the spectrum of PTFE. Its assignment is not possible, though it may be assumed that it is a Raman active band (symmetrical CF_2 stretch vibration, E_2) the selection rule of which is abolished by chain distortions. It is therefore called "amorphous band" and used for the characterization of the degree of crystallization of the PTFE [13 to 16]. It is another indication of the film to be amorphous. On the basis of the known chain structure of the PTFE molecule Liang and Krimm [13] and later Hannon et al. [11] have carried out the vibration analysis of the polymer.

In [11] dispersion curves of the PTFE helix are given which we used to assign the bands in the spectrum of the polymer film observed in the wave number range from 1500 to 300 cm^{-1}. It turned out that the bands may quite well be assigned to the vibrations occurring in the PTFE chain molecule. This encourages us to suggest that the polymer film consists of chains having a very similar structure.

In detail the following may be stated: According to the above mentioned authors the shoulder at 1450 cm^{-1} must be attributed to the asymmetrical stretch vibration (A_2) of the CF_2 group. At wave number 1221 cm^{-1} there is the E_1 form of the same vibration and at 1158 cm^{-1} the symmetrical stretch vibration (E_1) of the CF_2 group. In the wave number range between 1300 and 1100 cm^{-1} Hannon et al. mention another band at 1210 cm^{-1} which they assign to the stretch vibration of the chain structure (E_1). This band is apparent in the film spectrum at the utmost as a shoulder. Possibly it is overlapped by adjacent bands. The band at 650 cm^{-1} is assigned to the CF_2 wagging vibration (A_2). It is not clear how to assign the band at 532 cm^{-1}, as either the CF_2 wagging (E_1) or the CF_2 rocking vibrations (A_2) have to be considered.

From the mechanism of the film growth [1] we may expect that — as mentioned above — the growth of the $-(CF_2)_n-$ chains does not proceed in an undisturbed manner and that chain branching occurs because of charge carriers impinging on the growing film. In the infrared spectrum, therefore, weak bands occur at 1340 and 981 cm^{-1}, which Dube and Kriegsmann [17] interpret as C–F stretch vibration and "CF_3 vibration", respectively. Lo and Osborn [18] found these bands with high intensities in the infrared spectrum of liquid fluorocarbons which had been produced in a glow discharge between 150 and 800 Torr.

Cooper and Roy [19] as well as Gozzo and Bergomi [20] investigated the radiation polymerization of tetrafluoroethylene. They discussed the results on the basis of a radical polymerization mechanism. Probably radicals are included in the film during the growth of the polymer film in cases where further monomer molecules cannot be added to a chain for geometrical reasons. These $-\dot{C}F-$ or $-\dot{C}F_2$ radicals are also formed in the film volume by the impinging charge carriers originating from the gas discharge. Tamura and Oshima [21], for instance (see also [22]), found high concentrations of radicals in PTFE after the polymer had been irradiated with electrons and γ-radiation. In the presence [20] or after subsequent addition [21] of oxygen the radicals react with the O_2 molecules forming peroxide radicals.

These slowly disintegrate, possibly under the effect of water, and thereby give rise to the creation of various oxygen containing groups. Lo and Osborn [18] similarly discussed the findings of mass spectroscopic investigations of liquid fluorocarbons. The spectrum given by them also comprises the discussed bands.

Based on this conception, we assigned the band at 1850 cm^{-1} to the C=O stretch vibration of a carbonyl group the neighbourhood of which is unknown to us. The band at 1725 cm^{-1} has been identified as a C=O stretch vibration of an acid group. This assignment is substantiated by tests aiming at the chemical neutralization of the acid groups, in which the band disappeared, and by the band at 3555 cm^{-1} (OH stretch vibration). The band at 1783 cm^{-1} is the C=C stretch vibration of a perfluorinated unsaturated end-standing group, as bromine addition tests revealed.

A quantitative investigation of the oxygen proportion in the film from the IR spectrum seems to be not yet practicable.

5. Summary

The results of the X-ray diffraction method and of the IR spectroscopic method agree well with each other and may be summarized as follows:

1. In thin fluorocarbon polymer films chain molecules of the $-(CF_2)_n-$ form are present. The films are amorphous.

2. Disturbances of the chain structures are caused by chain branching and built-in foreign atoms. Chain distortions must be assumed.

Acknowledgements

The authors want to thank Dr. H.-U. Poll (Karl-Marx-Stadt) and Dr. Wagner (Dresden) for numerous useful hints and discussions.

References

[1] H.-U. Poll, Z. angew. Phys. **29**, 260 (1970).
[2] H.-U. Poll and U. Lippold, Wiss. Z. TH Karl-Marx-Stadt **12**, 463 (1970).
[3] L. U. Gregor, IBM J. Res. Developm. **12**, 140 (1968).
[4] H. G. Kilian and E. Jenckel, Z. Elektrochem. **63**, 308 (1959).
[5] K. Sagel, Tabellen zur Röntgenstrukturanalyse, Springer-Verlag, Berlin 1958.

[6] D. T. CROMER and J. T. WABER, Acta cryst. **18**, 104 (1965).
[7] Internat. Tables X-Ray Crystallogr., Kynoch Press, Birmingham 1959.
[8] N. B. HANNAY and C. D. SMITH, J. Amer. Chem. Soc. **68**, 171 (1946).
[9] C. H. DAUBEN and D. H. TEMPLETON, Acta cryst. **8**, 841 (1955).
[10] B. E. WARREN, H. KRUTTER, and O. MORNINGSTAR, J. Amer. Ceram. Soc. **19**, 202 (1936).
[11] M. J. HANNON, F. J. BOERIO, and J. L. KOENIG, J. chem. Phys. **50**, 2829 (1969).
[12] W. M. D. BRYANT, J. Polymer Sci. **56**, 277 (1962).
[13] C. Y. LIANG and S. KRIMM, J. chem. Phys. **25**, 563 (1956).
[14] R. G. MILLER and H. A. WILLIS, J. Polymer Sci. **19**, 485 (1956).
[15] R. G. BROWN, J. chem. Phys. **40**, 2900 (1964).
[16] G. DUBE and H. KRIEGSMANN, Z. Chem. **4**, 430 (1964).
[17] G. DUBE and H. KRIEGSMANN, Z. Chem. **5**, 421 (1965).
[18] E. S. LO and S. W. OSBORN, J. org. Chem. **35**, 935 (1970).
[19] R. COOPER and C. R. ROY, Austral. J. Chem. **23**, 939 (1970).
[20] F. GOZZO, A.BERGOMI, G. CAMAGGI, and V. OPRANDI, Europ. Polymer J. **6**, 219 (1970).
[21] N. TAMURA, Y. OSHIMA, H. K. YOTSUMOTO, and H. SUNAGA, Japan. J. appl. Phys. **9**, 1148 (1970).
[22] V. I. MUROMTSEV, R. A. ASATURYAN, and I. G. ACHVLEDIANI, Uspekhi Khim. **40**, 312 (1971).

(Received October 5, 1971)

phys. stat. sol. (a) 8, 521 (1971)

Subject classification: 3.1; 1; 22.2.2

Institute of Applied Physics, Moldavian Academy of Sciences, Kishinev

Growth Conditions and Morphological Pecularities of Gallium Phosphide Whiskers Obtained by Thermal Evaporation in Vacuum

By

P. O. BARABTARLO, YU. I. MAKSIMOV, and S. I. RADAUTSAN

Gallium phosphide whisker crystals were grown by thermal evaporation in vacuum in a closed volume. It is found that the crystals grow preferably along ⟨100⟩, ⟨110⟩, ⟨111⟩, and ⟨112⟩ directions. The crystallization mechanism from the gaseous phase through an intermediate liquid layer is observed when an excess of gallium is introduced. Different questions connected with possible growth mechanisms of gallium phosphide are discussed. Using an electron microscope the structure of the whiskers received was studied and a high degree of perfection was found.

Методом термической возгонки в замкнутом объеме получены нитевидные кристаллы фосфида галлия. Рассмотрены условия их образования. Установлено, что рост кристаллов происходит преимущественно вдоль направлений ⟨100⟩, ⟨110⟩, ⟨111⟩, ⟨112⟩. Показано, что при введении излишка галлия наблюдается механизм кристаллизации из газовой фазы через промежуточный слой жидкости. Рассматриваются вопросы, связанные с возможным механизмом роста нитевидных кристаллов фосфида галлия. Методом электронной микроскопии проведено исследование структуры полученных нитевидных кристаллов. Обнаружено их высокое совершенство.

1. Introduction

Gallium phosphide is one of the semiconductor materials, used for constructing instruments which are very effective at high temperatures and in agressive medium.

Many of the well-known techniques of synthesis and crystallization can be applied to receive it, but they become very complicated due to the high reactivity of gallium as well as to the considerable vapour tension of phosphorus at the equilibrium melting point. The methods of obtaining GaP from the vapour phase are most suitable as they allow to grow both epitaxial films and whisker crystals of different shape [1 to 3].

The growing interest in the examination of growth conditions, structure and properties can be explained by the nearly perfect structure, the specific character of dislocation orientation, the realization of the theoretical hardness, and by some other features which are of interest for scientific and practical applications.

2. Whiskers Growth and Morphology

In the present paper it is shown that the marked gallium phosphide evaporation in vacuum in the temperature range from 1000 to 1200 °C can be used to obtain perfect needle and ribbon crystals [4, 5]. Since dissociation of GaP is observed in the

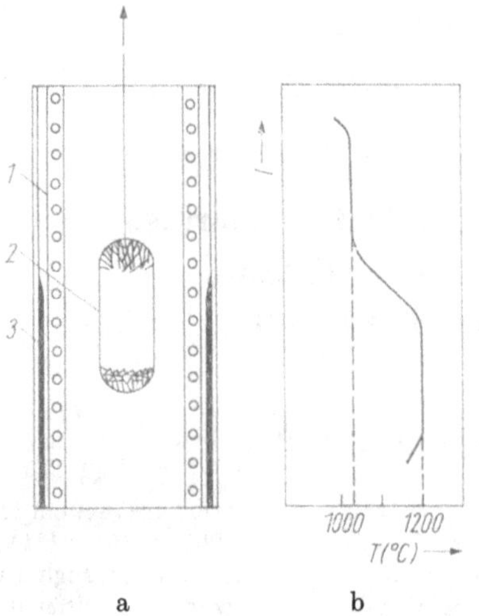

Fig. 1. a) Schematic diagram of the arrangement for growing gallium phosphide crystals; 1 heater, 2 quartz container, 3 semitransparent metal thermal reflector; b) dependence of temperature distribution along the heater

process of evaporation in vacuum (phosphorus predominates in the vapour phase as the more volatile component), the whole process was carried out in a limited volume to avoid stoichiometry violation. High temperatures and a small volume of the container prevented full dissociation and selected precipitation in the crystallization zone. Gallium phosphide crystal powder synthesized from a melt solution was used as starting material. The crystals were grown in an evacuated (10^{-6} Torr) ampoule of fused quartz previously thoroughly cleaned and degassed while it was moved along a heater with a temperature profile shown in Fig. 1. Combining the evaporation process in vacuum with the Bridgman method as described in [6] we succeeded in observing nucleation and further growth of needles and ribbons of GaP. In this case rather than in others the visual study of the growth conditions was possible. Note that nucleation and growth of good crystals took place in a certain temperature range, i.e. depend on the pressure in the ampoule and the evaporation rate. The effect of growth conditions on crystal shape and habitus is apparent. The observed variety of shapes depends strictly on temperature and velocity of moving the ampoule through the temperature zone and is determined by structure

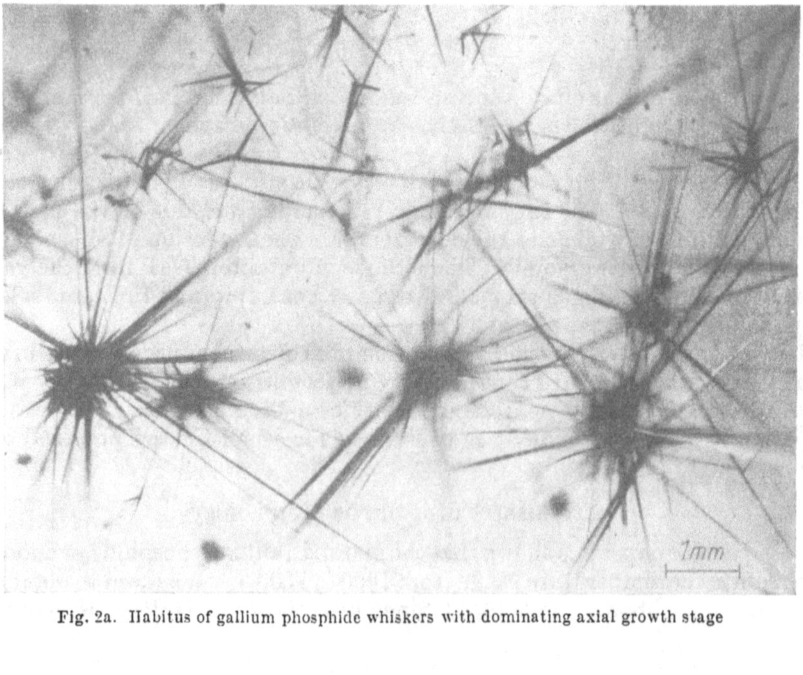

Fig. 2a. Habitus of gallium phosphide whiskers with dominating axial growth stage

transformations taking place during crystallization. Different morphological forms have characteristic growth temperatures, i.e. the crystals grow at an almost definite degree of oversaturation. The crystal forms changed from typical whisker crystals to polyhedral ones tending to an equilibrium shape, i.e. octahedron.

The external shape of crystals which grow in groups at the container walls is shown in Fig. 2. Some crystals are 30 mm or even longer depending on the ampoule dimensions, while their thickness varies from tenths of a micron to 1 mm. Among them ribbons can be distinguished which grew on needle crystals nucleating at protrusions observed on such ribbons. Not discussing details we must note that atoms or groups of atoms built in at growth steps have a large number of bonds and a better completeness of coordination of atoms is reached. The better the coordination the more probable is their connection to a given face and the higher must be the growth rate. Micrographs of GaP whiskers received by a scanning electron microscope are presented in Fig. 3. They show the importance of nucleation and motion of macrosteps in the process of crystal growth. The presence of a step-like surface can be explained by considering the crystal edge energy [7] as well as by the anisotropy of the surface energy, though beside the steps whose stability characterized by anisotropy a less stable form of density condensation of elementary steps (kinematic waves) is theoretically probable to exist.

By means of radiography it was stated that preferred growth directions of crystals are $\langle 100 \rangle$, $\langle 110 \rangle$, $\langle 111 \rangle$, $\langle 112 \rangle$. Various morphological types have quite different shapes and cross sections. Crystals oriented along $\langle 110 \rangle$ are most frequent. They look like additively pointed triangular pyramides thickened at the base or sometimes near the top (Fig. 4a, 2b).

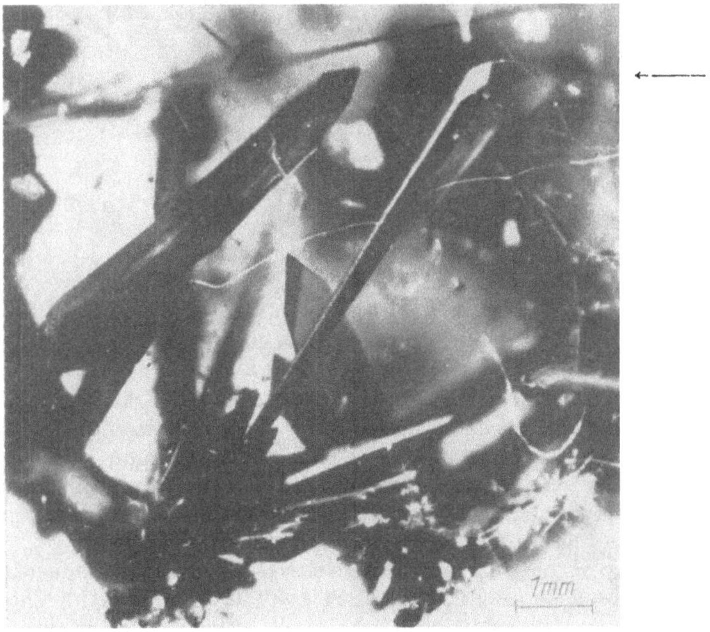

Fig. 2b. Habitus of gallium phosphide whiskers with dominating radial growth stage. The crystal thickening at the top is indicated by an arrow

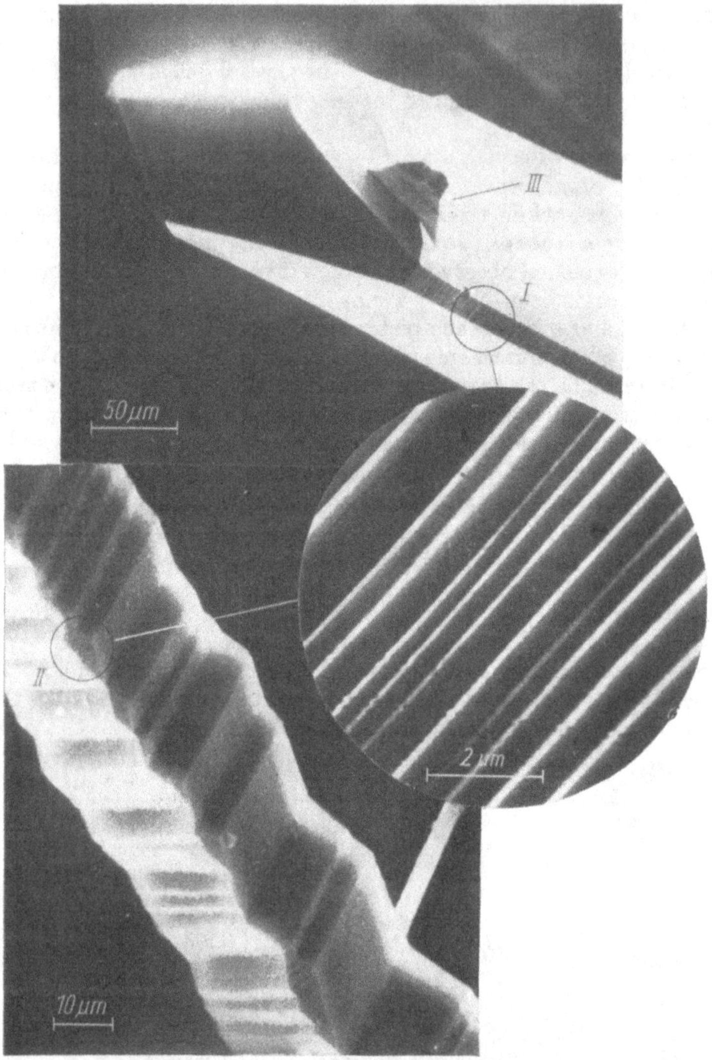

Fig. 3. Step growth of gallium phosphide whiskers; I, II characteristic growth steps, III nucleation of a new needle crystal

The thickening of whiskers can be determined by varying crystallization conditions and depends upon the increased concentration of precipitated substance at faces with a large area, at edges, and in entrance angles at the base. This results in a transformation from axial to radial growth which is characterized by a slow rate. During this period some layers appear on the crystal surface and gradually their high elastic properties disappear.

While the top of the growing whiskers moves its opposite end facets are limited (if it is possible) according to crystallographic laws of growth. The analysis of the process shows that slowing of the quick axial crystal growth is probably connected with the fact that the top gets into a region without strong oversatu-

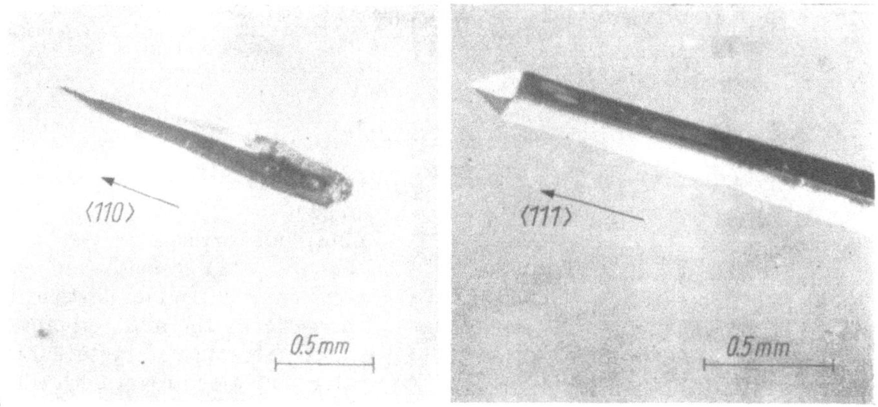

a b

Fig. 4. Characteristic habitus of needle crystals; a) ⟨110⟩ growth direction, b) ⟨111⟩ growth direction

ration. Thus, whisker growth can reduce the oversaturation and the effect of heat and mass transfer conditions on the crystal habitus.

Large needles in the form of triangular prisms with constant cross section described by the authors of [8, 9] are scarcely observed using the given method (Fig. 4b). Crystals in the form of prisms with hexagonal cross sections in ⟨111⟩ and ⟨110⟩ directions appear extremely rarely.

Almost perfect crystals with ⟨100⟩ growth direction are found. Their cross section has the form of a rectangular. In this case the crystal faces correspond to two (100) planes and two (001) planes.

Crystals in the form of perfect plates with faces oriented in ⟨100⟩ and ⟨112⟩ directions are of great practical interest. The high perfection of the crystalline plates is confirmed by the Kikuchi-lines on the electron diffraction patterns (Fig. 5).

Ribbon-like crystals growing in ⟨112⟩ direction are characterized by twinning in the (111) plane parallel to side faces of the crystal (Fig. 6). For gallium

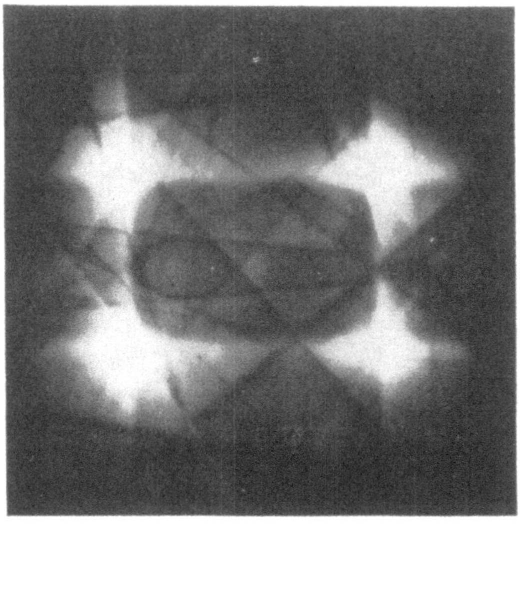

Fig. 5. Electron diffraction pattern of a gallium phosphide ribbon plate

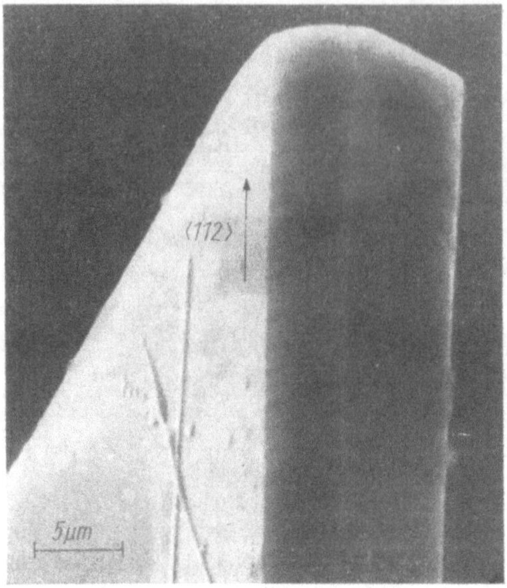

Fig. 6. Characteristic form of ribbon crystals growing in ⟨112⟩ direction (the twinning plane is parallel to the lateral face)

phosphide crystals crystallization in ⟨112⟩ direction of the face-centered cubic lattice is energetically the most advantageous . This type of crystal shape can be explained by growth taking place due to nucleation in entrance angles. Poorly developed {100} faces on the growth front of crystal is connected with crystal growth both by accumulation of single atoms and more complicated complexes of atoms.

Ribbons with ⟨100⟩ growth direction differ from crystals with ⟨112⟩ growth direction in the different degree of development of {100} and {111} faces. When ribbon crystals grow in ⟨100⟩ direction the twinning mechanism is impossible. This mechanism appears when the growth directions are in the twinning plane [10].

Ribbon crystals nucleate at the ampoule walls, at the tops and protrusions of the lateral faces of needle crystals (Fig. 7), and are themselves crystallization centres. Depending on the growth conditions the width of plates varies from tenths of microns to 1 mm while, sometimes their length reaches about 2 cm.

When metallic gallium ($\approx 1\%$) is introduced into the reaction ampoule, the growth becomes slower and one observed clearly a crystallization mechanism from vapour phase through an intermediate layer of liquid. But whiskers can grow according to this mechanism without special introduction of gallium. The resulting crystals are characterized by a stable cubic sphalerite structure with the above-mentioned growth directions and habitus. Whiskers in the form of the metastable hexagonal modification noted in [11] were not found.

Fig. 7. Intergrowth of gallium phosphide whisker

3. Examination of Gallium Phosphide Whiskers by Electron Microscopy

Examination of thin (up to 1000 Å) ribbon plates of GaP by means of electron microscopy showed that in most cases they had no dislocations. An electron microscope УЭМВ-100A with accelerating voltage of 75 kV was used. Micrograph of such whiskers is characterized by a lot of bent extinction contours as dark strips of constant width (Fig. 8). The bending of ribbons can be explained by randomly localized charged layers on their surfaces resulting in different surface tension on opposite planes of the crystals and thus the sample spontaneously bends. Whiskers can also bend due to thermal stress and mechanical effects during the preparation of samples.

On some thicker ribbons ($\approx$ 2000 Å) a break of bent extinction contours was observed which shows the presence of axial dislocations [12, 13]. The number of such crystals, however, is small in contrast to the total number of examined ribbons.

Most of the examined ribbon crystals had a high degree of perfection. This is shown by the Kikuchi-lines on the electron diffraction patterns. The occurrence of Kikuchi-lines is known to be associated both with samples thickness and perfection [14]. When crystals become thicker than some critical value multiple (dynamic) scattering effects appear. The critical thickness can be estimated [15] in the present case to 1500 Å. In addition the condition for Kikuchi-lines to appear should be as follows: the reflecting surfaces must be strictly parallel to each other, i.e. the crystal must be very perfect. A little plane disorientation makes these lines blurred , and at large disorientation angles they will disappear. In our experiments Kikuchi-lines in general were sharp (Fig. 5).

For indexing the observed bent extinction contours we carried out the dark field analysis [12, 16] of a gallium phosphide whisker. Passing a certain diffracted beam through an aperture we get in its dark field image the bright extinction contour. Each diffracted beam gives its indices to this contour thus determining the system of crystallographic planes. Diffraction at these planes causes the present extinction contour (Fig. 9). Hence it is possible to determine curvature radius, bend shape, and local orientation of any crystal area. With the help of the dark field image of some contours the crystal thickness can be estimated [12]. In our experiments the thickness of ribbons was estimated to be 700 to 1500 Å.

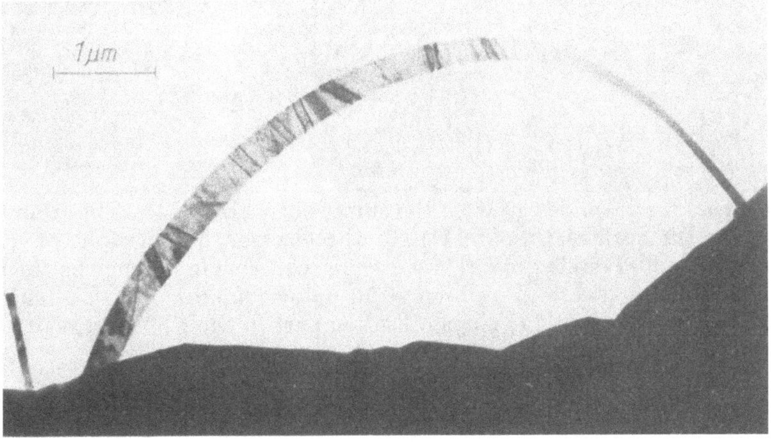

Fig. 8. GaP ribbon crystal without dislocations. Dark strips are bent extinction contours

a

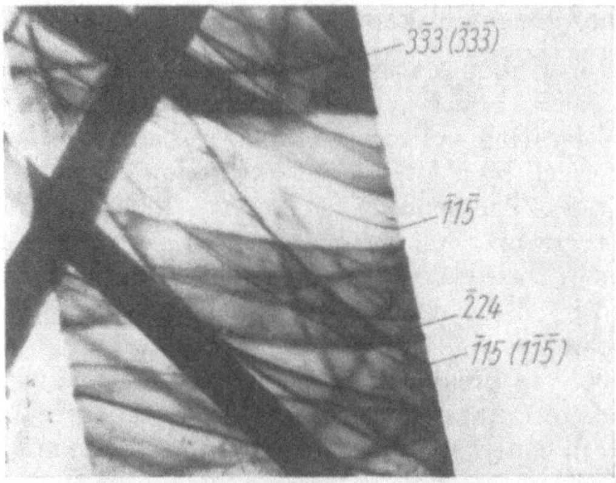

b

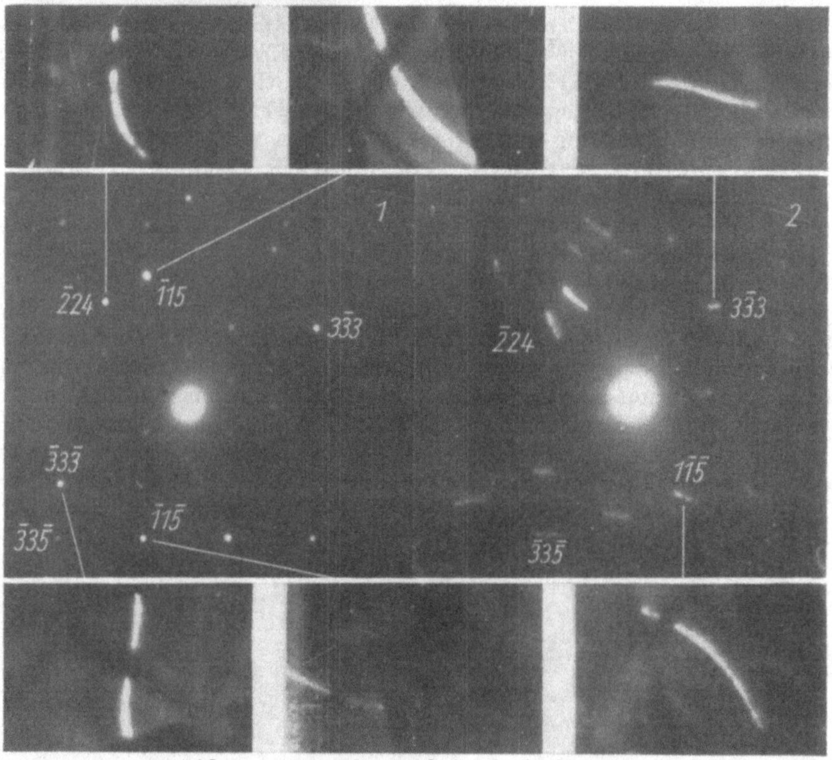

Fig. 9. Dark field analysis of GaP ribbon plate bent extinction contours. a) crystal area in light field, b) dark field image of extinction contours corresponding to different reflections of electron diffraction pattern. 1. focused electron diffraction pattern, 2. unfocused electron diffraction pattern

In some cases pyramides grew on the surface of whiskers starting from nuclei observed on the surface (Fig. 3, III). In the electron micrographs of gallium phosphide ribbon crystals growth layers were seen varying along the thickness. These growth layers were also observed in optical microscope. Evidently the layer growth mechanism [17] plays a leading part in the ribbon growth.

4. Conclusions

Analysing the results it is interesting to note that it is possible to get perfect gallium phosphide crystals by thermal evaporation in a limited volume. The whisker growth according to this method has specific features. It can take place by accumulation of single atoms as well as complexes of atoms. Though the main ideas about the whisker growth mechanism are still hypothesis and in general are based on the examination of the final crystal shape it is interesting that structure changes during the crystallization affect the shape and perfection of the crystals. In this connection it is necessary to note that crystals grow according to a vapour — liquid — solid crystal mechanism. A probable excess of phosphorus vapour does not exclude the possibility of forming gallium drops and nucleating of whiskers on them.

In the growth process whiskers were obtained growing preferably along ⟨100⟩, ⟨110⟩, ⟨111⟩, ⟨112⟩ directions. The micromorphological examination of whiskers at different stages of growth shows that imperfections of the lattice are mainly connected with the final slow growth period.

Ribbon crystals growing in ⟨100⟩ direction were probably observed for the first time.

The results of the electron microscopic examination of gallium phosphide whiskers show that crystals grown by means of this method have a high degree of perfection. The majority of examined ribbons are dislocation free. Only in some thicker ribbon crystals axial dislocations were observed. In ribbons growing in ⟨112⟩ direction twins were seen to occur both at old and at new crystallization centres. In the micrographs of some ribbon-like crystals growth layers with varying thickness were found. This indicates the possibility of a layer growth mechanism.

Acknowledgements

The authors are thankful to Yu. A. Simonov, V. N. Rozhanskii, and G. V. Berezhkova for their help in some experiments and for useful discussions.

References

[1] R. L. Barus and W. C. Ellis, J. appl. Phys. **36**, 2296 (1965).
[2] W. C. Ellis, C. J. Frosch, and R. B. Zetterstrom, J. Crystal Growth **2**, 61 (1968).
[3] S. I. Radautsan, Yu. I. Maksimov, V. V. Negreskul, and S. L. Pishkin, "Fosfid galliya", RIO Akad. Nauk Moldavian SSR, Kishinev 1970.
[4] Yu. I. Maksimov, A. N. Nateprov, Slozhnie poluprovodniki i ikh fizicheskie svoistva, RIO Akad. Nauk Moldaviam SSR, Kishinev 1971.
[5] G. P. Belousova, L. A. Bityutskaya, E. G. Goncharov, and Ya. A. Ugai, Fizika poluprovodnikov i mikroelektronika, Univ. Voronezh 1969 (p. 165).
[6] A. Rabenau, Philips tech. Rev. **26**, 117 (1965).
[7] S. A. Stroitelev, Mekhanizm i kinetika kristallizatsii, Izd. Nauka i Tekhnika, Minsk 1969 (p. 47).
[8] A. V. Lishina, S. A. Medvedev, A. Ya. Nashelskii, and B. A. Sakharov, Kristallografiya **9**, 436 (1964).
[9] M. Gershenzon and R. M. Mikulyak, J. Electrochem. Soc. **108**, 548 (1961).
[10] J. W. Faust and H. F. John, J. Phys. Chem. Solids **25**, 1407 (1964).
[11] J. E. Davey and T. Pankey, Appl. Phys. Letters **12**, 38 (1968).
[12] V. N. Rozhanskii and G. V. Berezhkova, phys. stat. sol. **6**, 185 (1964).
[13] G. V. Berezhkova, Nitevidnie kristalli, Izd. Nauka, Moscow 1969 (p. 64).
[14] S. A. Semiletov and Z. G. Pinsker, Kristallografiya **1**, 209 (1956).
[15] B. K. Vainshtein, Dokl. Akad. Nauk SSSR **104**, 537 (1955).
[16] K. A. Jackson and R. S. Wagner, J. appl. Phys. **36**, 2132 (1965).
[17] A. A. Chernov, Uspekhi fiz. Nauk **73**, 277 (1961).

(Received May 11, 1971)

phys. stat. sol. (a) 8, 531 (1971)

Subject classification: 1.2 and 10.1; 21; 21.1; 21.1.1

Department of Physics, Faculty of Science, Osaka City University

A New Concept on the Shape Memory Effect in Metals and Alloys

By

A. NAGASAWA

The shape memory effect is investigated on many kinds of materials undergoing the martensite transformation. The present experiment proves that such metals and alloys as Co, Ti, Zr, Co–Ni, Cu–Al, Fe–Ni and In–Tl exhibit the shape memory effect. From these results and considerations on the plastic deformation of material containing martensite phase, it is concluded that the shape memory effect is an inherent thermomechanical property of the martensite transformation.

Der Form-Gedächtniseffekt wurde in einer Anzahl von Materialien, die eine Martensitumwandlung aufweisen, untersucht. Die Experimente bestätigen, daß Metalle und Legierungen von Co, Ti, Zr, Co–Ni, Cu–Al, Fe–Ni und In–Tl den Gedächtniseffekt zeigen. Aus den Ergebnissen und den Betrachtungen der plastischen Verformung von Materialien, die die Martensitphase enthalten, wird gefolgert, daß der Gedächtniseffekt eine wesentliche thermomechanische Eigenschaft der Martensitumwandlung ist.

1. Introduction

The intermetallic compound NiTi exhibits a unique thermomechanical property known as the shape memory effect (hereafter abbreviated as the memory effect), which has been found by Buehler et al. [1]. The memory effect is now considered to be related with the martensite transformation of the alloy. That is, after a wire or sheet of the alloy is plastically deformed at a temperature below the martensitic transition point, it appears to revert back to its undeformed shape when heated to a temperature higher than the reverse transition point. It is also known that AuCd [2] and Cu–Al–Ni [3] alloys have similar properties. The unique nature of these alloys has been a subject of intense studies and, in order to explain it, proposals have so far been reported by de Lange et al. [4] and Otsuka et al. [3, 5]. The former, related only to the case of NiTi, however, is so simple that it cannot interpret a drastic shape recovery in the alloy and further it is meaningless for the cases of AuCd and Cu–Al–Ni. As to the latter, on the other hand, to realize the memory effect in NiTi, AuCd and Cu–Al–Ni the following conditions are required: 1. the structure must be ordered, 2. the martensite transformation is b.c.c. → h.c.p. type, ordering being disregarded, 3. lattice invariant strains in the martensites are not glide but twins, and 4. the martensite transformation is thermoelastic. It must be pointed out here that, though this proposal is based on a premiss that the memory effect is a peculiar property appearing only in several special alloys, it does not interpret the origin of peculiarity of the memory effect. Thus we can conclude that any convincing theory on the memory effect has not yet been established at this stage.

Recently, however, the present author [6, 7] has proposed briefly a hypothesis that the memory effect is not a peculiar property in such special alloys as NiTi, AuCd and Cu–Al–Ni, but it is principally expected to appear in every material undergoing the martensite transformation. From this point of view, the present author has investigated the memory effect and found it so far in such metals and alloys as Cu–Al [6], In–Tl [8], Fe–Ni [9], Co–Ni, Co, Ti and Zr. Enami et al. [10, 11] have also found the memory effect in Ni–Al and 304 type stainless steel. These facts strongly prove the present author's prediction to be reasonable. Consequently, it is necessary to reconsider the origin of the memory effect.

The present paper gives firstly experimental results on the memory effect. The present experiments have been done on the materials belonging to the following types of the martensite transformations; 1. b.c.c. → orthorhombic type (Cu–Al), 2. f.c.c. → f.c.t. type (In–Tl), 3. f.c.c. → b.c.c. type (Fe–Ni), 4. f.c.c. → h.c.p. type (Co and Co–Ni) and 5. b.c.c. → h.c.p. type (Ti and Zr). They contain almost all types of the martensite transformations. The paper presents secondly a new concept on the memory effect, which is deduced from both considerations on nature of plastic deformation in materials containing martensites and the obtained experimental results.

2. Experimental Procedures

In the present study, specimens except In–Tl were prepared by melting component metals, with purities listed in Table 1, in a plasma–arc furnace under argon atomosphere. Several melts were performed repeatedly for homogenization treatment. They were further treated as follows.

Table 1
Purities of component metals

metal	Al	Co	Cu	Fe	In	Ni	Ti	Tl	Zr
%	99.999	99.99	99.99	99.99	99.97	99.99	99.7	99.95	99.7

1. Cu–Al alloy: The specimen was cut from the prepared ingot and mechanically thinned into a sheet with about 0.2 mm thick. After chemical etching, it was annealed at 800 °C for 1 h in vacuum. The used alloy contains 25.0 at% Al.

2. In–Tl alloy: In– 21 at% Tl alloy was used. This was originally employed by Aoyagi and Sumino [12] for the investigation of the mechanical property. A sheet with about 0.2 mm thick was prepared by rolling the alloy at room temperature. It was annealed at 130 °C for 2 h in vacuum before observation of the memory effect.

3. Fe–Ni alloy: A thin sheet about 0.05 mm thick of Fe– 30.0 at% Ni alloy was made by cold rolling of an ingot. It was chemically etched and then deformed spirally at room temperature. The sheet was annealed at 870 °C for 1 h in vacuum and then cooled.

4. Co and Co–Ni alloy: Ingots of Co and Co– 25.0 at% Ni alloy were thinned to sheets about 0.1 mm thick by hot rolling. After chemically etching, they were severely deformed at room temperature. The deformation was done over limits of shape recoveries in order to prepare wavy shape at the high temperature state. They were annealed at 900 °C for 3 h in an evacuated quartz capsule, then quenched into brine.

5. Ti and Zr: Ti and Zr were cold rolled into thin sheets about 0.1 mm thick and deformed wavily for the same purpose as in the cases of Co and Co–Ni. After chemical etching, they were annealed at 950 °C for 3 h in quartz capsule enclosing argon gas, then quenched into brine.

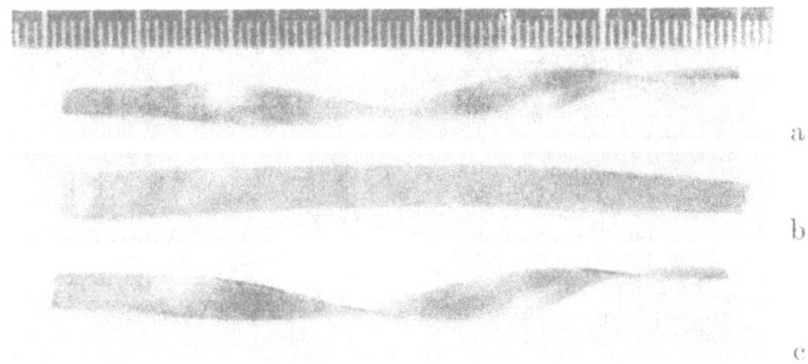

Fig. 1. Demonstrations of the memory effect in Cu–25.0 at% Al. Interpretations are given in the text

3. Experimental Results

3.1 The memory effect in Cu–Al (b.c.c. → orthorhombic type)

The alloy undergoes nearly completely the martensite transformation of a b.c.c. → orthorhombic type by quenching, ordering being disregarded; the M_s (martensite start) point is about 300 °C. The memory effect in the alloy is shown in Fig. 1. The flat sheet was twisted as a screw shape at about 600 °C and subsequently quenched to room temperature (Fig. 1a), at which sheet was deformed again into a flat shape (Fig. 1b). When the sheet was rapidly heated to about 450 °C, it recovered almost completely its original twisted shape (Fig. 1c); this is the memory effect. When the alloy is slowly heated, however, the memory effect cannot be observed, because decomposition of the martensite phase into the α and γ_2 phases occurs below the A_s (austenite start) point (about 400 °C).

3.2 The memory effect in In–Tl (f.c.c. → f.c.t. type)

The martensite transformation of the alloy, known as a single interface type, is a typical one of the f.c.c. → f.c.t. types. In the alloy, a phenomenon concerning the memory effect has been found by Basinski et al. [13] but its nature has not been clear.

At 130 °C, the prepared sheet was bent into a ring shape and held for 2 h. It was slowly cooled to room temperature so that it transformed completely into the f.c.t. martensite, the M_s point being about 60 °C (Fig. 2a). At room temperature, the ring was deformed into a flat shape (Fig. 2b). The memory effect was observed by reheating the sheet (Figs. 2c and d). It must be noted that annihilation of the deformation is realized through two steps. In the first step, a small shape recovery occurs below the A_s point (about 65 °C). Fig. 2c shows the shape at 55 °C. On the other hand, the second step of shape recovery bigins to occur on heating at the A_s point and ends

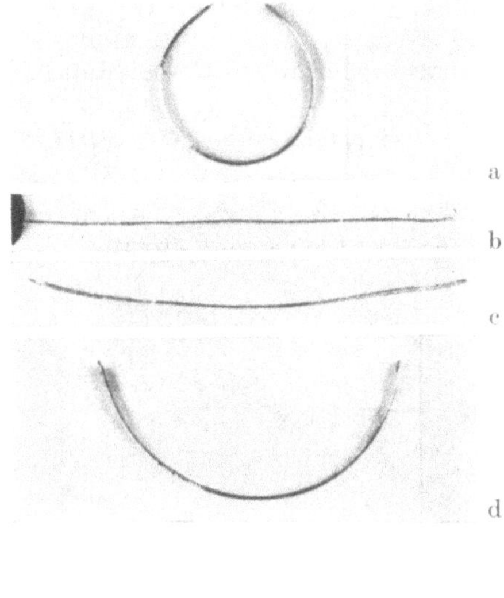

Fig. 2. Demonstration of the memory effect in In–21 at% Tl

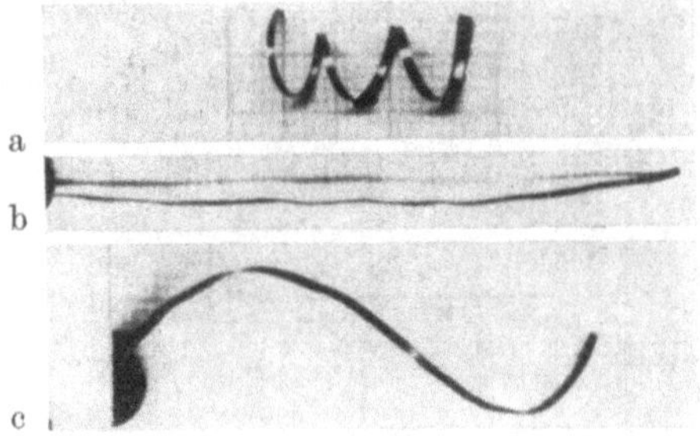

Fig. 3. Demonstration of the memory effect in Fe–30.0 at% Ni

at the A_f (austenite finish) point (about 95 °C), though it is incomplete (Fig. 2d).

3.3 The memory effect in Fe–Ni (f.c.c. → b.c.c. type)

The martensite transformation of the alloy is a typical f.c.c. → b.c.c. type, with a considerable temperature hysteresis. Fig. 3 shows the memory effect in the Fe–30.0 at% Ni alloy. The M_s and A_s points of the alloy are about −50° and 360 °C, respectively. The prepared sheet with spiral shape was sub-zero cooled to the liquid nitrogen temperature so that the martensite transformation occurred (Fig. 3a). It was deformed into nearly flat shape at room temperature (Fig. 3b). When the sheet was heated, shape recovery began to occur at the A_s point and ended at the A_f point (about 450 °C), though it was incomplete (Fig. 3c). It has been also confirmed that Fe–27.0 at% Ni alloy reveals the memory effect of incomplete type.

3.4 The memory effect in Co and Co–Ni (f.c.c. → h.c.p. type)

The f.c.c. → h.c.p. type martensite transformation is introduced by a stacking modulation of close packed layers. The transition modes of Co and Co–Ni alloy are of this type. Fig. 4 is an example demonstrating the memory effect in Co. The prepared wavy sheet was deformed into a flat shape at room temperature (Fig. 4a and b). When the sheet was heated up to about 500 °C, it regained the wavy shape (Fig. 4c); the A_f point is about 490 °C. Feature of the memory effect in the Co–25.0 at% Ni alloy is nearly the same as that in Co, though photographs are omitted. In this case the shape recovery is finished at about 300 °C; the A_f point of the alloy being about 290 °C. The memory effect has been also found in the Co–Ni alloys containing 19.0 and 30.0 at% Ni.

3.5 The memory effect in Ti and Zr (b.c.c. → h.c.p. type)

Ti and Zr undergo the b.c.c. → h.c.p. type martensite transformation with small hysteresis. The M_s points of them are about 865° and 860 °C, respectively. Fig. 5 shows the memory effect in Ti. The prepared sheet (Fig. 5a) was deformed

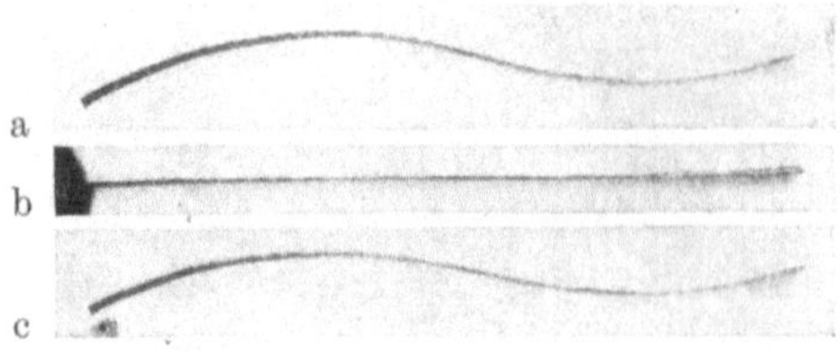

Fig. 4. Demonstration of the memory effect in Co

Fig. 5. Demonstration of the memory effect in Ti

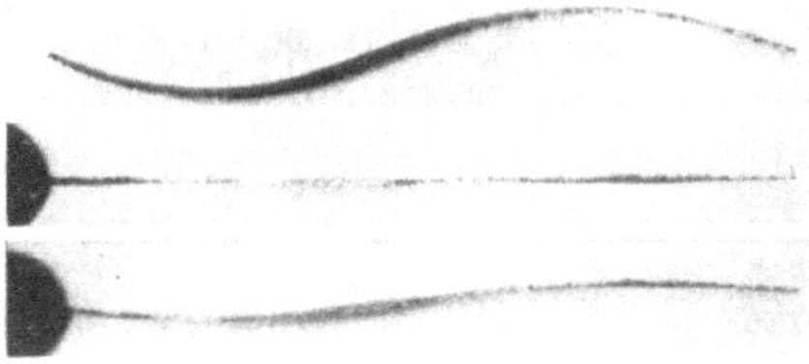

into a flat shape at room temperature (Fig. 5b). By heating up to about 900 °C, it recovered the curved shape at high temperature, though it was incomplete (Fig. 5c). The memory effect in Zr is similar to that in Ti, photographs being omitted.

4. A New Concept on the Memory Effect

Usually, plastic deformation of metals and alloys is due to slip or twinning resulting from the motion of line or surface dislocations. From a crystallographic point of view, the deformation mode of this kind can be regarded as a single phase process, because the deformation proceeds without inducing any phase transition in the materials. It is evident in principle that the single phase process is not responsible for the memory effect.

On the other hand, there is another mode of plastic deformation known as the martensitic shear. Besides cooling, an external stress field can induce the martensite transformation below the M_d (martensite start by deformation) point. This originates from the interaction of the stress field with the lattice strain involved in the transition process from austenite to martensite. If the interaction is favourable for the transition, formation of the martensite will relieve the external stress field. Thus it follows that the martensite transformation of this kind is an important mode of plastic deformation. Similarly, growth of martensites, twinning in martensites and others are also responsible for plastic deformation, if an external stress field is applied to a martensitically transformed material. Plastic deformations of these kinds can be en bloc regarded as martensitic shear. Evidently, this deformation mode is associated with a phase change process, which distinguishes it from slip or twinning.

The martensite transformation is a reversible reaction with hysteresis. The reverse transformation, from martensite to austenite, involves also a shear process, which is identical with, though opposite in sign, that occurring in the martensite transformation. This means that deformations associated with formation and/or growth of the martensites and others are annihilated through the reverse transformation, whether the martensites are produced by cooling or by external stress field. Namely, plastic deformation due to the martensitic shear is annihilated by heating up to the A_f point. This is just the same phenomenon as that of the memory effect. Consequently, we can conclude that the origin of the memory effect consists in the above mentioned nature of the martensite transformation with recoverable shearing process. Thus the memory effect is one of the chief characteristics of the martensite transformation. This conclusion leads necessarily a hypothesis that the memory effect is not a peculiar property in special materials such as NiTi, AuCd and Cu–Al–Ni, but it is expected to appear essentially in every material undergoing the martensite transformation.

Whether the hypothesis is reasonable or not should be determined by experiments. As stated in the previous section, the memory deffect has been found in many metals and alloys undergoing the martensite transformation. The fact justifies strongly the present author's hypothesis. Thus a new concept on the memory effect is established: the memory effect is an inherent thermomechanical property of the martensite transformation.

From this point of view, requirement as necessary and sufficient condition for the memory effect is that plastic deformation is possible to occur through the martensitic phase change process in a material. Such factors as thermoelasticity, mobility of twin boundaries and others proposed by Otsuka et al. [3, 5] are not essential conditions but only additional ones.

5. Discussions

From the present experiments it becomes clear that the memory effect in the respective materials exhibits different characteristics. These phenomena can be explained satisfactorily in terms of deformation modes as follows.

In Cu–Al the shape recovery is considerably large and nearly complete. By rapid quenching, the alloy is almost totally transformed to the martensites with layer structure expressed as the 18R by the Ramsdell's notation [14]. If the alloy is plastically deformed at room temperature, stacking faults are so easily introduced that the martensite is further changed to the 2H structure [15]. This means that, in order to release the external stress field, a preferential deformation is easily performed over nearly the whole range of the specimen through formation of the new martensites by stacking modulation. This mode leads to a large annihilable deformation.

The memory effect in In–Tl is characterized by a two step process. Below the M_s point, the alloy transforms completely into the f.c.t. martensite containing transformation twins. A stress field causes easily a preferential motion of twin boundaries, this being a main mode of plastic deformation of the alloy. On the other hand, the tetragonality of the martensite phase changes with temperature, i.e. when the martensite is heated the tetragonality decreases gradually. This lattice deformation may cause the twin boundary motion, which seems to be responsible for the first step recovery. The second step recovery results from the reverse transformation. By the transformation the twins involved in the martensite vanish and hence the associated deformation may disappear simultaneously, like in the case of Cu–Al–Ni [5]. It must be noted, however, that twinning is not responsible for the memory effect. For example, plastic deformations of In and Zn are mainly due to twinning but they do not exhibit the memory effect. Incompleteness of the memory effect in the present case may arise from an deformation exceeding a limit of the twinning mode in the martensite.

In Fe–Ni, there are umklapp and schiebung type martensite transformations. The former proceeds rapidly in the interior of the alloy containing more than about 25 at% Ni, while the latter produces only surface martensite in the alloy containing less than about 30 at % Ni. In the present case both transformations occur. According to Homma's investigation of the alloys containing 28 to 33 at% Ni [16], an external stress field induces only the schiebung type martensite in the alloys. Further, Krauss et al. [17] have found that, in the Fe–33 at% Ni alloy, twinning occurs in the umklapp type martensites by impingement. Thus it is possible to suppose that plastic deformation of the used alloy is due to such processes as formation of surface martensites, twinning in the umklapp type martensites and others, as well as slip in the matrix. It must noted here that deformation by twinning is responsible for the memory effect only when it occurs in martensites. In the present case, recoverable deformation is not dominant because the phase change process of plastic deformation is restricted to occur mainly at surfaces of the alloy and because the martensites do not grow so large. These circumstances lead to the incompleteness of the memory effect in the alloy.

The martensite transformation of Co or Co–Ni is only partial and occurs through stacking modulation from 3R to 2H, in which the lattice invariant shear is not necessary. When the transformed alloy is plastically deformed, formation and/or growth of martensites may occur with the stacking modulation. In fact, such mode has been found in Co–30.0 at% Ni [18]. There is no evidence on formation of transformation twins but stacking faults are frequently observed. These features are very similar to those in Cu–Al and, in fact, the shape recovery in Co and Co–Ni is nearly complete as we expect.

Rosi et al. [19] have reported that plastic deformation of the martensitically transformed Ti is mainly due to slip and twinning, the former occurring in a wide temperature range while the latter is confined to low temperatures. At room temperature slip seems to be dominant. This means that we cannot expect a large shape recovery in Ti. As shown in Fig. 5, the memory effect in Ti is a typical example of very small shape recovery. The case of Zr is considered to be analogous to that of Ti.

From the above discussions it is concluded that completeness or incompleteness of the memory effect is not essential, but caused by the relation between deformation mode and amount of deformation in the respective cases. Essential in the memory effect is, as stated in the previous section, that the recoverable modes are originally involved in the process of plastic deformation of metals and alloys undergoing the martensite transformation, if their temperatures are below the respective M_d points.

Acknowledgements

The present author wishes to thank Mr. K. Kawachi for his considerable assistance in the experiments, and to Dr. K. Sumino for providing the sample of In–Tl alloy.

References

[1] W. J. Buehler, J. V. Gilfrich, and R. C. Wiley, J. appl. Phys. **34**, 1475 (1963).
[2] L. C. Chang and T. A. Read, Trans. AIME **189**, 47 (1951).
[3] K. Otsuka and K. Shimizu, Scripta metall. **4**, 469 (1970).
[4] R. G. de Lange and J. A. Zijderveld, J. appl. Phys. **39**, 2195 (1968).
[5] K. Otsuka, Japan. J. appl. Phys. **10**, 571 (1971).
[6] A. Nagasawa and K. Kawachi, J. Phys. Soc. Japan **30**, 296 (1971).
[7] A. Nagasawa, J. Phys. Soc. Japan **31**, 136 (1971).
[8] A. Nagasawa, J. Phys. Soc. Japan **30**, 1200 (1971).
[9] A. Nagasawa, J. Phys. Soc. Japan **30**, 1505 (1971).
[10] K. Enami and S. Nenno, Metall. Trans. **2**, 1487 (1971).
[11] K. Enami, S. Nenno, and Y. Minato, Scripta metall. **5**, 663 (1971).
[12] T. Aoyagi and K. Sumino, phys. stat. sol. **33**, 317 (1969).

[13] Z. B. BASINSKI and J. W. CHRISTIAN, Acta metall. **2**, 148 (1954).
[14] L. S. RAMSDELL, Amer. Mineralogist **32**, 64 (1947).
[15] S. KAJIWARA, J. Phys. Soc. Japan **27**, 712 (1969).
[16] T. HOMMA, J. Japan Inst. Metals **21**, 263 (1957).
[17] G. KRAUSS and W. PITSCH, Acta metall. **12**, 278 (1964).
[18] S. TAKEUCHI and T. HOMMA, Sci. Rep. Res. Inst. Tohoku Univ. **A9**, 492 (1957).
[19] F. D. ROSI, F. C. PERKINS, and L. L. SEIGLE, J. Metals **8**, 115 (1956).

(Received October 20, 1971)

phys. stat. sol. (a) 8, 539 (1971)

Subject classification: 10.2; 22.5.3

Jena, Carl-Zeiß-Straße 1

On the Dynamics of Photostimulated Processes in Additively Coloured CaF_2:La Crystals

By

HJ. BERNHARDT

Crystals doped with Y^{3+}, La^{3+}, Ce^{3+}, Gd^{3+}, Tb^{3+}, or Lu^{3+} ions show after an additive coloration very similar absorption spectra and a characteristic photosensitive behaviour. Light irradiation makes the electrons leave their traps and shift to another ones. Three different mechanisms of such transfer are described in the present paper. Mathematical relations describing such mechanisms are applied to experimentally recorded progresses of the dynamical photosensitive behaviour. The theoretical models are compared with the experimental observations. This practice showed, that irradiation of different wavelengths (410 and 650 nm) induces different types of dynamical photosensitive behaviour. This let us assume different mechanisms of electron transfer, too. The models proposed in [13, 15, 16] assuming a double-model which consists of an impurity ion and an F-centre are supported. On the contrary to the model of the authors of references [17 to 19] this model traps only *one* electron. Trapping of *two* electrons should result in the lost of the photosensibility. Following our model the photochromic process is identical with the transfer of one electron between the partners of the above-mentioned double model. The present investigations are interpreted by assuming an 410 nm-induced electron transfer from the F_g-centre to the $La^{3+}_{trig.}$ with possible participation of the conduction band. 650 nm-irradiation induces the reverse process but without the conduction band.

Werden MeF_2-Kristalle, die mit Y^{3+}-, La^{3+}-, Ce^{3+}-, Gd^{3+}-, Tb^{3+}- oder Lu^{3+}-Ionen dotiert sind, additiv verfärbt, so besitzen alle diese Kristalle untereinander verwandte Absorptionsspektren und ein charakteristisches, photosensibles Verhalten. Unter Einwirkung von Licht können Elektronen aus ihren Haftstellen gelöst und auf andere Plätze gebracht werden. Drei verschiedene Mechanismen einer solchen Übertragung werden in der vorliegenden Arbeit beschrieben. Mathematisch erhaltene Beziehungen, die solche Mechanismen beschreiben, werden auf experimentell bestimmte Verläufe des dynamischen photosensiblen Verhaltens angewandt. Es wird verglichen, welcher theoretische Prozeß am besten die Beobachtungen beschreibt. Auf diese Weise wurde herausgefunden, daß bei Einstrahlung unterschiedlicher Wellenlängen (410 bzw. 650 nm) unterschiedliche Typen des dynamischen photosensiblen Verhaltens auftreten. Aus diesem Grunde wird angenommen, daß auch verschiedene Elektronenübertragungsmechanismen vorkommen. Die in [13, 15, 16] dargelegten Vorstellungen über ein Doppelmodell, die eine Vereinigung des Dotierungsions mit dem F-Zentrum annehmen, werden gestützt. Im Gegensatz zu den Autoren der Arbeiten [17 bis 19] wird nach unseren Vorstellungen dieses Zentrum mit nur einem Elektron besetzt. Der photoinduzierte Prozeß besteht nach unserer Vorstellung in der Übertragung eines Elektrons zwischen den Partnern des oben genannten Doppelzentrums. Die vorliegenden Untersuchungen werden dahingehend gedeutet, bei 410 nm-Anregung eine Übertragung vom F_g-Zentrum zum $La^{3+}_{trig.}$-Zentrum bei möglicher Beteiligung des Leitungsbandes und bei 650 nm-Einstrahlung den umgekehrten Prozeß ohne Leitungsband anzunehmen.

1. Introduction

Many authors investigated the nature of the processes observed on Y-, La-, Ce-, Gd-, Tb-, or Lu-doped MeF_2 crystals after additive coloration. The authors of references [17 to 19] state the model of a PC centre consisting of a fluorine defect and a La^{3+} ion binding together two electrons. The photochromic process consists in the transfer of one of these electrons to an isolated La^{3+} ion in a cubic environment. According to our model [13, 15, 16] this centre traps only one electron, the change of the absorption spectrum is due to the transfer of this electron between the partners F_g and $La^{3+}_{trig.}$ within one kind of centre. The present paper will give further arguments dealing with these problems. The mechanism of photochromic transfer will be investigated.

While the authors of references [5, 7, 11, 17, 18, 19] suppose exclusively an electron transfer via the conduction band, we already in [13, 15, 16] proposed the possibility of transfer excluding the conduction band.

This paper presents investigations of the dynamical photochromic behaviour of CaF_2:La crystals, taking CaF_2:La as a substance which is representative for the afore-mentioned class of photochromic systems. We observed characteristic deviations of the photochromic behaviour of two main absorption bands which are, according to our model, due to the partners F_g and $La^{3+}_{trig.}$, respectively. It will be pointed out, that our observations can be understood in terms of the proposed double-model. As the double-model, proposed in [13, 15, 16] is non-stoichiometric we cannot exclude that this centre forms higher aggregates, e.g. with F_i^- ions.

2. Mathematical Considerations

In the discussion of experiments it is useful to consider three important processes:

a) The transfer of an electron within a double centre excluding retrapping processes. This model has been supposed in a former paper [16].

b) The transfer of an electron between different centres with low concentrations via the conduction band. This case describes e.g. the PC centre mechanism of Kiss and co-workers.

c) The excitation of electrons into the conduction band from centres with very high concentration and trapping of the electrons on centres with low concentration.

2.1 Electron transfer between locally separate centres with participation of the conduction band

While the previous paper [16] is dealing with the process a) we here consider at once the point b). Fig. 1 points out, that two main processes have to be expected in this mechanism: the transfer of electrons from the centre A to the centre B across the conduction band or retrapping of the electron in a centre A which has been ionized before. Therefore we must regard the probabilities

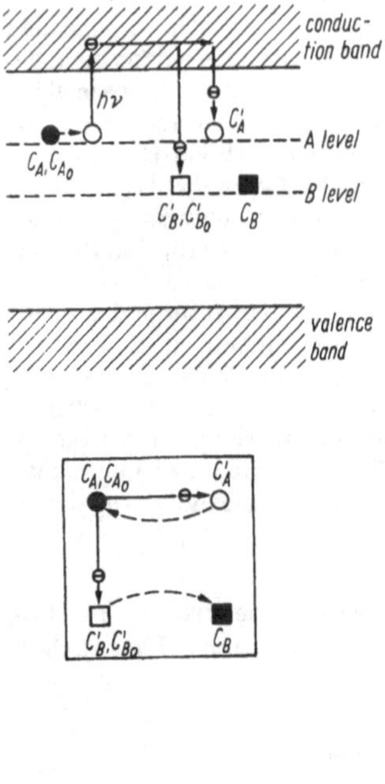

Fig. 1. Energy band model for the electrontransfer between centres A (● with and ○ without an electron) and B (■ with and □ without electron)

of the centres A to absorb photons and of the electrons to be trapped in the centres B. We want to point out, that the word "low" in process b) has the meaning, that the concentration of centres A decreases measureably during the irradiation process.

We denote the ionization probability of centres A by $\sigma_A \varrho_A C_A$, the probability of centres B to trap electrons by $\sigma_{B'} \varrho_{B'} C_{B'}$, and the probability of empty centres A to trap electrons by $\sigma_{A'} \varrho_{A'} C_{A'}$.

Taking into account the above-mentioned we get the following differential equation for the trapping of electrons in centres B:

$$dC_B = (\sigma_A \varrho_A C_A)(\sigma_{B'} \varrho_{B'} C_{B'}) \left(\frac{1}{\sigma_{A'} \varrho_{A'} C_{A'}}\right) dt = \Omega \frac{C_A C_{B'}}{C_{A'}} dt \tag{1}$$

with σ trapping cross section, ϱ quantum efficiency, and C number of centres.

Postulating the conservation of electrons and centres A and B enables us to state the following secondary conditions:

1. Conservation of centres A: $C_A + C_{A'} = C_{A_0}$.
2. Conservation of centres B: $C_B + C_{B'} = C_{B_0'}$.
3. Conservation of the electrons: $C_A + C_B = C_{A_0}$.

From 1. and 2. follows

$$C_{A'} = C_B\,. \tag{2}$$

Inserting this into the differential equation (1) it follows

$$dC_B = \frac{\Omega}{C_B} (C_{A_0} - C_B)(C_{B_0'} - C_B)\, dt, \tag{3}$$

and regarding the initial condition $C_B(t = 0) = 0$ we get after integration

$$\frac{1}{2} \ln \left| \frac{C_B^2 - C_B (C_{A_0} + C_{B_0'}) + C_{A_0} C_{B_0'}}{C_{A_0} C_{B_0'}} \right| + \frac{C_{A_0} + C_{B_0'}}{2\, C_{A_0} - C_{B_0'}} \times$$

$$\times \ln \left| \frac{2\, C_B - (C_{A_0} + C_{B_0'}) - C_{A_0} + C_{B_0'}}{2\, C_B - (C_{A_0} + C_{B_0'}) + C_{A_0} - C_{B_0'}} \frac{C_{B_0'}}{C_{A_0}} \right| = \Omega\, t\,. \tag{4}$$

This relation has been derived using very general assumptions. Let us consider the special case of a double-centre, that is the local aggregation of centres A and B. This leads to the following equations:

$$C_{A_0} = C_{B_0'}\,, \tag{5}$$

$$dC_B = \Omega \frac{(C_{A_0} - C_B)(C_{A_0} - C_B)}{C_B} dt\,. \tag{6}$$

After integration we get

$$\frac{C_B}{C_{A_0} - C_B} + \ln \left| \frac{C_{A_0} - C_B}{C_{A_0}} \right| = \Omega\, t\,. \tag{7}$$

C_{A_0} is the measurable quantity of the process.

2.2 *Excitation of electrons from centres with high concentration and trapping by centres of low concentration with participation of the conduction band*

This point considers case c) of Section 2. Let us first state some presumptions:
1. The absorbed fraction of the incident light shall be considered constant.
2. The electrons are transferred via the conduction band.
3. The electrons can be trapped in centres A already ionized.
4. The transferred electrons are not retrapped.
5. The incident photon densities are not too large.

Concerning with 1. to 5. we suppose the following processes to take place: An exited electron is shifting in the condution band, an oxidized centre A remains, the electron gets trapped by a centre B.

The shifting electron will be retrapped by an oxidized centre A and reduces it again, this centre is able to get ionized once more.

Observing the reduction of the centres we get the reduction probability by forming the quotient of the number of "favourable" and possible trapping acts. Considering the "favourable" acts we must take into account the number of empty acceptors C_A, the initial number of empty acceptors C_{A_0} minus the number of reduced ones $\sum_j W_j$ up to the moment of observation.

The possible cases include the initial number of acceptors C_{A_0} minus the number of reduced ones as favourable acts and further retrapping, the number of donators (occupied centres A), which have been oxidized up to the moment of observation, and exited electrons as not favourable acts.

We use the denotations of [16]:
m number of photons in an irradiated photon layer, k number of irradiated photon layers, B_0 incident light intensity, $\sum_k W_k$ number of absorption acts after irradiating k photon layers.

According to the above-mentioned we get

$$W_k = m \frac{C_{A_0} - \sum_{j=1}^{k-1} W_j}{C_{A_0} - \sum_j W_j + \sum_j W_j + 1} = m \frac{C_{A_0} - \sum_{j=1}^{k-1} W_j}{C_{A_0} + 1} . \tag{8}$$

This induces a relation between the number of transferred electrons and irradiated photon layers:

$$\left.\begin{aligned} W_1 &= \mu\, C_{A_0} , \\ W_2 &= C_{A_0} (\mu - \mu^2) , \\ W_3 &= C_{A_0} (\mu - 2\,\mu^2 + \mu^3) , \\ &\vdots \\ W_k &= \mu\, C_{A_0} (1 - \mu)^{k-1} . \end{aligned}\right\} \tag{9}$$

For $1 \gg \mu$ we get

$$W_k \approx \mu\, C_{A_0}\, e^{-\mu k} . \tag{10}$$

The sum of all transfers from the beginning to the k-th photon layer is

$$W_{\Sigma k} = \frac{\mu\, C_{A_0}}{\ln (1 - \mu)} [e^{(k-1) \ln |1-\mu|} - 1] , \tag{11}$$

$$W_{\Sigma k} \approx C_{A_0} [1 - e^{-\mu k}] . \tag{12}$$

In a former paper [16] we showed that

$$\mu = \frac{m}{C_{A_0} + 1}, \quad m = \left[\frac{B_{0\lambda} \lambda}{h c^2}\right]^{2/3}, \quad k = \frac{c t}{[h c^2/B_{0\lambda} \lambda]^{1/3}}. \tag{13}$$

Inserting this into (12) we get

$$W_{\Sigma k} = C_{A_0} \left[1 - \exp\left(- \frac{B_{0\lambda} \lambda / h c}{C_{A_0} + 1} t\right)\right]. \tag{14}$$

3. Experimental Investigations

The dynamical progressions of the described photochromic reactions are recorded continuously. The sample is irradiated simultaneously with measuring and bleaching light. The measuring light has been kept at low intensity. In this paper investigations of photochromic processes in additively coloured CaF_2:La crystals are presented. Equations (7), (13), and (6″, reference [16]) have been applied to the observed processes. The following cases have been investigated:

a) $\lambda_i = 410$ nm, $\lambda_m = 650$ nm;
b) $\lambda_i = 410$ nm, $\lambda_m = 410$ nm;
c) $\lambda_i = 650$ nm, $\lambda_m = 410$ nm;
d) $\lambda_i = 650$ nm, $\lambda_m = 650$ nm;

λ_i irradiation wavelength,
λ_m measuring wavelength.

We observed two different types of dynamical photostimulated behaviour. As the same centres take part in all the four described cases the type of the observed behaviour must be determined by the irradiation wavelength. Therefore the cases a) and b) or c) and d) are observed to have the same types. Applying relation (7) to the experimental results of a) and b) one gets an ideal linear progress after a very small curved one, see Fig. 2 (a). Application of the same formula to the case c) results in a throughout curved behaviour, see Fig. 2 (b). Applying formula (6″, reference [16]) to the experimental results of

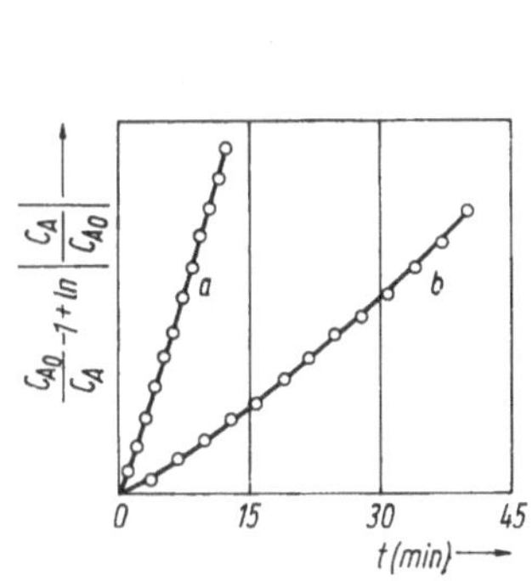

Fig. 2. Dynamical photosensitive behaviour of additively coloured CaF_2:La at $T_m = 300$ K. (a) $\lambda_i = \lambda_m = 410$ nm; (b) $\lambda_i = \lambda_m = 650$ nm

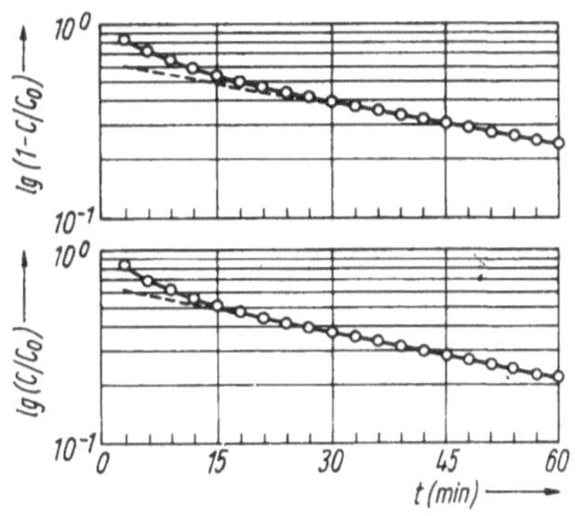

Fig. 3. Dynamical photosensitive behaviour of additively coloured CaF_2:La at 300 K. a) $\lambda_i = 650$ nm, $\lambda_m = 410$ nm; b) $\lambda_i = \lambda_m = 650$ nm

c) and d) we get an initially curved dependence soon changing into a linear one. The initially curved dependence may be due to an effect of the crystal thickness or an photosensitive spectroscopic background, see [16] and Fig. 3.

According to our observations two of the four observed processes are described very well by a common relation. This could be interpreted such that the assumed models are correlated very well with the observed processes.

We can conclude that the processes are different whether we irradiate with light of 410 or 650 nm wavelength.

4. Conclusions

The experimentally observed dynamical photochromic behaviour of the 410 nm absorption band of additively coloured CaF_2:La crystals showed a surprising conformity with the theoretically expected one. The relation (7), which has been applied to our experimental results, has been calculated under the assumption of an electron transfer possibly across the conduction band and equal concentrations of donator and acceptor centres. This is satisfied if both centres A and B are aggregated. This is the case when assuming a double-model as proposed in [13, 15, 16].

The behaviour of the 650 nm band at light irradiation in the same band satisfies the supposition of an electron transfer within a F_g–$La^{3+}_{trig.}$ double centre without participation of the conduction band and excluding retrapping processes.

Our results did not contradict the proposed double model and the supposition that the photochromic process is an electron transfer within such centres. Short-wavelength irradiation should induce an electron transfer with possible participation of the conduction band, long-wavelength irradiation without conduction band.

References

[1] J. R. O'Connor and P. R. Hilton, Appl. Phys. Letters **7**, 531 (1962).
[2] U. T. Höchli, Phys. Rev. **162**, 262 (1967).
[3] W. Hayes and J. W. Twidell, Proc. Roy. Soc. **82**, 330 (1963).
[4] Z. J. Kiss and Y. Yocom, Phys. Rev. **84**, 520 (1964).
[5] D. L. Staebler, S. E. Schnatterly, and W. Zernik, IEEE J. Quantum Electronics **10**, 575 (1968).
[6] F. S. Chen and J. R. O'Connor, Bull. Amer. Phys. Soc. **8**, 231 (1963).
[7] Z. J. Kiss and D. L. Staebler, Phys. Rev. Letters **14**, 691 (1965).
[8] J. R. O'Connor and F. S. Chen, Phys. Rev. **130**, 1790 (1963).
[9] P. M. Heyman, Appl. Phys. Letters **14**, 81 (1969).
[10] D. L. Staebler and Z. J. Kiss, Appl. Phys. Letters **14**, 93 (1969).
[11] D. L. Staebler and Z. J. Kiss, Bull. Amer. Phys. Soc. **12**, 670 (1967).
[12] G. Kötitz, Thesis, Friedrich-Schiller-Universität, Jena 1971.
[13] Hj. Bernhardt, Diploma Work, Friedrich-Schiller-Universität, Jena 1970.
[14] G. Schwotzer, Diploma Work, Friedrich-Schiller-Universität, Jena 1969.
[15] Hj. Bernhardt, P. Görlich, and G. Kötitz, phys. stat. sol. (a) **6**, 479 (1971).
[16] Hj. Bernhardt, phys. stat. sol. (a) **7**, 167 (1971).
[17] D. L. Staebler and S. E. Schnatterly, Phys. Rev. **B3**, 516 (1971).
[18] C. H. Anderson and E. S. Sabisky, Phys. Rev. **B3**, 527 (1971).
[19] R. C. Alig, Phys. Rev. **B3**, 536 (1971).

(Received October 13, 1971)

phys. stat. sol. (a) 8, 545 (1971)

Subject classification: 14.3.2; 13.3; 22.1.2

Centre Electronique Horloger S.A., Neuchâtel

Charge Pumping and Low-Frequency Noise in MOS Structures

By

F. LEUENBERGER

Interface state density in MOS structures determined from charge pumping measurements correlates well with the level of low-frequency excess noise. Small-signal charge pumping current, low-level diode forward current, and excess noise peak at the same value of surface potential. Exposure to a dose of 10^7 rd of γ-radiation increases the pumped current by a factor of 30. In a gold-doped n-channel MOS transistor, the typical generation–recombination noise spectrum has been observed. The charge pumping current vs. frequency plot changes slope at the frequency corresponding to the transition from a flat to a ω^{-2}-spectrum.

Die mit Hilfe von charge pumping-Messungen bestimmten Dichten von Grenzflächenzuständen in MOS Strukturen korrelieren gut mit der Intensität des niederfrequenten Rauschens. Kleinsignal-charge pumping-Strom, Diodenstrom in Leitrichtung bei kleinen Stromdichten und Rauschen durchlaufen Maxima beim gleichen Wert des Oberflächenpotentials. Der charge pumping-Strom nimmt um einen Faktor 30 zu, wenn z. B. eine Diode mit Feldelektrode einer γ-Strahlendosis von 10^7 rd ausgesetzt wird. Im Falle eines golddotierten n-Kanal-MOS-Transistors wurde das typische Generations–Rekombinations-Rauschspektrum beobachtet. Der als Funktion der Frequenz gemessene charge pumping-Strom ändert seine Neigung bei jener Frequenz bei der das Rauschspektrum von $S(\omega)$ = const. in $S(\omega) \sim \omega^{-2}$ übergeht.

1. Introduction

Charge pumping in MOS devices has been described by Brugler and Jespers [1]. For gate voltages V_G exceeding the threshold voltage V_T of a MOS transistor, these authors give the following expression for the pumped substrate current:

$$I_{Sub.} = A_G f [- \alpha C_0 (V_G - V_T) + q N_{st}] , \tag{1}$$

where A gate surface [cm^2], f frequency of signal applied to the gate [Hz], C_0 oxide capacitance per cm^2 [F cm^{-2}], q electronic charge [C], and N_{st} total fast-surface-state density contributing to the effect [cm^{-2}].

The factor α depends strongly on device geometry. Therefore the first term in the square bracket is called "geometric component" of the charge pumping current. It has been shown experimentally [1] that using a sawtooth gate signal leads to a negligible small value of the geometric component. In this case, the net injected charge Q_T is proportional to the trap density (we assume a uniform distribution in energy) times the peak-to-peak bending of the bands $\Delta\psi_s$:

$$Q_T = q N_{ss} \Delta\psi_s . \tag{2}$$

With $N_{st} = N_{ss} \Delta\psi_s$ the substrate current will be given by

$$I_{Sub.} = f A_G q N_{ss} \Delta\psi_s . \tag{3}$$

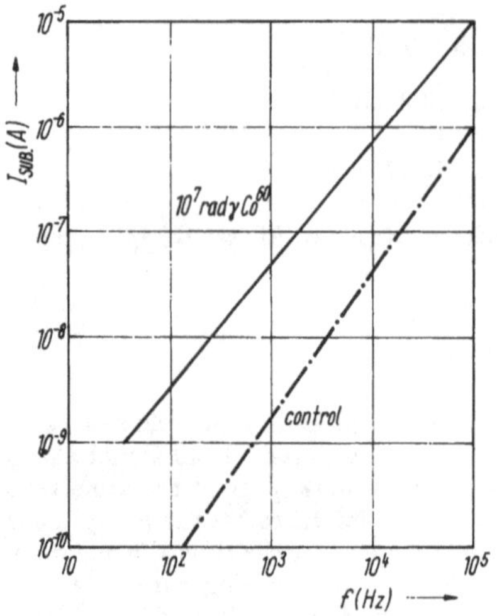

Fig. 1. Large signal charge pumping current $I_{SUB.}$ (substrate current) as a function of frequency of the sawtooth signal applied to the gate of a p^+–n-gate-controlled diode which had been exposed to 10^7 rd γ-radiation from a Co^{60} source. Also shown is the $I_{SUB.}(f)$ plot for an otherwise identical control device

Golder and Baldinger [2] derived a quantitative small-signal theory of the charge pumping effect. The determination of the energy distribution of the interface states is feasible in principle but the nonlinear relationship between substrate current and band bending amplitude renders the procedure quite impractical.

Even neglecting the fact that interface state distribution is not uniform and that surface potential determination should account for the presence of interface states, large-signal charge pumping measurements permit global determination of the interface state densities by using equation (3).

2. Charge Pumping in Devices with Radiation-Induced Interface States

The most important effects occuring when MIS structures are exposed to ionizing radiation [3] are a) the introduction of a fixed positive oxide space charge by the occupation of pre-existing charge trapping sites in the insulator and b) the creation of new interface states at the semiconductor–insulator interface. The saturated large-signal charge pumping current depends linearly on frequency of the gate signal. In Fig. 1 we have plotted the substrate current as a function of frequency for a gate-controlled p^+–n diode which had been exposed to a dose of 10^7 rd of γ-radiation (Co^{60}). Also shown is the substrate current of an otherwise identical control device. The gate signal was a sawtooth with an amplitude of ± 8 V. According to (3) the current ratio is proportional to the ratio of interface state densities. The ratio of 30 measured at 1 kHz has been confirmed by MOS conductance measurements [4]. The difference of slope is presumably due to differences in the spatial distribution of the traps within the oxide layer.

3. Charge Pumping, Low-Level Forward Current and Noise in a Gate-Controlled Diode

Measurements of reverse current or low-level forward current in gate-controlled diodes have shown [5] that these currents peak when the semiconductor surface is depleted of carriers. It has also been shown [6 to 8], that the low-frequency noise vs. gate voltage characteristic has a maximum value at a gate voltage corresponding to surface depletion. It seems therefore of interest to look for possible correlation between these characteristics and charge pumping. Two kinds of charge pumping measurements have been performed. In the case of large-signal charge pumping, a triangular waveform is applied to the gate electrode. The point $U_G = -3.5$ V for instance, corresponds to a peak

Fig. 2. Large-signal charge pumping current $I_{SUB.}$ at $f = 1$ kHz, small-signal charge pumping current $I_{sub.}$ and equivalent noise resistance $R_{eq.}$ at $f = 4$ kHz as well as diode forward current I_f at $V_f = 0.25$ V, all as function of gate voltage. The gate-controlled p^+–n diode is made on (111) orientation silicon with $N_D = 5 \times 10^{14}$ cm^{-3}. Oxide thickness is (1000 ± 100) Å

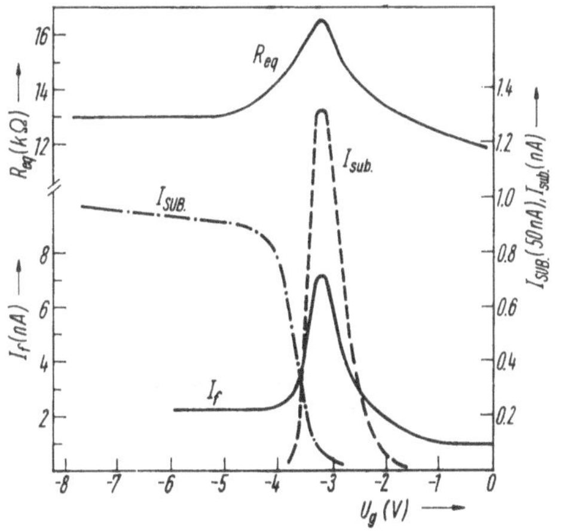

signal voltage of ±3.5 V. The charge pumping current $I_{SUB.}$ saturates more or less when the surface goes into heavy inversion, because surface potential as a function of applied gate voltage is nearly constant under that condition. In order to measure the small-signal charge pumping current $I_{sub.}$, a triangular signal of ±50 mV is superimposed on a variable bias corresponding to the gate voltage indicated in Fig. 2. As would be expected, the small-signal charge pumping current, after passing through a maximum, tends to become very small when the surface is heavily inverted. In Fig. 2 are shown measurements of diode forward current (at $V_f = 0.25$ V), large- and small-signal charge pumping currents as well as equivalent noise resistance all as functions of gate substrate bias. It can be seen, that the small-signal charge pumping current peaks at the same value of gate voltage and thus at the same value of surface potential, as do the low-level diode forward current and the equivalent noise resistance. We therefore conclude that the same set of interface states gives rise to the peaks of the three characteristics. At small-bias conditions the forward characteristic is essentially the same as that of the reverse current. The height of the reverse current peak is related to the surface recombination velocity s_0 by means of [5]

$$I_R = q\, n_i\, s_0\, A_G\,, \tag{4}$$

where

$$s_0 = \frac{\pi}{2} (\sigma_{sn}\, \sigma_{sp})^{1/2}\, v_{th.}\, (kT\, N_{ss})\,. \tag{4a}$$

Assuming the effective capture cross section $(\sigma_{sn}\, \sigma_{sp})^{1/2}$ to be independent of temperature over a limited range, the interface state density N_{ss} may be determined from measurements of the reverse current peak at different temperatures.

The ratio of the surface of the metallurgical junction plus the field-induced junction to that of the metallurgical junction alone is 2.6 for the diodes used in these measurements. The ratio of current with the surface inverted to that with the surface accumulated is 2.2. This difference might be due to different densities of recombination centres within the depletion regions of the metallurgical junction and the field-induced junction, respectively. We suggest an

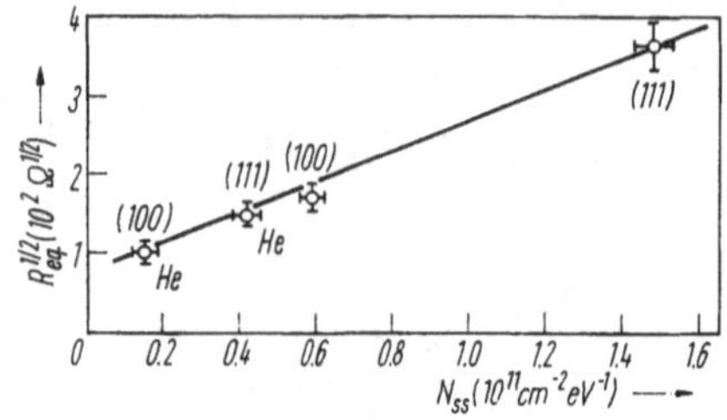

Fig. 3. Correlation between square root of equivalent noise resistance, $R_{eq.}^{1/2}$, measured at 4 kHz and interface state density determined from measurement of large-signal charge pumping current. The circular geometry p-channel devices have a gate surface of 1.5×10^{-3} cm². The noise measurements were performed in saturation at 0.1 mA of drain current

alternative interpretation. The inversion layers lateral resistance causes an IR drop which leads to a junction bias voltage which decreases with increasing radial distance from the diode centre. For low reverse bias, the current ratios observed correspond closely to the surface ratios, presumably because the IR drop is negligible at these very-low-current levels.

4. Correlation between 1/f Noise and Interface State Density

The existence of such correlation has been demonstrated by Sah and Hielscher [9] and by Abowitz et al. [10]. The authors of reference [10] have shown that the noise of n-channel MOS transistors measured at 10 kHz is proportional to the interface state density determined from channel conductance of devices with non-specified geometry.

In our investigation we used circular devices with a channel length of 60 μm. The four parameters investigated are the combination of (100) and (111) substrate orientations with dry-oxygen oxidation and dry-oxygen oxidation followed by helium annealing at the oxidation temperature of 1200 °C. Using equation (3) the interface state density was determined by measuring the saturated charge pumping current when a sawtooth signal of 10 kHz was applied to the gate. A linear relationship (Fig. 3) was found between the RMS value of the input noise voltage and the interface state density. Noise was measured at 4 kHz. For the best devices: (100) orientation and He annealing, noise levels off at approximately 8 kHz.

5. Influence of Deep-Level Recombination Centres on Excess Noise and Charge Pumping Characteristics

If a MOS transistor is doped with an impurity giving rise to a discrete energy level within the gap of the semiconductor substrate, the typical generation–recombination type of noise spectrum will be observed:

$$S(\omega) \sim \frac{\tau}{1 + (\omega\,\tau)^2}. \tag{5}$$

The noise is independent of frequency for $\omega \ll 1/\tau$ and falls off with ω^{-2} for $\omega \gg 1/\tau$. One would expect this time constant to show up as a change of slope of the charge pumping current vs. frequency characteristic. Fig. 4 shows that this is indeed the case. The slope changes at the same frequency for which the generation–recombination noise spectrum changes slope. The n-channel MOS transistor used in this experiment had undergone a gold diffusion step at 1100 °C. The solid solubility of gold in silicon [11] at this temperature is

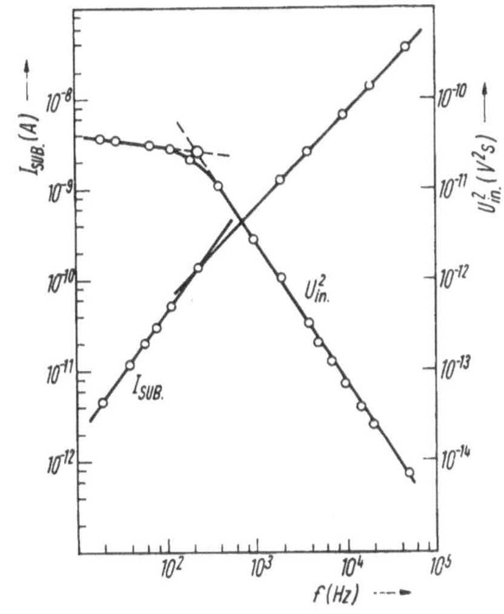

Fig. 4. Mean square input noise voltage per 1 Hz bandwidth and large-signal charge pumping current $I_{SUB.}$ as a function of frequency of a gold-doped n-channel MOS transistor

3×10^{16} cm^{-3} but non-uniform concentration of the gold due to preferential segregation at the silicon–silicon dioxide interface might have to be taken in consideration. The presence of an accidental and unknown concentration of deep level impurities, like gold should be carefully taken into account when charge pumping measurements are used to determine interface state densities.

6. Conclusions

Our experimental results may be summarized as follows:

a) Small-signal charge pumping current, low-level diode forward current, and equivalent noise resistance of a gate-controlled diode peak at the same value of surface potential. The three characteristics are therefore governed by the same set of interface states.

b) The $1/f$ noise of p-channel MOS transistors operated in saturation correlates well with the interface state density determined by charge pumping measurements. Lowest interface state density and excess noise is obtained on transistors on (100) substrates with the gate oxide annealed in He *in situ* at the oxide growth temperature (1200 °C).

c) Traps introduced by exposing a gate-controlled diode to a dose of 10^7 rd γ-radiation from a Co^{60} source have lead to an increase of a factor of 30 in charge pumping current for the particular device tested.

d) The presence of a discrete energy level within the gap produces a change of slope of the charge pumping current vs. frequency plot. Correlation between this behaviour and the well-known shape of generation–recombination spectra has been demonstrated. Charge pumping provides a very simple way of determining the effective interface state density, but the interpretation of such measurements becomes ambiguous in the presence of a non-negligible concentration of deep-level bulk recombination centres.

References

[1] J. S. Brugler and P. G. A. Jespers, IEEE Trans. Electron Devices **16**, 297 (1969).
[2] J. Golder and E. Baldinger, Helv. phys. Acta **44**, 387 (1971).
[3] K. H. Zaininger and A. G. Holmes-Siedle, RCA Rev. **28**, 208 (1967).
[4] E. H. Nicollian and A. Goetzberger, Bell Syst. tech. J. **46**, 1055 (1967).
[5] A. S. Grove and D. J. Fitzgerald, Solid State Electronics **9**, 783 (1966).
[6] F. Leuenberger, Verh. DPG (6) **2**, 74 (1967).

[7] S. T. HSU et al., Appl. Phys. Letters **12**, 287 (1968).
[8] F. LEUENBERGER, Electronics Letters **3**, 280 (1968).
[9] C. T. SAH and F. H. HIELSCHER, Phys. Rev. Letters **17**, 956 (1966).
[10] G. ABOWITZ, et al., IEEE Trans. Electron Devices **14**, 775 (1967).
[11] F. A. TRUMBORE, Bell Syst. tech. J. **39**, 205 (1960).

(Received October 22, 1971)

phys. stat. sol. (a) 8, 551 (1971)

Subject classification: 10; 22.5.2

Institute of Crystallography, Academy of Sciences of the USSR, Moscow

Electron Microscopy Investigation of the Surface Relief after Concentrated Load Deformation of NaCl Crystals

By

V. N. ROZHANSKII and M. A. VELEDNITSKAYA

The investigation by electron microscopy of the relief change after indentation reveales new peculiarities of the plastic deformation in NaCl crystals. The formation of hillocks near the indentation and on a lateral surface was observed. The hillocks are oriented in ⟨110⟩ directions, coinciding with the directions of chains of similar ions in the crystal. The formation of hillocks may be explained by the motion of interstitial ions by means of crowdions, their coagulation into prismatic dislocation loops, and their emergence at the surface by means of the joint action of prismatic slip and climb.

Eine elektronenmikroskopische Untersuchung der Reliefänderungen nach Eindruck zeigt neue Besonderheiten der plastischen Deformation in NaCl-Kristallen. Die Bildung von Erhebungen in der Nähe des Eindrucks und auf einer Seitenfläche wurden beobachtet. Die Erhebungen sind in ⟨110⟩-Richtung angeordnet, in Übereinstimmung mit den Richtungen der Ketten von ähnlichen Ionen im Kristall. Die Bildung der Erhebungen kann durch die Wanderung von Zwischengitterionen durch „crowdions", ihre Koagulation in prismatische Versetzungsschleifen und ihr Auftreten an der Oberfläche durch die gemeinsame Wirkung von prismatischem Gleiten und Klettern erklärt werden.

1. Introduction

The present work was undertaken with the aim of obtaining information on the crowdion plasticity in NaCl crystals. The development of crowdion plasticity is expected to occur at impact loading and non-uniform straining with steep gradients of the normal strains. It has been assumed that this mechanism can also work under the conditions of uniform deformation [1]. It is expected that crowdions formed during deformation move mainly along close packed directions in metals or along the directions of chains of similar ions in ionic crystals.

Crowdion plasticity has been found for the first time in corundum crystals [2]. Further evidence of the occurrence of this mechanism was obtained in tungsten [3] and cesium iodide [4].

However, no information has been obtained concerning the directional anisotropy of crowdion motion, the rates of their displacement and the path length as a function of the loading conditions. This is primarily due to the absensce of direct techniques for studying crowdion plasticity.

2. Experimental Technique

The decoration technique of atomic steps on crystal surfaces [5, 6] allows the observation of the motion of dislocations with Burgers vectors non-parallel to the surface, and the emergence of point defect clusters on the surface [7 to 9]. It was hoped that the high resolution of this techniques would allow its application for the study of crowdion plasticity.

Serious experimental difficulties arise from the combination of electron microscopy and concentrated load deformation. The impressions caused by microhardness indents and the resulting deformation prove to be too large for observation by electron microscopy and their search and identification are tedious. On the other hand, brushes with rows of germanium whiskers [10] were found to be suitable for making indentations on crystal surfaces. The regulation of the growth conditions of these whiskers allowed to obtain an optimal arrangement, radius of curvature of the tip and crystal length. The small differences in crystal height resulted in the simultaneous formation of many indentations with different sizes on a sufficiently large area. A whisker density of $\approx 6 \times 10^4$ cm^{-2}, a radius at the whisker tip of ≈ 1000 Å, and a whisker length ≈ 10 μm proved to be most convenient. Smooth indentations were made by carefully pressing the brush by hand and impact indentations by lightly striking the reverse side of the brush placed on the crystal surface with a hard object. The indentations were made on freshly cleaved surfaces of sodium chloride crystals which were then coated with a layer of gold ≈ 10 to 20 Å thick. The evaporation was performed in vacuum, the temperature of the crystal being 100 to 150 °C; subsequently a thin layer of carbon was evaporated. The film was separated from the crystal by immersion in water and picked up with a specimen support grid.

The crystals were cleaved and deformed in air. In order to verify whether the water adsorbed by the crystal surface does not appreciably influence the surface relief appearing during indentation the experiments were partly carried out by means of the following technique. The crystals were annealed for two hours at a temperature of 400 °C. Then the temperature was decreased to 150 °C and cleavage, indentation, and decoration of the crystals were performed. Thus the absorption of water at the crystal surface was excepted. The structure of the cleaved surface allowed to draw conclusions about the absence of water on the surface; pointed cleavage steps typical for crystals cleaved in vacuum were observed. The relief near the indentation was really the same in both cases. Therefore it was possible to postulate that the water influence on the discussed features may be neglected.

The decoration patterns of the surface being deformed were often compared with those on the adjacent lateral side. For this purpose gold and carbon were evaporated onto both surfaces simultaneously. We succeeded in separating the film from the substrate without breaking it at the boundary between these two faces. The boundary between the two crystal surfaces could be readily identified since it was marked by zones of different transparency and/or a sudden change in the size of the gold particles.

3. Observations on the Upper Surface

Electron microscope observation of the areas, surrounding the indentation, revealed the emergence of hillocks in ⟨110⟩ directions surrounded by terraces of elemental steps (Fig. 1). The early stages of the formation of these hillocks may be seen near small indentations as shown in Fig. 2 where the formation of three hillocks is apparent, a fourth one being only outlined. A well formed hillock is shown in Fig. 3. Aside from terrace-like steps, Fig. 3 also shows small loops which are presumably connected with the emergence on the surface of flat clusters of interstitial ions.

Fig. 1. Anisotropy of the arrangement of elemental steps surrounding the hillocks formed near the spot of concentrated load application

Some attention should be given to the reasons that the systems of concentric elemental steps indicate the presence of hillocks and not depressions. It is known that during indentation of crystal surfaces, material is forced upward by the indent, giving rise to hillocks [11]. The profile of these hillocks, established

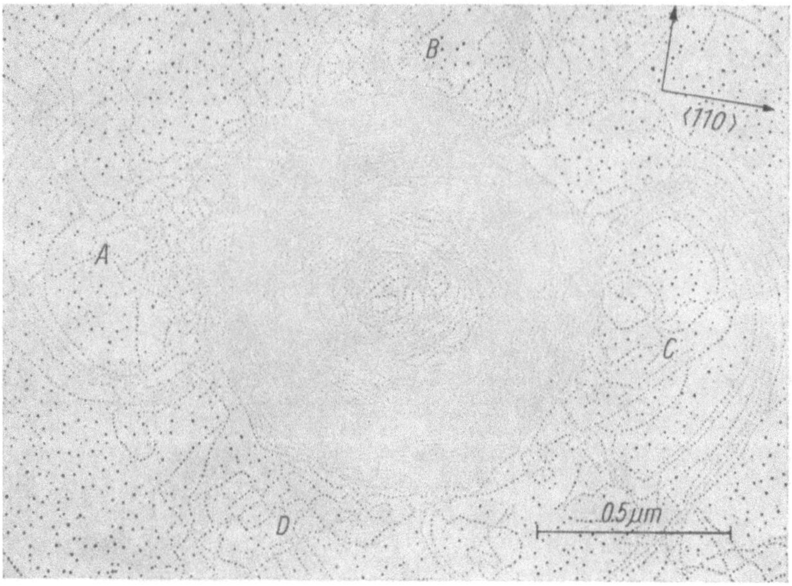

Fig. 2. Initial stage of hillock formation near a small indentation. The hillocks A, B, and C are already partly formed. The hillock D is only outlined

by interferometric measurements of the surface relief near diamond indenter traces [11], completely corresponds to the profile of elemental steps arranged around the hillock in Fig. 4. An increase of the height and base length of the hillocks is caused only by a monotonous growth of the number of elemental steps of which the hillocks are composed. The specific profile of the hillocks does not depend on the size of the indentations. The ratio of distance of hill top from the indentation edge to the base length of the hillock is approximately the same for all indentations. In our case this distance was 0.2 to 3 μm.

The arrangement of hillocks in ⟨110⟩ direction is most clearly observed after smooth deformation. In the instance of impact deformation this effect is somewhat blurred. For indentation sizes of less than 0.5 μm, the hillocks are not yet developed although a considerable number of loops, surrounding the indentation, is observed (Fig. 5).

The mechanism of hillock formation is often well characterized near ≈1 μm indentations after smooth deformation (Fig. 6). It consists of screw dislocations moving along arc lines running around the periphery of the future hillock. This motion seems to be the result of the joint effect of climb and prismatic slip of flat clusters of interstitial ions toward the surface. The edge arms of dislocation rosettes [12] are not developed by the decoration because edge dislocation motion is not accompanied by the emergence of steps at the crystal surface. However, they may be revealed by the appearence of etch pits after treatment in acetic acid. Etch pits are usually located further from the indentation than hillocks. As a rule, the edge arms also appear near the smallest indentations. Dislocation motion in the screw arms starts to be visible only near indentations with more than 3 to 5 μm diameter.

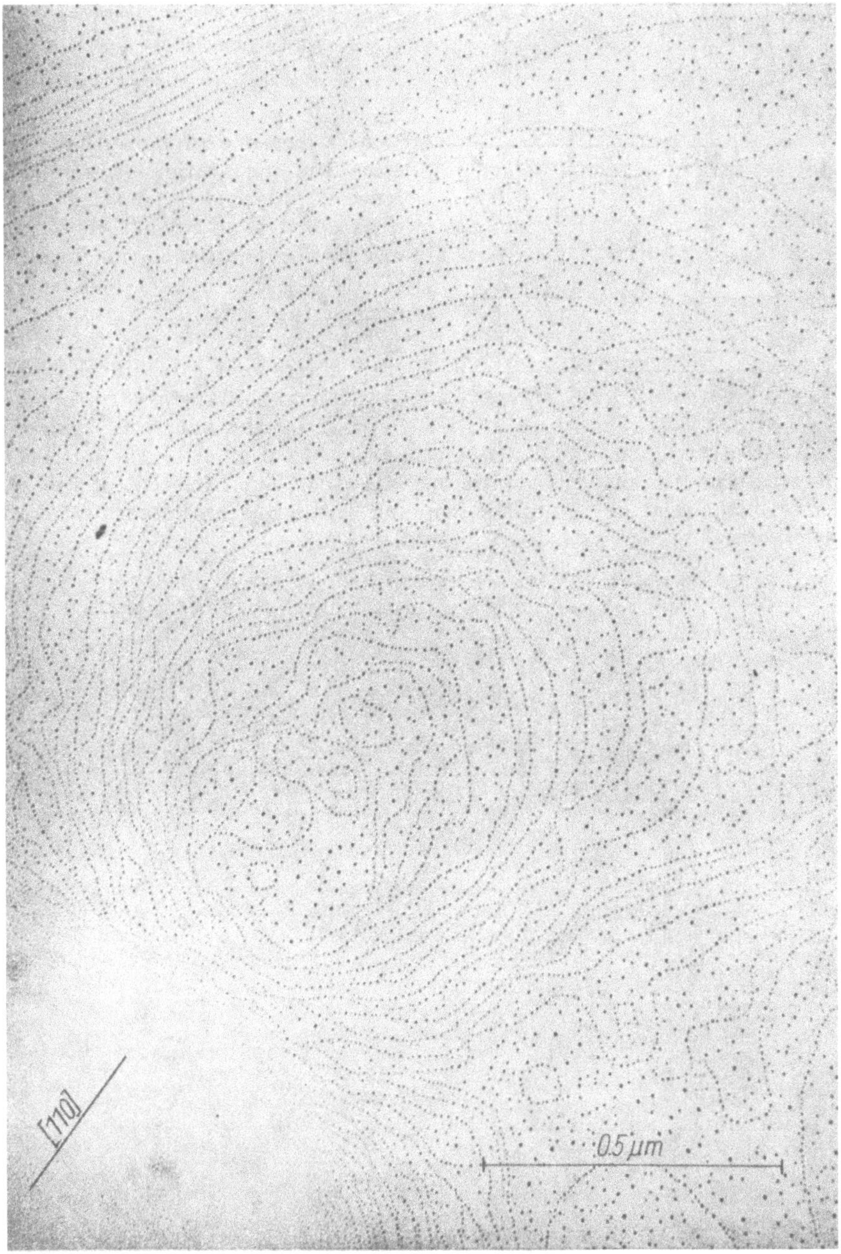

Fig. 3. Well formed hillock. The small loops are connected with the emergence of the separate planar clusters of interstitial ions onto the surface

During usual slip in NaCl crystals the screw dislocations move mainly along the {110} planes, leaving behind monoatomic surface steps stretching in ⟨100⟩ directions. Such slip is often observed at the periphery of the hillocks and near large indentations.

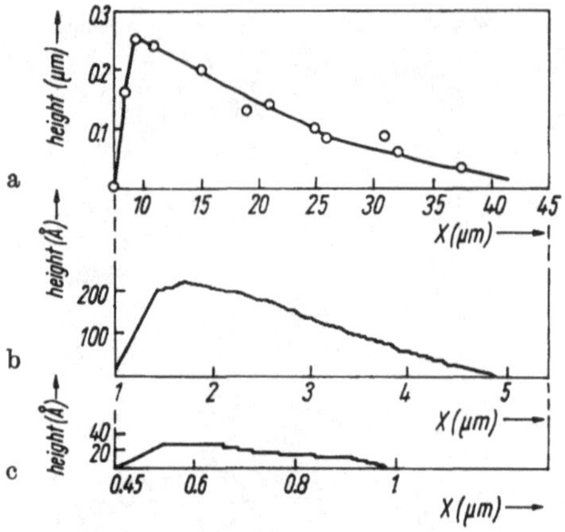

Fig. 4. Profile of the hillocks forming near the indentation. X distance from the centre of the indentation. a) Profile established by interferometric measurements [11], b) and c) profiles established from the arrangement of elemental steps around hillocks

At a distance of more than 10 µm from the indentation, a number of loops is observed the density of which does not depend upon their azimuth orientation relative to the impression (Fig. 7). Under impact load the loop density increases.

The sign of the loops, i.e. whether they correspond to a hillock or a depression on the crystal surface, may be defined by the nature of their intersection with a step connected to a moving screw dislocation. After the intersection the loop assumes the shape of a spiral curl (Fig. 8). The direction of the spiral depends on the relative sign of the loop and of the Burgers vector of the inter-

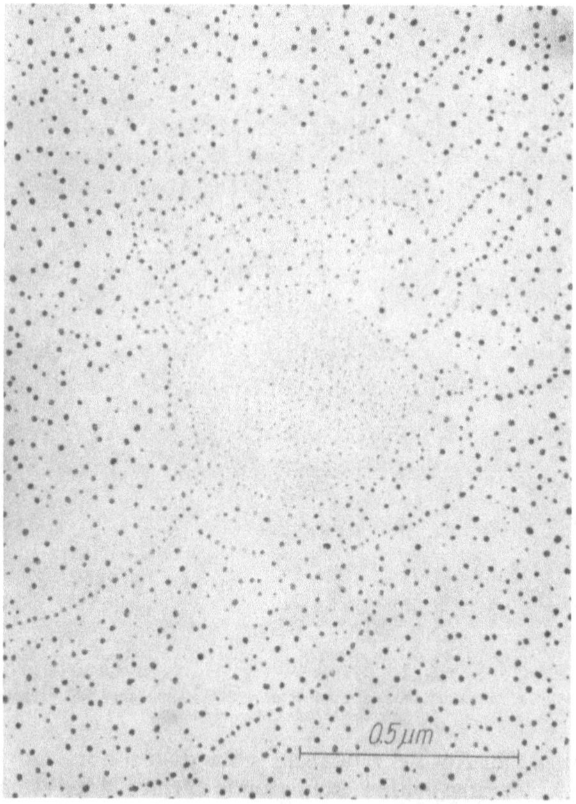

Fig. 5. Very small indentation made by impact load. Loops are revealed in its vicinity, but hillocks are not yet formed

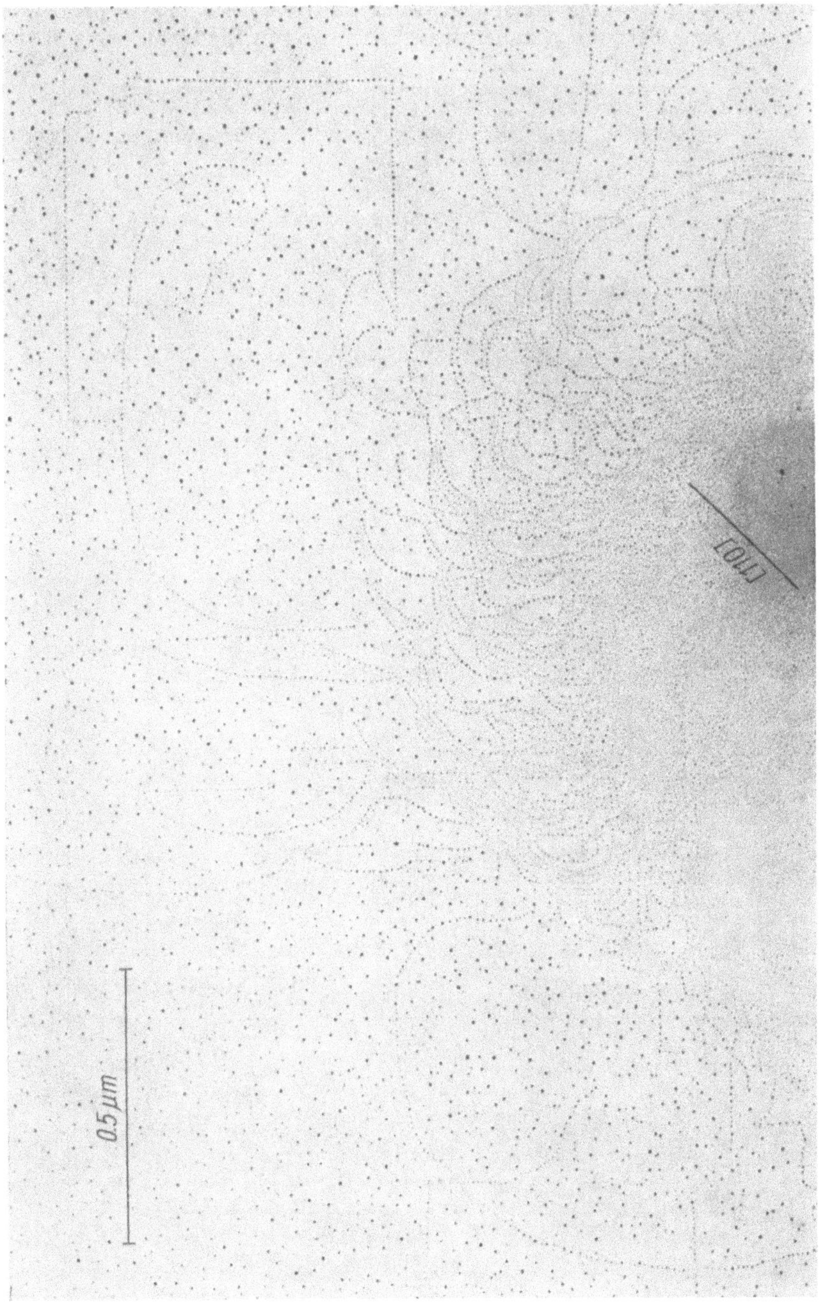

Fig. 6. Mechanism of hillock formation during smooth indentation. Screw dislocations run around the hillock, moving along arc-shaped lines. Straight sections are formed during slip of several dislocations which apparently arose after load removal

secting dislocation. The sign of the Burgers vector of the dislocations moving in the screw arms of the dislocation rosette is known since it has to provide a drop in surface level between parallel dislocation rows in the screw arms of the

Fig. 7. Rise of the diffuse distribution of loops along the periphery of the indentation. The number of loops does not depend upon their azimuth arrangement relative to the indentation

Fig. 8. Formation of spiral curl at the intersection of a loop with a step connected to a screw dislocation

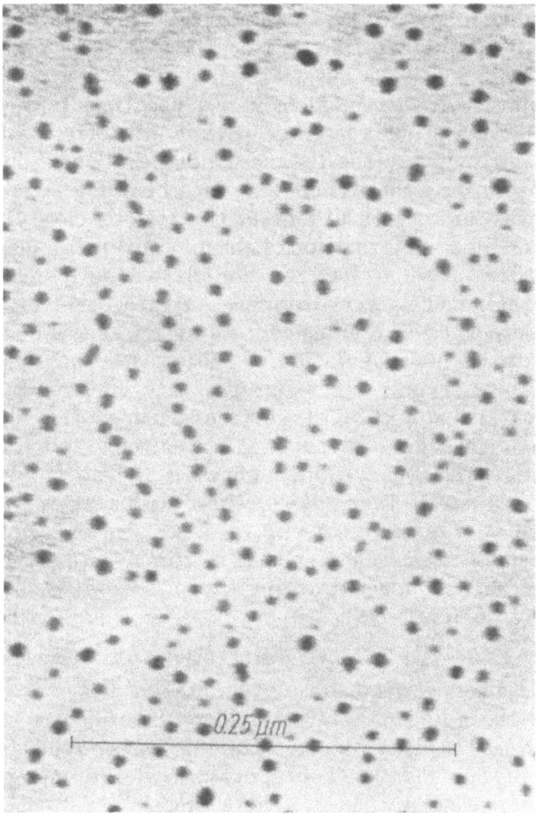

rosette. It was thus possible to conclude that the loop corresponded to a single hillock on the crystal surface and therefore was formed as a result of interstitial dislocation loop emergence on the surface.

The absence of anisotropy in the arrangement of these loops allows to conclude that interstitial ions from which they were formed were displaced to the periphery by a diffusion process from zones of high concentration near the indentation. The area around indentation with a diameter of about 20 µm contains a small number of these loops; this effect seems to be connected with the absorption of interstitial ions by the rosette dislocations and also with their being used largely for hillock formation near the indentation in ⟨110⟩ directions.

The formation of hillocks is associated with the accumulation of interstitial ions in these areas. It is possible to postulate two mechanisms by which the interstitial ions reach these zones from the region under the indent: by crowdions or by a system of prismatic dislocation rings. The latter seems to be less probable for the following reasons:

1. Loops of prismatic slip near the indentation in NaCl have not been reported previously, although they should be revealed by the emergence of a great number of flat-bottomed etch pits.

2. During observation by transmission electron microscopy of indented corundum crystals [2], interstitial loops are formed around the indentation only after considerable heating. Also, the loops do not move from the indentation, but arise at some distance from it and gradually increase in size. In other words the aggregation of interstitial ions into flat discs is observed. Thus, at least in corundum, the transport of interstitial ions from the region under the indenter takes place not by prismatic slip but by crowdions.

Nevertheless, in plastic crystals the question probably should not be raised whether the crowdions move individually or coalesce to form prismatic dislocation loops. During mass formation and motion of crowdions, in general, both possibilities can occur.

The following mechanism of hillock formation may be postulated:

During indentation the material from the region under the indent is transported mainly by crowdions scattering in ⟨110⟩ directions. The removal of material is partly due to the motion of dislocations forming dislocation rosettes near the indentation. The average path length of crowdions is not large and probably slightly exceeds the distance from the indentation edge to the top of the hillock. It depends on the extent and rate of deformation. During smooth loading the excess in interstitial ions of the regions close to the indentation is larger; this promotes the formation of sharply outlined hillocks in ⟨110⟩ directions. During impact loading the crowdion path length increases and the interstitial ions travel over larger distances. Aggregation into interstitial dislocation discs takes place in regions containing an excess of interstitial ions. Furthermore, growth and coagulation of these discs and their simultaneous motion to the surface occurs. During growth and emergence of interstitial discs screw dislocations appear on the surface. Their motion along the periphery of the hillock leads to a mass transfer to the surface and to a reduction of the number of interstitial ions within the crystal and ultimately to the formation of hillocks. The separate segments of these dislocations can move into the screw planes and later on move by slip. But, as already mentioned, the contribution of slip is small compared with that of climb. In particular, as seen in Fig. 6, straight sections are formed during slip of several dislocations which apparently arose after load removal.

4. Observations on the Lateral Surface

Defects formed during indentation near the edge of the crystal can emerge at the adjacent lateral surface. The emergence of screw and two edge slip planes [12] leaving corresponding steps on the lateral surface is shown schematically in Fig. 9. The spots where the emergence of crowdions is expected are marked by rings. These areas were investigated by electron microscopy with special care.

Fig. 10 is a micrograph showing the place of the emergence of a dislocation rosette edge arm corresponding to 1 in Fig. 9 on the lateral surface. The position of the edge between the lateral and upper crystal surfaces is clearly evident in the lower part of the micrograph. The presence of a large number of wavy steps surrounding the hillock at the place of emergence of the edge arms of dislocation rosettes on the lateral surface is highly typical. Depending on the size of the indentation and the distance to the edge, one or several hillocks both round and elongated may arise (Fig. 10 and 11).

In Fig. 11 an elongated hillock with several tops is shown. The system of wavy steps seems to be the result of the emergence of edge arm dislocations on the lateral surface and the flow of crowdions. The wavy shape of the steps is connected with the interaction between dislocations and point defects. The absorption of interstitial ions corresponds to the elongation of the extra half-planes of edge arm dislocations and hence to their displacement from the axial plane between rosette arms to the periphery. In virtue of the

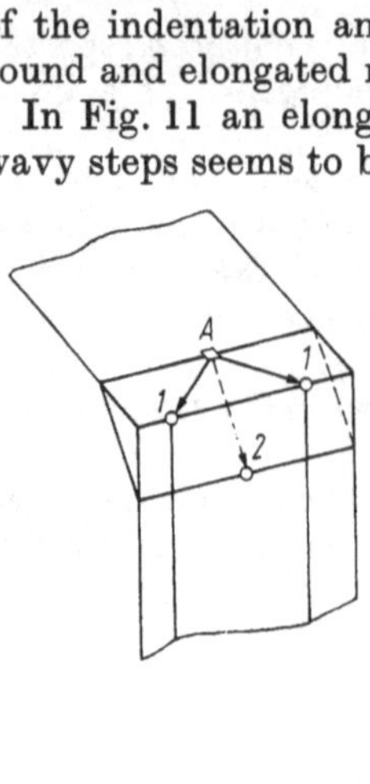

Fig. 9. Scheme of screw and two edge slip planes emerging into the lateral surface. The places of expected crowdion emergence are marked by rings; A is the application spot of concentrated load

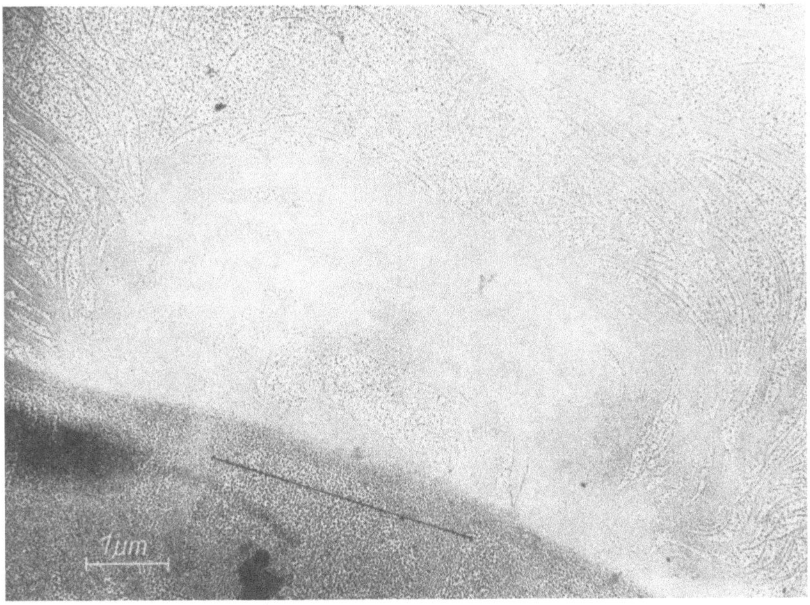

Fig. 10. Rise of a hillock at the place of edge arm emergence on the lateral surface (ring 1 in Fig. 9). The edge position is marked by a line

symmetry peculiarities in the bends of elemental steps with opposite signs, close to the surface some decrease of interstitial supersaturation occurs in connection with the emergence of interstitial ions.

A more rounded hillock is shown in Fig. 12. This shape appears to be connected with the predominance of crowdions in its formation on the lateral surface.

Observation at the place of emergence marked by 2 in Fig. 9, also revealed a great number of hillocks; however, they usually were not very high and surrounded by a small number of elemental steps (Fig. 13).

5. Discussion

In order to explain the surface microtopography resulting from the indentation of NaCl crystals it is necessary to postulate the transport of interstitial ions from the region under the indent by crowdions or by a system of prismatic dislocation rings. It is difficult to decide between these two possibilities in a given case, since the results of both the crowdion mechanism and prismatic slip are alike. In either case, the differences in surface relief caused by concentrated arrays of crowdions or by a system of prismatic rings, are not very clear. Nevertheless, on the basis of the considerations mentioned above it seems to us that the crowdion mechanism is the most probable.

The following interpretation of the observed phenomena is the most suitable at the present time.

During concentrated load deformation of a crystal, a great number of crowdions arises and moves from the indentation in $\langle 110 \rangle$ directions. The crowdion path length is several times the size of the indentation and strongly increases with the rate of impact loading, although the main moving force is created by

Fig. 11. Elongated hillock on the lateral surface at the place of edge arm emergence as a result of the joint emergence of crowdions and dislocation beam

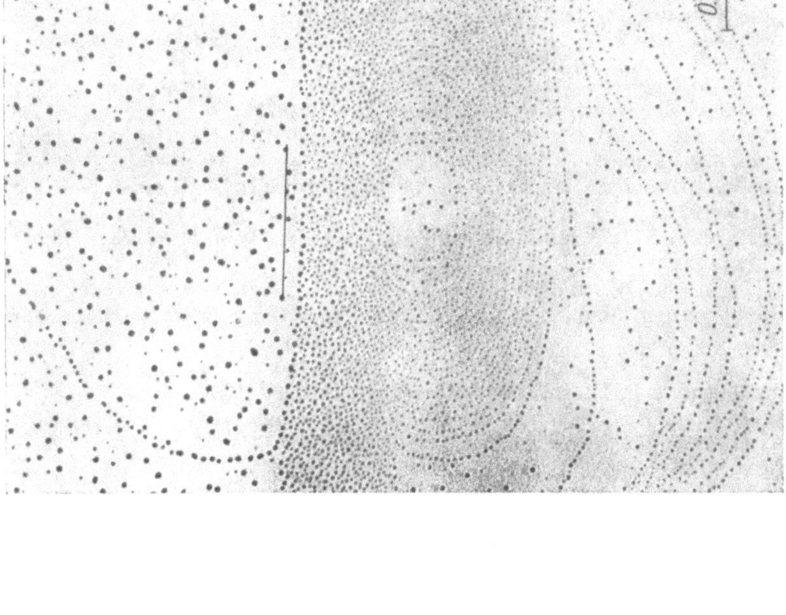

Fig. 12. Round hillock observed in the place marked by 1 in Fig. 9. The edge position is marked by a line

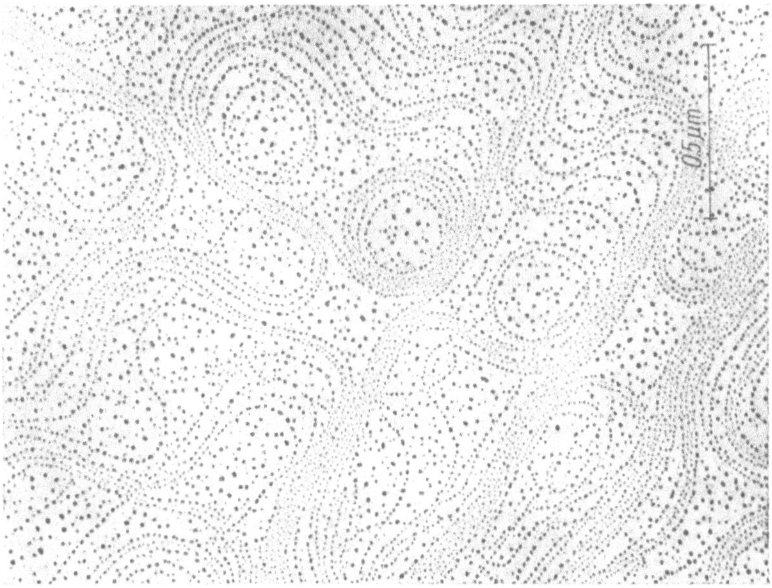

Fig. 13. Hillocks formed at the spot marked by 2 in Fig. 9

the non-uniform field of normal strains near the indentation. Part of the material transport is due to dislocation motion under the effect of the indenter. The dislocations take an active part in the interaction with interstitial ions supersaturating the zone around the indentation. The formation of dislocation rosettes which are developed by etching [12] or by volume decoration [4] is due to both the dislocations and the interstitial ions carried out from the region under the indent by the crowdions.

It is very important to estimate how exactly the surface relief describes the deformation processes proceeding within the crystal. One cannot exclude the possibility of microtopographic changes during decoration which is carried out at 100 to 150 °C, although this heating is relatively short (about 30 min). In particular, interstitial ion clusters appear to emerge onto the crystal surface during heating in view of diffusionally distributed peripheric loops.

On the other hand, the relief in the vicinity of the hillocks near the indentation probably changes very little since it does not appreciably depend on temperature and duration of the decoration process. Therefore, small changes in the arrangement of elemental steps on the crystal surface during the decoration process can be neglected.

The second cause for the distortion of the surface relief is the effect of water rounding off the surface relief steps. As it was mentioned a special examination was performed under conditions which excluded the effect of water vapour. One can be sure that the round form of elemental steps surrounding the hillocks arises as a result of the indentation of the crystal and is not connected with the effect of water. The round form is connected with screw dislocation motion along arc lines around the hillocks resulting in their successive increase during the deformation process.

Acknowledgements

The authors express their sincere appreciation to E. I. Givargizov for making germanium brushes available to them and to E. V. Parvova for her assistance in the course of this work. The authors thank Dr. G. I. Distler, Dr. V. Schmidt, and Dr. U. Messerschmidt for helpful discussions.

References

[1] V. L. INDENBOM, Zh. eksper. teor. Fiz., Pisma **12**, 526 (1970).
[2] V. N. ROZHANSKII, M. P. NAZAROVA, I. L. SVETLOV, and L. K. KALASHNIKOVA, phys. stat. sol. **41**, 579 (1970).
[3] G. I. DRANOVA, A. M. DYACHENKO, and I. M. MICHAILOVSKII, Zh. eksper. teor. Fiz., Pisma **13**, 40 (1971).
[4] V. N. ROZHANSKII, N. L. SIZOVA, and A. A. URUSOVSKAYA, Solid State Phys. **13**, 411 (1971).
[5] G. BASSET, Phil. Mag. **3**, 1042 (1958).
[6] C. SELLA, P. CONJEAND, and J. J. TRILLAT, IV. Internat. Congr. Electron Microscopy, Berlin 1958.
[7] H. BETHGE, Surface Sci. **3**, 33 (1964).
[8] G. I. DISTLER and V. P. VLASOV, Solid State Phys. **11**, 2226 (1969).
[9] G. I. NAZVLISHVILI, N. G. POLITOV, and S. F. GETZ, Kristallografiya, to be published.
[10] E. I. GIVARGIZOV and N. N. SHEFTAL, J. Crystal Growth **9**, 326 (1971).
[11] A. A. SHPOONT, Solid State Phys. **4**, 718 (1962).
[12] A. A. PREDVODITELEV, V. N. ROZHANSKII, and V. M. STEPANOVA, Kristallografiya **7**, 418 (1962).

(Received September 6, 1971)

phys. stat. sol. (a) 8, 565 (1971)

Subject classification: 14.3; 1.2; 22.5

Research Laboratory for Chemical Structures of the Hungarian Academy of Sciences, Budapest

The Electrical Conductivity and the Phase Transformation of Thallous Iodide Single Crystals

By

Z. MORLIN

The temperature-dependent specific electrical conductivity of zone-refined TlI single crystals was measured in the temperature range between 400 and 600 °K. Repeated measurements of the same crystal samples resulted in annealing effects in the course of which metallic-grey precipitates were formed. The conductivity in the transformation range can be explained by the formal kinetics of a diffusion-controlled nucleation and growth process. An estimation of the activation energy of the formation of the second-phase nuclei in the parent phase by the evaluation of TTT curves yields energy values of 0.2 to 0.4 eV.

Es wurde die temperaturabhängige spezifische elektrische Leitfähigkeit zonengeschmolzener, polymorpher TlJ-Einkristalle im Temperaturbereich 400 bis 600 °K gemessen. Wiederholte Messungen an denselben Kristallen kommen einer Wärmebehandlung gleich, deren Ergebnis die Bildung metallisch-grauer Prezipitate in der Mutterphase ist. Die Leitfähigkeit im Temperaturbereich des Phasenübergangs kann mit der formellen Theorie der diffusionsgesteuerten Keimbildung und des Wachstums gedeutet werden. Eine Schätzung der Aktivierungsenergie der Keimbildung mittels TTT-Kurven ergab die Werte 0,2 bis 0,4 eV.

1. Introduction

The thallous halides crystallize in the cubic structure as stable AB type crystals except TlI which has a B_{33} double-layered orthorhombic structure (D_{2h}^{17} — Cmcm) with four molecules per unit cell. This ordinary, yellow, low-temperature modification transforms on heating above 170 °C into a B_2, CsCl-type structure which has a bright red colour [1]. Not much work has been done so far to clarify the details of the $B_{33} \rightleftarrows B_2$ tranformation. Zahner and Drickhamer have shown that a pressure-induced transformation occurs also at room temperature at about 5 kbar [4]. More recently Samara et al. have investigated the polymorphism, compressibility, and thermal expansion of this crystal [9]. Since experiments with CsCl have shown that the measurement of the temperature dependence of electrical condutivity is an extremely sensitive method to investigate the phase transformation [7, 10 to 13], this method has been applied to study the behaviour of TlI in the temperature range of 400 to 600 °K.

2. Experimental

Single crystals were grown with the zone melting technique from British Drug House pro analysi material. The samples were cut from the pure ends of the zone-molten slabs, and their conductivity was measured in the usual way at a stabilized voltage of 1 V. The samples covered with a thin graphite layer were clamped between platinum electrodes, and heated at a rate of 1 deg/min in an electric furnace of stabilized current which could be regulated with a toroid transformer. The temperature of the specimen was measured with a Fe–Co thermocouple and a Norma millivoltmeter. The results could be represented by the well-known function

$$\sigma = A \exp\left(-\frac{B}{T}\right), \tag{1}$$

where σ is the specific conductivity in $\Omega^{-1}\,\mathrm{cm}^{-1}$, T is the temperature in °K, A and B are constants.

3. Results

Fig. 1 shows a typical example of the conductivity measurements in $\ln \sigma$ vs. $1/T$ representation. The graph consists of three straight lines, the low-temperature section up to 453 °K, the high-temperature line, and the range of the phase transformation indicated by a considerable increase of the conductivity. The transformation shows a well developed hysteresis, when the sample was measured on heating as well as on cooling.

The temperature-dependent specific conductivity values can be represented by the equations

$$\sigma_{B_{33}} = 0.037 \exp\left(-\frac{6200}{T}\right) \tag{2}$$

and

$$\sigma_{B_2} = 48.2 \exp\left(-\frac{8500}{T}\right). \tag{3}$$

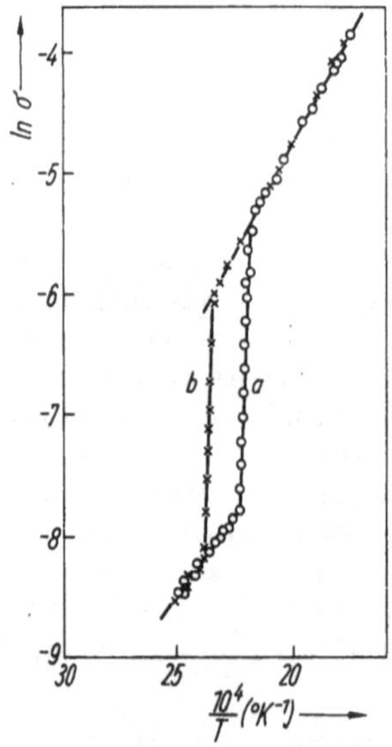

Fig. 1. ln σ vs. $1/T$ representation of the specific conductivity of TlI. (a) Transformation on heating; (b) transformation on cooling

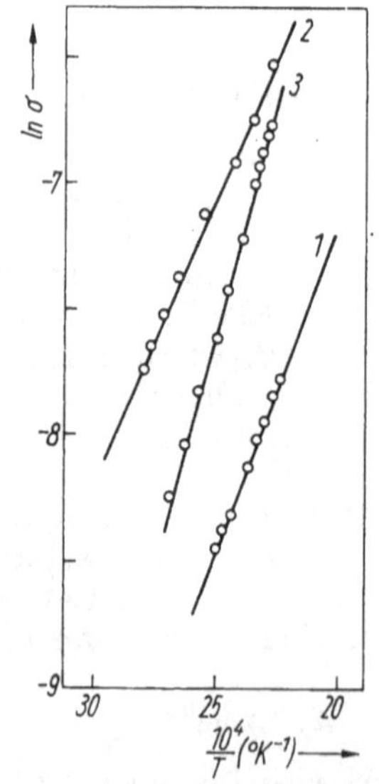

Fig. 2. The change of the low-temperature slope of the ln σ vs. $1/T$ lines on annealing by repeated measurements. The numbers refer to repeated measurements

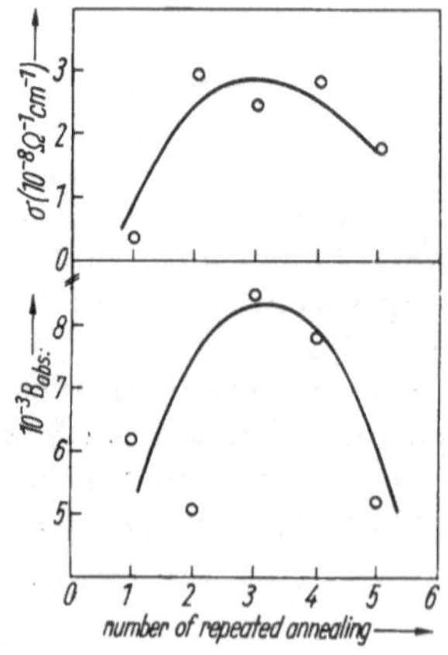

Fig. 3. The annealing dependence of σ and B. The numbers on the abscissa indicate the repeated measurements

The electrical conductivity proved to be extremely structure-sensitive. If in the same sample the σ-, A-, and B-values were measured several times, also the hysteresis loop showed marked changes. This is demonstrated in Fig. 2 and 3. Since the repeated measurements correspond to a repeated annealing, the figures show annealing effects. The changes may be due to a change of the chemical composition of the material. In fact, at the end of several runs, metallic-grey particles were found in the samples, which differed greatly from the bright yellow colour of the B_{33}-type TlI crystals. Though this effect was not studied in detail it seems to be probable that during the repeated measurements Tl_2O has formed in TlI, all the more since the measurements were carried out in air. In order to avoid this structure sensitivity the kinetics of the transformation was investigated only by the results of the first measurements of the samples.

4. Kinetics of the Phase Transformation

In this section the experimental data of the conductivity in the temperature range of the transformation are presented in terms of the formal theory of phase transformation developed by Avrami [2] and Johnson and Mehl [3], and treated exhaustively by Christian [5].

In order to study the kinetics of transformation the experimental conductivity data must be expressed in terms of the transformed fraction [14 to 16] by either of the corrected formulae used previously for CsCl. The result is shown in Fig. 4. The curve can be analysed in two ways:

1. An $\ln 1/(1-x)$ vs. T (or t) plot enables to test the general kinetic formula

$$1 - x = \exp\,(k\,t^n)\,. \tag{4}$$

This has been done in Fig. 5 which shows that the transformation clearly consists of two stages. The first one up to $x = 0.8$ is parabolic ($n \neq 1$), and the second one linear ($n = 1$), which leads to the conclusion that in this latter stage the growth of the equilibrium-phase grains is restricted to one direction only,

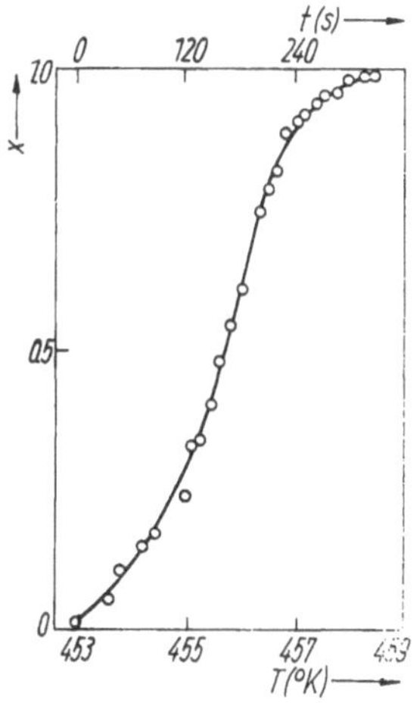

Fig. 4. The corrected x vs. T diagram as obtained from conductivity measurements. x is the transformed fraction

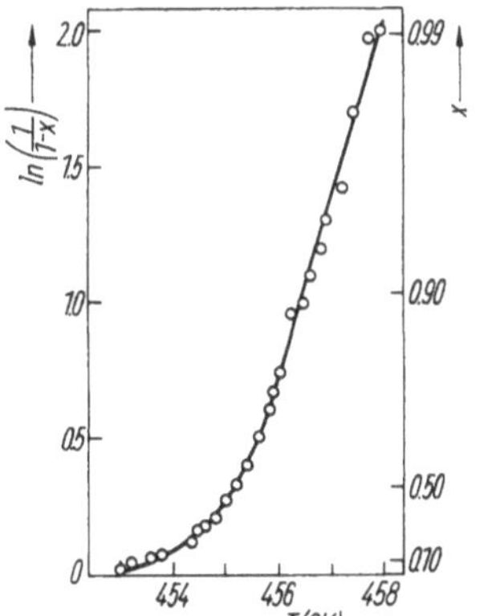

Fig. 5. ln $1/(1-x)$ vs. T as evaluated from Fig. 4

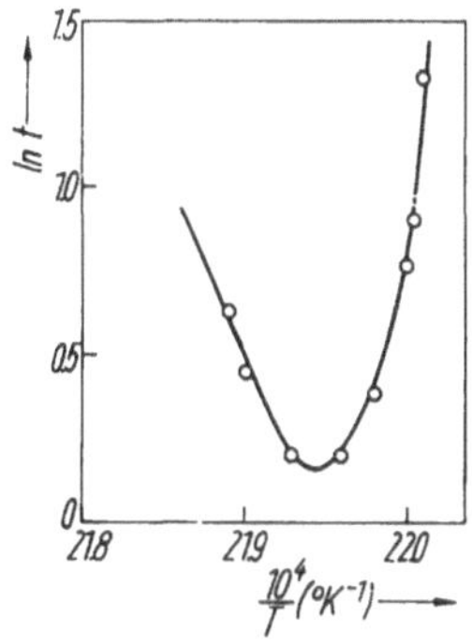

Fig. 6. TTT diagram as evaluated from Fig. 4

where new nuclei are formed, and the driving force is constant. An alternative possibility would be — as has been pointed out by Weijma [16] — to assume that the transformation is in this stage a first-order process (the probability of the existence of a high-temperature phase is proportional to the transformed volume at the time $|t_x|$, when $x = 0.80$). Since at $t_{0.8}$ still 20% of the material is untransformed, the first possibility seems to be more probable.

A numerical test of the first (parabolic) stage has shown that for $0 < x < < 0.80$ $n \approx 2$. This means that the nucleation rate is a rapidly decreasing function of time due to the decreasing grain boundary area of the orthorhombic phase, and that the nuclei grow along the edges of the new phase [5].

2. The second type of analysis of the experimental curve (Fig. 4) consists in constructing a TTT (time, temperature, transformation) curve [6]. This can be done if the transformation time of always the same fraction plotted versus temperature yields C-shaped curves. Actually $\ln t_x$ vs. $1/T$ was plotted as shown in Fig. 6. The diagram represents the relation

$$\ln t_x = \text{const} \frac{1}{T} + \frac{W}{kT} - \ln A \,, \tag{5}$$

where W is the activation energy of the formation of a critical nucleus. By applying the graphical method developed by Burke [6] to estimate W the values 0.2 to 0.4 eV were obtained. This seems to be reasonable compared with our experimental B-values expressed in terms of energy, since the activation energy of the grain boundary migration was found to be equal or lower than that of the vacancy migration [8]. Of course, in this case one must assume that the conductivity of TlI is purely ionic.

5. Conclusion

From the results of the electrical conductivity experiments one may conclude that the phase transformation of polymorphic TlI is dominated by a diffusion-controlled process. The transformation consists of two stages, the first of which follows a parabolic law, whereas the second one points to a uniform growth of the new phase with a constant driving force. The experimental results can also be represented in the form of TTT curves. This enables the estimation of the activation energy of nucleus formation.

Acknowledgement

The author wishes to thank Dr. R. Voszka, Research Group for Crystal Physics of the Hungarian Academy of Sciences, for growing the TlI single crystals.

References

[1] L. HELMHOLTZ, Z. Krist. **95**, 129 (1936).
[2] M. AVRAMI, J. chem. Phys. **7**, 1103 (1939).
[3] W. A. JOHNSON and R. F. MEHL, Trans. AIME **135**, 416 (1939).
[4] J. C. ZAHNER and H. G. DRICKHAMER, J. Phys. Chem. Solids **11**, 92 (1959).
[5] J. W. CHRISTIAN, The Theory of Transformation in Metals and Alloys, Pergamon Press, Oxford 1965 (pp. XVI, 973).
[6] J. BURKE, The Kinetics of Phase Transformation in Metals, Pergamon Press, Oxford 1965 (pp. VIII, 226).
[7] Z. MORLIN, Acta phys. Hungar. **21**, 137 (1966).

[8] P. Gordon and R. A. Vandermeer, in: Recrystallization, Grain Growth, and Textures, Ed. H. Margolin, Amer. Soc. for Metals, Metals Park (Ohio) 1966 (pp. 205 to 266).
[9] G. A. Samara, L. C. Walter, and D. A. Northrop, J. Phys. Chem. Solids **28**, 1875 (1967).
[10] J. Arends and H. Nijboer, Solid State Commun. **5**, 163 (1967).
[11] P. Süptitz and J. Teltow, phys. stat. sol. **23**, 9 (1967).
[12] P. J. Harvey and J. M. Hoodless, Phil. Mag. **16**, 543 (1967).
[13] J. Arends and H. Nijboer, phys. stat. sol. **26**, 537 (1968).
[14] Z. Morlin, Acta phys. Hungar. **24**, 277 (1968).
[15] I. Gaál, Z. Morlin, and I. Tarján, Acta phys. Hungar. **27**, 25 (1969).
[16] H. Weijma, Some Aspects of the Phase Transformation in Solid CsCl, V.R.B. Offsetdrukkerij, Groningen 1971 (pp. VI, 85).

(Received July 1, 1971)

phys. stat. sol. (a) 8, 571 (1971)

Subject classification: 10.1 and 18.2; 21.1.1

Division of Engineering, Brown University, Providence, Rhode Island

X-Ray Topographic Observations of Dislocation–Ferromagnetic Domain Interaction in Fe–3% Si Crystals

By

C. Cm. Wu[1]) and B. Roessler

Ferromagnetic domains as well as dislocations in cube-oriented Fe–3% Si crystals grown by secondary recrystallization are observed in X-ray anomalous transmission topographs. The arrangement of surface closure domains is influenced by the presence of dislocations. Nodes in the dislocation network frequently coincide with the topographic images of 90° walls which bound the closure domains. A determination of the Burgers vectors of the dislocations suggests that the strain fields of the dislocations influence the domain arrangement through magnetostrictive interaction.

Ferromagnetische Domänen sowie Versetzungen in kubisch orientierten Fe–3% Si-Einkristallen, die durch sekundäre Rekristallisation gezüchtet wurden, werden in den Topogrammen anomaler Röntgenstrahlentransmission beobachtet. Die Anordnung der Oberflächenabschlußdomänen wird von der Anwesenheit der Versetzungen beeinflußt. Häufig treffen Knoten des Versetzungsnetzwerkes mit den topographischen Bildern von 90°-Wänden zusammen, die die Abschlußdomänen begrenzen. Aus der Bestimmung der Burgersvektoren der Versetzungen wird geschlossen, daß die Spannungsfelder der Versetzungen die Domänenanordnung durch magnetostriktive Wechselwirkungen beeinflussen.

1. Introduction

It is possible, by employing anomalous transmission X-ray topography, to observe both ferromagnetic domains and dislocations in Fe–3% Si crystals [1 to 4]. Dislocation images occur in the X-ray topograph if the strain field of the dislocation distorts the diffracting plane of atoms; the images appear as regions of reduced X-ray intensity. X-ray images of ferromagnetic domains and domain boundaries can also be observed in the same topograph in which dislocation images appear if suitable diffracting planes are employed. The contrast effects associated with these domain images are quite complex [4] and will be described separately [5]. The purposes of the present paper are to present direct X-ray topographic observations showing that the ferromagnetic domain arrangement can be influenced by the strain fields of the dislocations present in a crystal and to suggest that this interaction arises because of magnetostriction.

2. Experimental Procedure

An Fe–3% Si crystal about 1 in. square grown by secondary recrystallization was cut by spark machining from a large-grained polycrystalline sheet having a thickness of 0.018 in. The [010] and [001] crystallographic directions were within approximately 0.5 and 2 degrees, respectively, of the crystal surface. The surfaces of the crystal were prepared by standard metallographic polishing

[1]) Present address: Department of Mechanical Engineering, University of Maryland, College Park, Maryland.

Fig. 1. The $(01\bar{1})$ anomalous transmission topograph showing dislocation network at RECTANGL

with four abrasive papers and 6 μm and 1 μm diamond paste and were then chemically polished in a solution of 50% by volume of phosphoric acid–30% hydrogen peroxide for a time sufficient to remove the surface damage caused by mechanical polishing. The final thickness of the crystal was 0.0137 in. Anomalous transmission topographs were recorded with CoKα X-rays using the experimental arrangement shown in Fig. 1 of reference [1].

3. Results and Discussion

3.1 Dislocation-domain interaction

Fig. 1 shows an X-ray topograph employing the $(01\bar{1})$ reflecting plane. The dislocation images appear as white lines, corresponding to reduced X-ray intensity. They are part of a dislocation network which extends into the crystal and is visible to a depth of about 100 μm. The rectangular dislocation array at RECTANGL in Fig. 1 actually consists of pieces of several different dislocations, as may be seen by comparing Fig. 1 with the (011) reflection in Fig. 2. In Fig. 2 the dislocations forming certain portions of the rectangle are out of contrast. For example, the images in Fig. 1 of the dislocations EC, CT, NG, GL, and LR are invisible in the (011) reflection (Fig. 2). The vanishing of the dislocation image will be taken to imply that the Burgers vector of the dislocation lies in the diffracting plane [6].

Chevron-shaped images of fir-tree closure domains are seen in the (002) reflection, Fig. 3. These closure domains, which are present at the surface of the crystal [7], extend several microns into the crystal at V and taper to a point at H. Their images do not appear in the $(01\bar{1})$ and (011) reflections under the diffraction conditions employed in Fig. 1 and 2. It can be seen in Fig. 3 that the

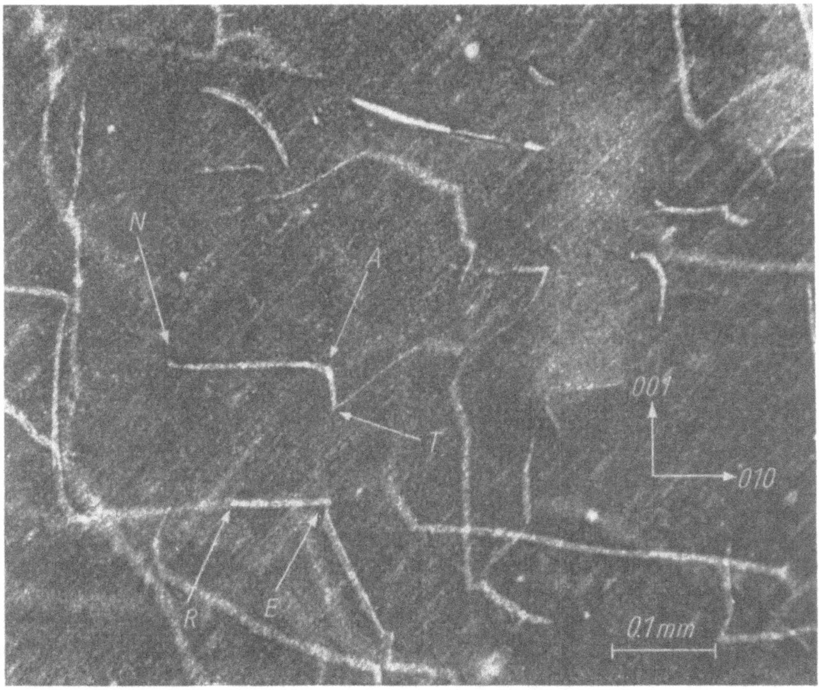

Fig. 2. The (011) topograph from the same region of the crystal as shown in Fig. 1. Dislocation segments EC, CT, NG, GL, and LR are out of contrast

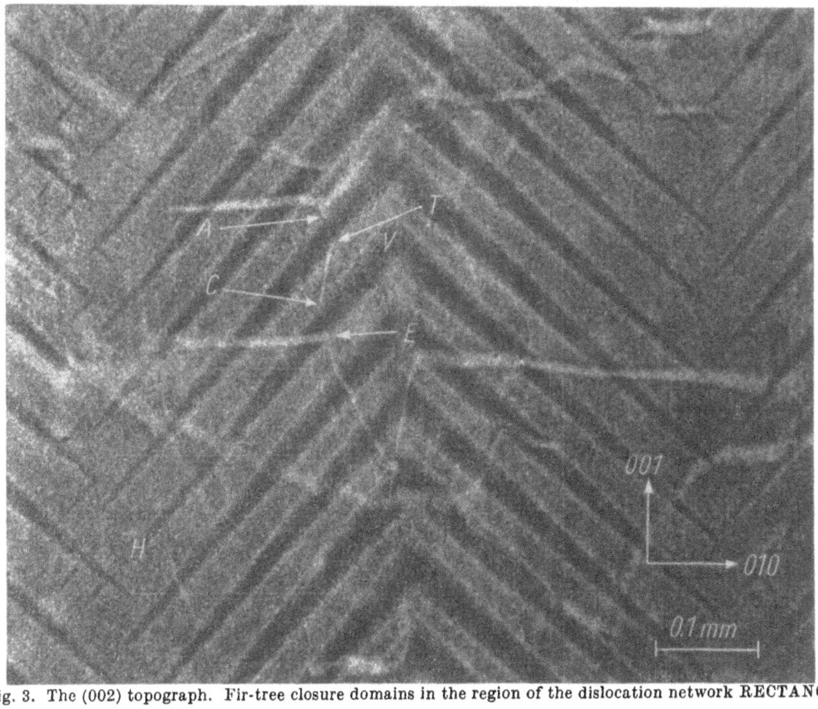

Fig. 3. The (002) topograph. Fir-tree closure domains in the region of the dislocation network RECTANGL

images of the two edges of the fir-tree closure domain HV coincide with the dislocation nodes C and E, and that the nodes T and A coincide with the images of the 90° walls bordering the adjacent closure domain. These coincidences can be seen much more clearly when the negatives of the photographs shown in Fig. 1 to 3 are superposed and viewed with transmitted light. They can also be seen by making a tracing of the images seen in one reflection and superposing it upon the other photograph. It seems clear, therefore, that the arrangement of the ferromagnetic closure domains is affected by the presence of dislocations and dislocation networks in the crystal.

3.2 Burgers vectors of interacting dislocations

An additional example of dislocation–domain interaction in which it has been possible to determine the Burgers vectors of the dislocations influencing the domain arrangement is presented in Fig. 4 to 6, which show respectively the $(01\bar{1})$, (011) and (002) reflections. It can be seen in the (002) reflection, Fig. 6, that the nodes at the ends of dislocation segment 21 again coincide with the 90° walls of a closure domain.

The Burgers vectors of several of the dislocations in Fig. 4 to 6 were determined by employing the criterion that if the Burgers vector is parallel to the diffracting plane the dislocation image vanishes [6]. If the dislocation image vanishes in two different reflections, the Burgers vector is determined as the line of intersection of these two reflecting planes. In addition to the reflections shown in Fig. 4 to 6, this same area of the crystal was also examined with (101), $(\bar{1}\bar{1}0)$, and $(0\bar{2}0)$ reflections; i.e., six different reflecting planes were used.

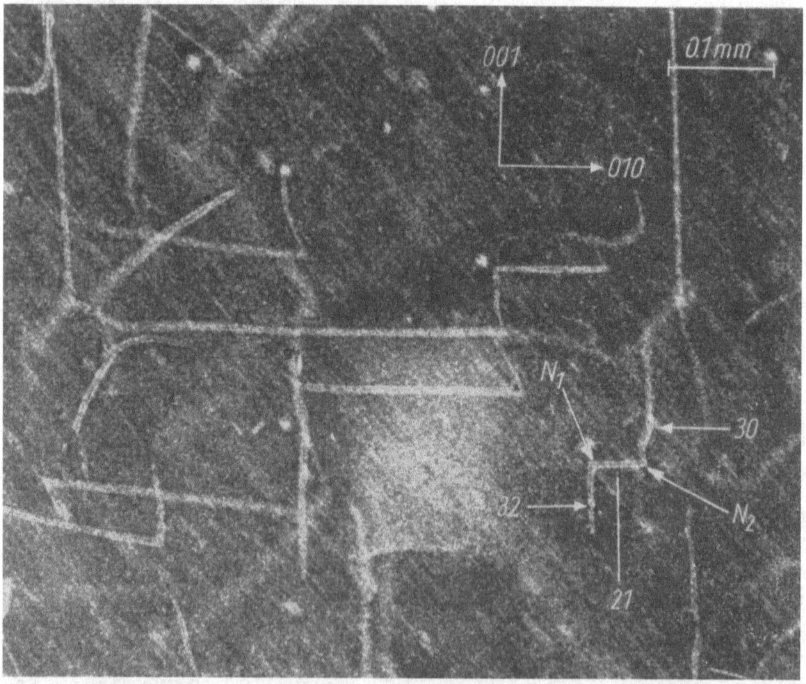

Fig. 4. The $(01\bar{1})$ topograph from another region of the crystal. Dislocation segments 21, 30, and 32 are visible segments 29 and 31 are not

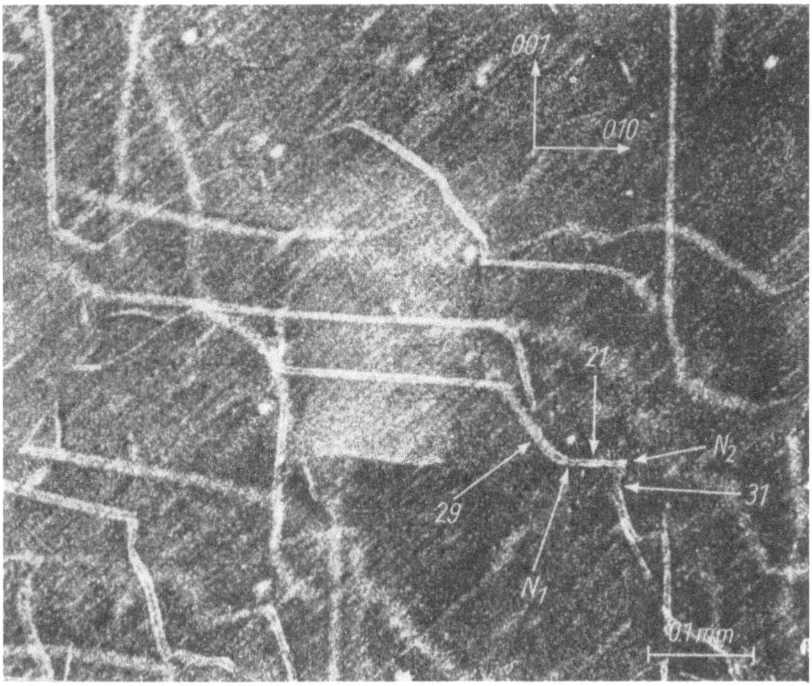

Fig. 5. The (011) topograph, from same region as in Fig. 4. Dislocation segments 21, 29, and 31 are visible, segments 30 and 32 are not

Fig. 6. The (002) topograph. Closure domains in the region shown in Fig. 4 and 5

It is a feature of this anomalous transmission technique that the clarity of the image of a particular dislocation differs in each reflection. While this is sometimes a result of experimental factors which are difficult to control, in most cases it is due to inherent differences in the diffraction process when different reflecting planes are employed. An example of an inherent limitation is the fact that as the depth of the dislocation beneath the crystal surface increases, the width of the dislocation image increases [8] if the dislocation line is perpendicular to the plane of incidence [9]. Moreover, the width is inherently greater in {020} reflections than in (011) and (01$\bar{1}$) reflections. In order to take this variation in the "quality" of the image into account in determining whether the image of a given dislocation was present or absent in a reflection, a reliability factor was assigned to each such determination. The results of this analysis are collected in Table 1. The reliability factors were assigned to account as well as possible both for varying quality of the topograph and for inherent diffraction effects. It has not been possible to determine the sense of the Burgers vector, and therefore the Burgers vectors can be those listed in Table 1 or their negatives.

Dislocations 29, 30, 31, and 32 are, apparently, slip dislocations with Burgers vectors of the $a/2\ \langle 111 \rangle$ type, whereas dislocation 21, which has a [001] Burgers vector, appears to have formed by the combination of two slip dislocations along the line of intersection of their slip planes according to the following reaction:

$$\boldsymbol{b}_{29} + \boldsymbol{b}_{32} = \boldsymbol{b}_{21}\,,$$

$$\frac{a}{2}\,[\bar{1}11] + \frac{a}{2}\,[1\bar{1}1] = a\,[001]\,. \tag{1}$$

This reaction and the X-ray topographic evidence for it are discussed more fully elsewhere [6]. It is energetically favorable on the basis of Frank's criterion [10], since it reduces the elastic energy by a factor of 1.5 to 1.

Table 1

Analysis of dislocation images in Fig. 4 to 6 and Burgers vector determinations

dislocation	reflecting plane						Burgers vector direction
	(01$\bar{1}$)	(011)	(0$\bar{2}$0)*)	(002)	($\bar{1}\bar{1}$0)*)	(101)*)	
30	P – 100	A – 90	P – 80	P – 100	A – 70	P – 10	[1$\bar{1}$1]
32	P – 100	A – 80	P – 95	P – 40	A – 70	P – 10	[1$\bar{1}$1]
31	A – 90	P – 100	P – 45	P – 55	A – 80	A – 60	[$\bar{1}$11]
29	A – 90	P – 100	P – 85	P – 95	A – 80	A – 75	[$\bar{1}$11]
21	P – 100	P – 100	A – 95	P – 95	A – 40	P – 20	[001]

Key: P indicates that the dislocation image was present, A indicates that the dislocation image was absent. The number following P or A is an estimate of the reliability of the determination (100 indicates 100% reliable).

*) (0$\bar{2}$0), ($\bar{1}\bar{1}$0), and (101) reflections are not shown in this paper.

3.3 Discussion

The evidence that dislocations interact with ferromagnetic domains by affecting the positions of closure domains is perhaps most clearly seen in Fig. 1 to 3. The additional example shown in Fig. 4 to 6, in which the Burgers vectors are

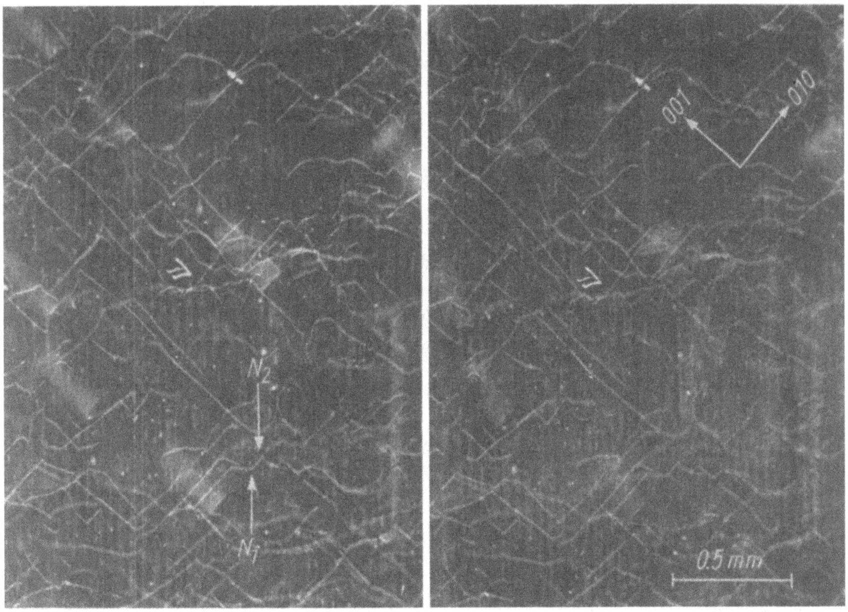

Fig. 7. A (011) stereoccopic pair from the region shown in Fig. 4 to 6

known, strongly suggests that this interaction is an inverse magnetostrictive effect [11] caused by the elastic strain field of the dislocations. Dislocation 21 appears in the stereoscopic photographs shown in Fig. 7 to lie approximately parallel to the crystal surface. In addition, the constant width of the image also suggests that the tangent vector of the dislocation is [010]. Dislocation 21 is thus a pure edge dislocation and has an elastic strain field corresponding either to compression or tension in the [001] direction in the region above the (100) plane on which it lies (depending on the sense of the Burgers vector). The arrangement of dislocations and domains is shown in the schematic diagram of Fig. 8. If the region of tension associated with dislocation 21 is, in fact, above the (100) plane, and if, in addition, the dislocation lies at a depth beneath the crystal surface about equal to the depth of the closure domain in that region, it is reasonable to expect that the arrangement of closure domains will be influenced by the dislocation strain field. The magnetization, m_1, in the closure domain, whose walls RU and US coincide with the dislocation nodes N_1 and N_2 at the ends of

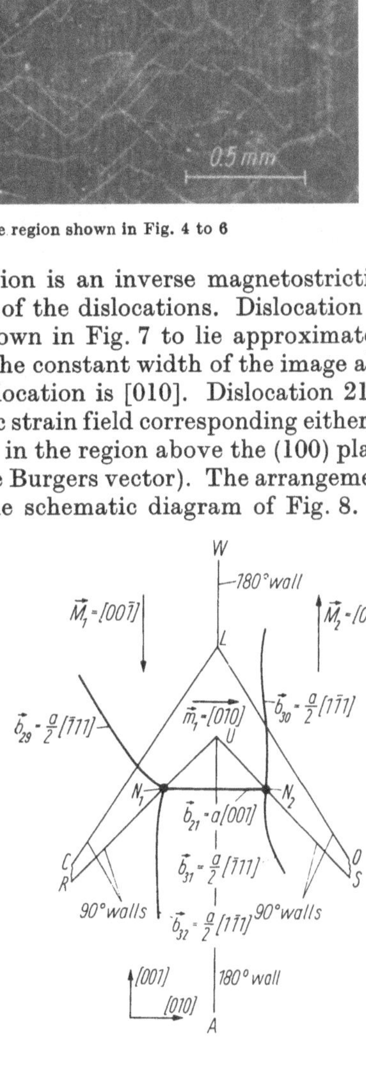

Fig. 8. A schematic drawing of the arrangement of domains and dislocations shown in Fig. 4 to 6. Closure domain at CLOSUR. 180° wall at WA is perpendicular to the crystal surface. Dislocation nodes at N_1 and N_2. Burgers vectors assigned as in Table 1. $\boldsymbol{M}_1$ and $\boldsymbol{M}_2$ indicate magnetization in adjacent main domains, $\boldsymbol{m}_1$ indicates magnetization in closure domain. Dislocation nodes N_1 and N_2 coincide with 90° walls along RU and US

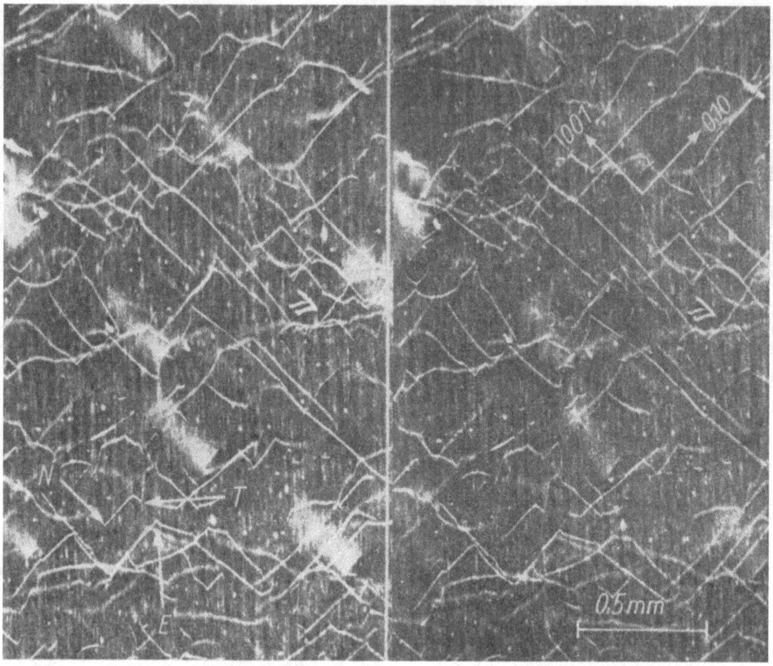

Fig. 9. A (011) stereoscopic pair from the region shown in Fig. 1 to 3

dislocation 21, lies in the [010] direction. As a result of magnetostriction, the atomic unit cell within the closure domain is slightly elongated in the [010] direction and is slightly contracted in the [100] and [001] directions. If the closure domain is arranged so as to avoid the region of elastic tension in the region above dislocation 21, the total energy will be lower than if this arrangement did not occur. It seems likely, therefore, that this type of interation accounts in a qualitative way for the arrangement of domains in the regions near the dislocations discussed in Fig. 4 to 6.

Fig. 9 is a stereoscopic pair of the region shown in Fig. 1 to 3. Although portions of the rectangle are out of contrast, it is apparent that the dislocations lie at varying depths beneath the crystal surface. In those places where the dislocations are close enough to the surface so that their strain fields are still appreciable compared to the magnetostrictive strain, an interaction sufficient to affect the closure domain arrangement can occur. If, on the other hand, the dislocation lies much deeper in the crystal, its strain field will not influence the closure domain arrangement.

One would expect to find that the strain field of the dislocation can cause a domain either to avoid (as in Fig. 4 to 6) or to prefer a region of the lattice. Although it has not been possible in the experiments reported here to determine completely the Burgers vectors for all dislocation–domain interactions observed, this should in principle be possible. Although an unequivocal determination of the Burgers vector for dislocation TA in Fig. 1 was not possible, the evidence available suggests that it is a [010]. If the tensile region for dislocation TA is towards the crystal surface this should be a favorable region for the closure domain which lies there.

The frequent coincidence of domain walls with dislocation nodes is probably due to the fact that at such nodes the combining dislocations change their depth in the crystal. This causes a rapid change in the elastic strain field near the closure domain, since the two reacting dislocations generally run away from the node in different directions.

Finally, it should also be mentioned that the domain walls are not always arranged in the same fashion with respect to the dislocation nodes and strain fields. X-ray topographs have been recorded over a period of many months; during this period, as a result of changes in applied magnetic fields or by application of elastic stress, all the domains have been swept out of the crystal or at least the domain arrangement has been altered substantially. When these influences are removed, the magnetic domains rearrange themselves in essentially the same manner as before. Slight differences in the location of both 180° and 90° domain walls, however, do occur. It is our opinion, therefore, that, in Fe–Si crystals, this domain-dislocation interaction acts to control the details of the closure domain arrangement. It does not play a major role in determining the gross features of the domain arrangement; these are determined by other factors [7].

Acknowledgements

The crystals used in this investigation were grown by Prof. J. J. Kramer of the University of Delaware while at the Westinghouse Electric Corporation. The authors are extremely grateful to Prof. J. J. Kramer and to Westinghouse for supplying the crystals. The capable assistance of Mr. W. Oates with the photographic work is also appreciated. The financial support of the Advanced Research Projects Agency, Contract SD-86, and of the National Science Foundation, under Grant GK-4123, is gratefully acknowledged.

References

[1] B. Roessler, J. J. Kramer, and M. Kuriyama, phys. stat. sol. **11**, 117 (1965).
[2] B. Roessler, phys. stat. sol. **20**, 713 (1967).
[3] C. Cm. Wu and B. Roessler, J. appl. Phys. 42, 1814 (1971).
[4] C. Cm. Wu, Ph.D. Thesis, Brown University, Providence, Rhode Island 1970.
[5] C. Cm. Wu and B. Roessler, to be published.
[6] C. Cm. Wu and B. Roessler, to be published.
[7] R. Carey and E. D. Isaac, Magnetic Domains and Techniques for Their Observation, Academic Press, New York 1966.
[8] F. Meier and V. Gerold, Direct Observation of Imperfections in Crystals, Interscience Publishers, New York 1962 (p. 509).
[9] R. W. James, Solid State Phys. **15**, 53 (1963).
[10] F. C. Frank, Physica (Utrecht) **15**, 131 (1949).
[11] S. Chikazumi, Physics of Magnetism, John Wiley & Sons, New York 1964 (p. 182).

(Received September 20, 1971)

phys. stat. sol. (a) 8, 581 (1971)

Subject classification: 14.4; 22.9

Department of Physics, University of Saugar, Sagar, India

Effect of Polarizing Field on Sealing Wax Thermoelectret Charge

By

R. NATH and D. R. BHAWALKAR

The surface charge characteristics of sealing wax thermoelectrets, prepared at different temperatures by varying polarizing field in steps of 2 up to 20 kV/cm, show that the retained charg eincreases with the polarizing field. The surface charge characteristics are explained in terms of Gross's two charge theory and the three layer capacitor model of Wiseman and Feaster and confirm Gross's views on the nature of the hetero- and homocharges also in the case of a dielectric which is a mixture of organic and inorganic substances.

Die Oberflächenladungscharakteristiken von Dichtwachs-Thermoelektreten, die durch Änderung des Polarisationsfeldes in Stufen von 2 bis 20 kV/cm bei verschiedenen Temperaturen präpariert wurden, zeigen, daß die Restladung mit dem Polarisationsfeld zunimmt. Die Oberflächenladungscharakteristiken werden mit der Zwei-Ladungs-Theorie von Gross und dem Dreischichten-Kondensator-Modell von Wiseman und Feaster erklärt und bestätigen die Ansichten von Gross über die Natur der Hetero- und Homo-Ladungen auch im Fall eines Dielektrikums, das eine Mischung aus organischen und anorganischen Substanzen ist.

1. Introduction

Studies on the dependence of thermoelectret charge on the strength of the polarizing field have been carried out on various types of thermoelectret materials by various workers. Polarizing fields as low as 100 V/cm and also 10 kV/cm or more have been used to prepare thermoelectrets of carnauba wax [1] and the important difference between the electrets thus prepared has been reported [2]. Thiessen et al. [3] have observed heterocharge only up to field strengths of 10 kV/cm, the magnitude of which is proportional to the intensity of the polarizing field, and a characteristic reversal of hetero- to homocharge for field strengths above 10 kV/cm. Similar results but at different values of polarizing fields have been observed for plastics [4]. Mikola [5] has reported the influence of polarizing fields on the sign of the thermoelectret charge the magnitude of which is proportional to the intensity of the polarizing field up to 30 kV/cm.

Besides these, studies of the authors on the effect of the temperature during the preparation on the thermoelectret charge of sealing wax thermoelectrets, which has been reported elsewhere, show clearly that the thermoelectret charge is not a function of temperature alone but depends on the polarizing field as well. In view of these results the systematic study of the dependence of the thermoelectret charge of sealing wax on the strength of the polarizing fields has been undertaken.

2. Experimental

Method of preparation of sealing wax thermoelectrets has been developed by the authors in this laboratory [6]. Samples of sealing wax were clamped in an electret cassette, designed and prepared locally in the workshop, and then subjected to an electric field in a thermostat for the formation of the thermoelectret. Each thermoelectret was preserved by keeping the short-circuited thermoelectret between cotton pads inside a desiccator. The surface charge has been measured on the anode surface of the thermoelectret by the lifted electrode method using a Lindemann electrometer.

Four series of thermoelectrets were prepared by varying the polarizing fields in steps of 2 kV/cm between 2 kV/cm and 20 kV/cm at the preparation temperatures of 35, 55, 75, and 95 °C. The thickness of the thermoelectrets was 2 mm for the range 2 to 10 kV/cm and 1 mm for the range 12 to 20 kV/cm. Each thermoelectret was subjected to the polarizing field for a total time of 9 h, i.e. 3 h of heating and 6 h of cooling. The electrode used was a tin metal foil for all thermoelectrets.

On the figures the thermoelectrets are designated by their preparation conditions in terms of the polarizing field in kV/cm right of the capital letter E and preparation temperature in °C left of E.

3. Results and Discussion

The anode side of each thermoelectret has been used for the measurement of the surface charge for this study. The surface charge of each thermoelectret has been measured just after 9 h, the time during which field has been applied. Then the charge has been measured daily for a period of one month. These measurements thus give the decay and the surface charge retained within a period of one month for each thermoelectret.

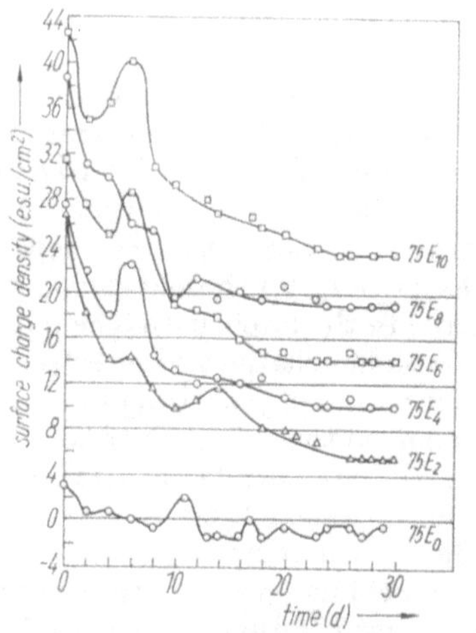

Fig. 1. Decay mode of sealing wax thermoelectret prepared at 75 °C with different polarizing fields in the range of 0 and 10 kV/cm. The origin for the field strengths 2, 4, 6, 8, and 10 kV/cm has been shifted to 4, 8, 12, 16, and 20 on the ordinate, respectively. The horizontal line just below the decay curve shows the position of abscissa for that curve

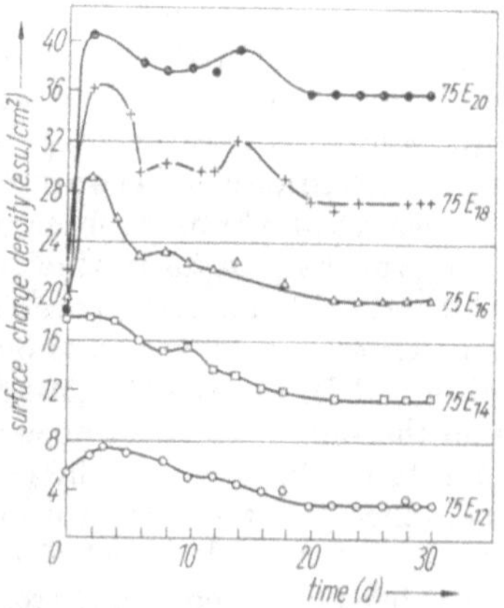

Fig. 2. Decay mode of sealing wax thermoelectret prepared at 75 °C with different polarizing fields in the range of 12 and 20 kV/cm. The origin for the field strengths 14, 16, 18, and 20 kV/cm has been shifted to 8, 16, 24, and 32 on the ordinate, respectively. The horizontal line just below the decay curve shows the position of abscissa for that curve

The decay has been represented by plotting the curves of surface charge densities in e.s.u./cm² versus time in days. The decay curves for each series of thermo-electrets prepared are represented in two figures one showing the field range of 0 to 10 kV/cm and the other of 12 to 20 kV/cm. This has been done in view of the different thicknesses in these two ranges of field. These decay curves of a representative series are shown in Fig. 1 and 2.

These curves show that the surface charge density varies irregularly and after nearly 20 days it has a tendency to persist with time as is shown by the flat portion of the curve. The initially observed charge on the surface of thermo-electrets is shown in Fig. 3. Though in most of the thermoelectrets the initially observed charge is a homocharge, in the case of thermoelectrets $95E_{14}$, $75E_{18}$, $75E_{20}$, $55E_{18}$, $55E_{20}$, $35E_8$, $35E_{10}$, $35E_{12}$, and $35E_{18}$ the initially observed charge is a heterocharge. This heterocharge has a tendency to rapid decay.

The results of the variation of surface charge of the thermoelectrets with field strength are shown in Fig. 4. In this figure the variation of observed retained charge with polarizing field is plotted for each series of thermoelectrets. The left-hand and right-hand sides of the figure show the results for the field ranges 2 to 10 kV/cm and 12 to 20 kV/cm, respectively. The general study of the nature of these curves shows a tendency of the surface charge to increase with polarizing field at temperatures above room temperature. At 35 °C and in the field range 12 to 20 kV/cm the curve is flat up to 16 kV/cm and shows no charge in the case of 18 and 20 kV/cm fields. Apparently there is no appreciable influence of a change of the field stength by 2 kV/cm in the range 12 to 20 kV/cm at 35 °C in the case of thermoelectrets of 1 mm thickness. This is also observed in the range of 6 to 10 kV/cm for the thermoelectrets of 2 mm thickness at the same temperature of 35 °C.

The two charge theory of Gross [7, 8] is most widely used in attempting the explanation of the observed results in thermoelectrets. It is quite often found to provide satisfactory explanation of the results. It has been extended to the three layer capacitor model for the behaviour and formation of thermoelectrets by

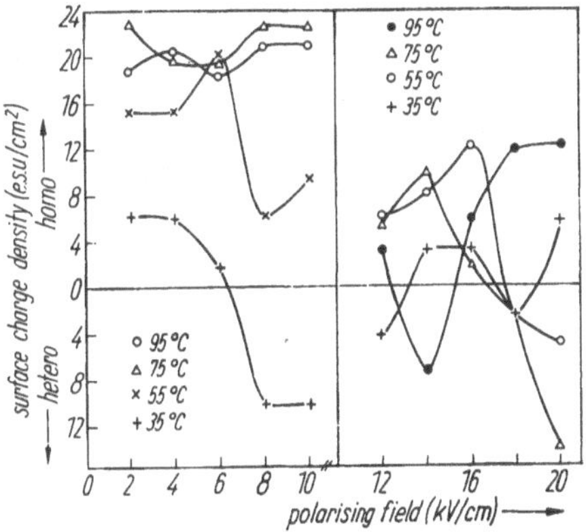

Fig. 3. Initial charge of sealing wax thermoelectret versus polarizing field at different temperatures

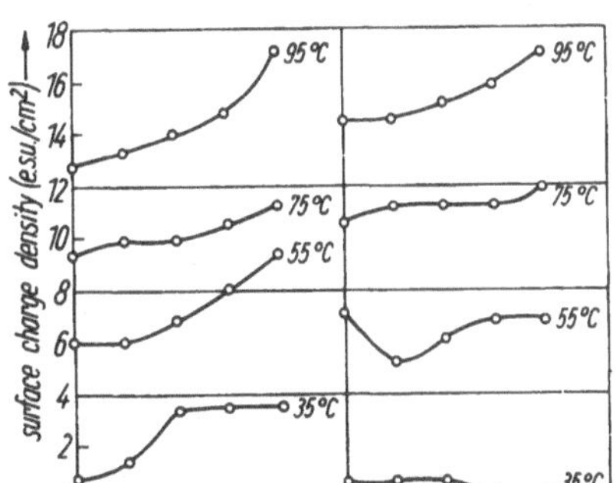

Fig. 4. Variation of final charge of sealing wax thermoelectrets with polarizing field at different temperatures of preparation. The origin for the series prepared at 55, 75, and 95 °C has been shifted respectively to 4, 8, and 12 on the ordinate and is shown by horizontal lines below the curves

Wiseman and Feaster [9]. The ideas put forward by these authors together with the results of some other workers will be used here in an attempt to explain the results obtained in these investigations. According to Gross [7, 8] the nature of hetero- and homocharges is different and they co-exist on the surface of thermoelectrets so that the measured surface charge is the algebraic sum of two charges. The heterocharge decays quite rapidly initially and then the decay rates of homo- and heterocharges are different. The observed nature of the decay curves of the thermoelectrets studied here is understandable in the light of the foregoing. The flat portion of the curve is due to the persistence of polarization of thermoelectrets.

The homocharge observed initially in the majority of thermoelectrets can be explained by the rapid decay of the heterocharge in initial stages so that it seems to have decayed before the measurements are made, such that its influence has been completely masked by the homocharge formed on the surface as the measured charge is the combined effect of the two. Besides other reasons the dependence of the time of persistence of heterocharge on the time elapsing between the end of polarization and removal of electrodes as observed by Handerek and Piech [10] may also be one of the causes resulting in the initially observed heterocharge in some of the thermoelectrets.

In some of the thermoelectrets prepared at low field strengths the initially observed homocharge does not appear to be quite reasonable in terms of Gross's views on the formation of the homocharge [7, 8] as the lower fields may not be sufficient to bring about dielectric breakdown in the air at the dielectric–electrode interface. This may, however, be explained in terms of the constituents of the sealing wax used for these studies. This wax is made of rosin, an organic amorphous material, and barytes ($BaSO_4$) and TiO_2, inorganic crystalline materials as revealed by X-ray analysis reported earlier [11]. According to Wiseman and Feaster [9] the rosin gives positive moulding charges on stripping the metal foil electrodes and thus the presence of rosin in sealing wax may contribute to such positive charges. Besides this the presence of inorganic crystalline constituents in the amorphous organic insulating material rosin is expected to increase the value of the dielectric constant in view of the improved dielectric properties according to Herman and Minne [12]. These large values of the dielectric constant of sealing wax may be considered to decrease the effective electrical thickness of the thermoelectrets forming the central layer of the three-layer capacitor [9], so that even low values of applied potential differences are able to cause breakdown in the dielectric–electrode interface, resulting in the formation of a homocharge.

The positive charge observed in the case of unpolarized thermoelectrets can be explained in terms of the presence of the rosin in the sealing wax. The negative charge appearing on the surface of such thermoelectrets may, however, be attributed to the induced volume polarization resulting from the shielding of the thermoelectrets during storage. Such induced volume polarization, however, suffers thermal decay as the shielding reduces the internal field.

The variation of the finally retained surface charge density with polarizing field observed here can be explained by the formation of more homocharge, increasing in the polarizing field, whatever may be the origin of homocharge in these thermoelectrets. The slight deviations observed in the case of the series prepared at 35 °C can be due to the low value of the preparation temperature as the observed thermoelectret effect is less intense at low temperatures.

The results of these studies on the thermoelectret effect in sealing wax clearly confirm the ideas of Gross [7,8] on the nature of hetero- and homocharges, but, however, the origin of these charges in thermoelectrets like sealing wax, a mixture of organic and inorganic substances, cannot be understood completely with these studies alone. So, in order to understand the origin of these charges in sealing wax thermoelectrets the authors have extended their studies to the effect of the electrode material on the thermoelectret charge of the sealing wax thermoelectrets.

Acknowledgements

The authors are grateful to the Ministry of Education, Government of India, for the award of Senior Research Training Scholarship to one of them (R. Nath) which made this work possible. The authors are also thankful to the Director of the National Physical Laboratories, New Delhi (India), for providing library facilities.

References

[1] O. J. Johonson and P. H. Carr, Phys. Rev. **42**, 912 (1932).
[2] C. S. Bhatnagar and D. R. Bhawalkar, Proc. Indian Acad. Sci. **36**, 170 (1952).
[3] P. A. Thiessen, A. Winkel, and K. Herrmann, Phys. Z. **37**, 511 (1936).
[4] H. H. Wieder and S. Kaufman, J. appl. Phys. **24**, 156 (1953).
[5] S. Mikola, Z. Phys. **32**, 476 (1925).
[6] R. Nath, Ph.D. Thesis, Univ. Saugar, Sagar (India) 1965.
[7] B. Gross, J. chem. Phys. **17**, 866 (1949).
[8] B. Gross, Brit. J. appl. Phys. **1**, 259 (1950).
[9] G. G. Wiseman and G. R. Feaster, J. chem. Phys. **26**, 521 (1957).
[10] J. Handerek and T. Piech, Acta phys. Polon. **16**, 339 (1963).
[11] R. Nath and D. R. Bhawalkar, Indian J. pure appl. Phys. **9**, 10 (1971).
[12] G. Herman and R. Minne, Revue E. (Belgium) **3**, 339 (1962).

(Received August 16, 1971)

phys. stat. sol. (a) 8, 587 (1971)

Subject classification: 9; 1.2; 10.1; 21.1

Physics Department, Millikin University, Decatur, Illinois

Tracer Diffusion in the Chromium–Nickel System[1])

By

J. ASKILL

Self diffusion of Cr^{51} in pure chromium, Cr + 20 wt% Ni, Cr + 44 wt% Ni, and Cr + + 65 wt% Ni has been measured over the temperature ranges of 1000 to 1900, 1000 to 1525, and 900 to 1350 °C, respectively. Over the temperature range of 1000 to 1750 °C the Arrhenius plot for the self diffusion of Cr^{51} in pure Cr is linear and gave the relation $D = 1.6 \exp(-81.0/RT)$ cm²/s. The penetration plots for the experiments on pure chromium and the Cr rich alloys within ≈ 100 deg of the melting point were non linear, yielding two diffusion coefficients. By comparison with results from a single crystal specimen at high temperatures, the higher diffusion coefficients are shown to be due to some type of high temperature grain boundary diffusion process, possibly due to local melting at the grain boundaries. In addition to this, the Arrhenius plots for these systems all exhibit upwards curvature within ≈ 100 deg of the melting point. This is associated with a divacancy contribution to the diffusion process. The penetration plots and Arrhenius plot for the diffusion of Cr^{51} and Ni^{63} in the f.c.c. Cr + 65 wt% Ni alloy were all linear, typical of a normal face centered cubic system. No evidence was found for a high temperature structure of chromium other than b.c.c., either from the diffusion data or from associated X-ray experiments.

Die Selbstdiffusion von Cr^{51} in reinem Chrom, Cr + 20 Gew % Ni, Cr + 44 Gew % Ni und Cr + 65 Gew % Ni wurde in den Temperaturbereichen von 1000 bis 1900, 1000 bis 1525 bzw. 900 bis 1350 °C untersucht. Innerhalb des Temperaturbereiches von 1000 bis 1750 °C ist die Arrheniuskurve für die Selbstdiffusion von Cr^{51} in reinem Cr linear und ergab die Beziehung $D = 1{,}6 \exp(-81{,}0/RT)$ cm²/s. Die Eindringkurven für die Untersuchungen in reinem Chrom und den Chrom-reichen Legierungen waren bis zu ungefähr 100 Grad vom Schmelzpunkt nichtlinear und ergaben zwei Diffusionskoeffizienten. Durch Vergleich mit Ergebnissen von Einkristallproben bei hohen Temperaturen wurde gezeigt, daß die höheren Diffusionkoeffizienten durch eine Art von Hochtemperatur-Korngrenzendiffusionsprozess verursacht werden, wahrscheinlich durch lokales Schmelzen an Korngrenzen. Zusätzlich dazu zeigen die Arrheniuskurven dieser Systeme alle aufwärts gerichtete Krümmung bis etwa 100 Grad vom Schmelzpunkt. Dies ist verbunden mit dem Beitrag von Doppelleerstellen zum Diffusionsprozeß. Die Eindring- und Arrheniuskurven für die Diffusion von Cr^{51} und Ni^{63} in der k.f.z. Legierung Cr + 65 Gew% Ni waren alle linear und typisch für ein normales k.f.z.-System. Es wurden keine Hinweise für eine andere Hochtemperaturstruktur als die k.r.z.-Struktur für Chrom gefunden, weder aus den Diffusionswerten noch aus damit verbundenen Röntgenstrahlenuntersuchungen.

1. Introduction

Chromium has attracted considerable interest because of its refractory nature, the possibility of a high temperature phase change to a structure other than b.c.c.,

[1]) Work carried out at the Physics Department, Reading University, Reading, Berkshire, England (research sponsored by the British Ministry of Aviation), and Metals and Ceramics Division, Oak Ridge National Laboratory, Oak Ridge, Tennessee, (research sponsored by the U.S. Atomic Energy Commission under contract with Union Carbide Corporation).

and also because of the possibility that the activation energy for self diffusion may be temperature dependent. Numerous papers [1 to 4] have discussed the structure of this high temperature phase since the work of Bloom and Grant [5] in 1951, who first suggested the possibility of a high temperature f.c.c. phase in pure Cr.

The author has previously shown [6] that the activation energy for self diffusion in chromium from 1000 to 1600 °C is not temperature dependent as suggested by earlier self diffusion data. The primary purpose of this investigation was to extend previous self diffusion in Cr data [6] to 1900 °C, the melting point of chromium. In addition, if a high temperature phase change does exist, one would expect a noticeable change, probably a discontinuity, in the self diffusion coefficient at the transition temperature. It was also hoped to understand more fully the resulting high temperature diffusion data by studying the diffusion of Cr^{51} into some chromium rich Cr–Ni alloys.

2. Experimental Procedures

Diffusion specimens 1 cm in diameter and 8 mm long were prepared from rods of 99.98% pure chromium, and 20 wt% Ni, 44 wt% Ni and 65 wt% Ni alloys. The main impurity in the pure Cr was Fe (135 ppm). Preanneals at about nine-tenths of the melting point of these materials resulted in grain sizes of 2 to 3 mm. However, in the case of one of the pure chromium samples, a single crystal resulted. This was extremely fortunate as single crystals of chromium 1 cm in diameter were not available at the time of these experiments, and this enabled vital comparisons, of single crystal and polycrystalline penetration plots at high and low temperatures, to be performed.

The radioisotope Cr^{51} was applied to the prepared samples in the form of $CrCl_2$ in HCl solution and dried under a heat lamp. For the 65% Ni alloy the radioisotopes Cr^{51} and Ni^{63} were deposited electrolytically using a bath of saturated ammonium oxalate. In this case the thickness of the plated layer was about 1 μm for Ni^{63} and 0.1 μm for Cr^{51}. All of the diffusion anneals were carried out in a tantalum element furnace under an absolute pressure of two atmospheres of high purity argon, except for those on the 65% Ni alloy which were carried out in vacuo. A typical analysis of the argon used gave the impurity content as 1 ppm O_2 and 24 ppm N_2 + CO.

For the experiments using Cr^{51} in the 65 wt% Ni alloy, the samples were analysed by the standard serial sectioning method using a lathe. Before sectioning, the diameters of the samples were reduced by about 1 mm to eliminate the possibility of any surface effects. Between 10 and 15 sections were taken, the turning weighed, dissolved in acid, and assayed using a well type scintillation counter and a single channel analyser. Linear penetration plots were obtained extending over 2 to 3 orders of magnitude in concentration.

Autoradiography, essentially as described by Graham and Tomlin [7] (for the diffusion of Fe^{55} in Fe) was employed for the analysis of samples using the low energy Ni^{63} beta emitter.

The pure chromium, 20 wt% Ni, and 44 wt% Ni samples were also analysed by the serial sectioning method, although hand grinding was used as the method of layer removal. Details of this procedure have been reported previously [6], although they were refined for these experiments by using a balance capable of measuring ± 1 μg. This extended the range of measurable diffusion coefficients down to 10^{-13} to 10^{-14} cm²/s.

3. Experimental Results and Discussion

Self diffusion measurements in b.c.c. pure chromium have previously been reported by the author from 1000 to 1550 °C [6]. Over this temperature range the self diffusion Arrhenius plot was shown to be linear. The possibility of a phase change in chromium was first suggested by Bloom and Grant [5] in 1951. On the basis of metallographic and X-ray studies on various chromium rich Ni–Cr alloys they concluded that there was a phase change in pure b.c.c. chromium, to a f.c.c. structure at high temperatures. This suggestion seemed to be substantiated by the works of Vasyutinskii et al. [4] and Grigorev et al. [2, 3]. If there is, in fact, a phase change between 1650 °C (the highest self diffusion measurements previously made) and the melting point of chromium (1900 °C) one might expect an abrupt change in the diffusion coefficient at the transition temperature. In iron, for instance, there is a considerable discontinuity in the diffusion coefficient at both the b.c.c. to f.c.c. phase changes.

Over the temperature range of 1000 to 1750 °C, normal penetration plots were found. Even as high as 1450 °C, considerable grain boundary diffusion was observed for the polycrystalline samples. A typical penetration plot for a polycrystalline sample annealed at 1440 °C is shown in Fig. 1. Three anneals were carried out at lower temperatures (1200, 1079, and 1014 °C) using a single crystal sample. Good linear penetration plots were observed covering two or more orders of magnitude in concentration. The calculated diffusion coefficients were somewhat lower than those measured previously [6]. This is probably due to the fact that at temperatures below 1200 °C it is difficult to analyse accurately non-linear distance squared penetration plots into volume and grain boundary contributions, particularly when the grain boundary contribution is so important.

At temperatures above 1700 °C, an interesting phenomenon was observed [8]. The penetration plots for polycrystalline samples were found to be non-linear in a way similar to that observed at lower temperatures (< 1400 °C) where the non-linearity is associated with normal grain boundary diffusion. Comparisons were made of the penetration plots resulting from two anneals at

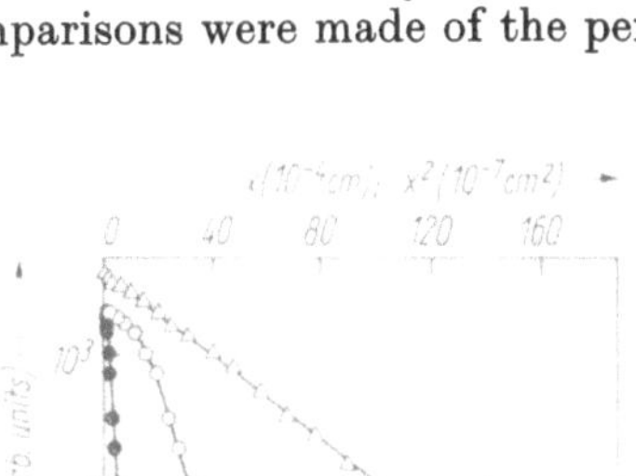

Fig. 1. Penetration plot for self diffusion in chromium at 1440 °C. △ x^2 plot (upper abscissa) after background subtraction (activity ×2); ○ x plot; ● x^2 plot (lower abscissa)

Fig. 2. Penetration plot for self diffusion in chromium at 1824 °C. Single crystal and polycrystal specimens

1824 °C, one with a polycrystalline sample, and the other with a single crystal. The result is shown in Fig. 2.

The curved penetration plot for the polycrystalline sample was analysed into two straight line regions. The steeper dotted line, formed by subtraction of the extrapolated deeper penetration was essentially parallel to the single crystal penetration plot, and thus gave the same value of the diffusion coefficient within 5%. Considerable activity of Cr^{51} was present at much greater depths in the polycrystalline samples than in the single crystal sample. This is probably due to some form of enhanced diffusion in the grain boundaries. The high temperature polycrystalline data are well fitted by a straight line on an x plot as Fisher's [9] analysis suggests for normal grain boundary diffusion. However, it was found impossible to distinguish between a linear x and a linear x^2 plot for the grain boundary regions of the penetration plots for the anneals at temperatures between 1750 and 1900 °C.

One possible explanation of this effect is that there is considerable segregation of impurities to the grain boundaries at high temperatures causing local melting in the grain boundaries and enhanced grain boundary diffusion. The vapor pressure of chromium at these temperatures is very high. Curved penetration plots, similar to those described here, have been reported by Kucera [10] in 1962, for the diffusion of Fe^{55} and Cr^{51} in polycrystalline samples of a multi-component Ni–Cr–Fe alloy at temperatures within 50 °C of the melting point.

However, a more important observation from the high temperature data is that the Arrhenius plot for the self diffusion coefficients that are associated with volume diffusion (i.e. smaller D values) was found to be non-linear. Upwards curvature was observed in the Arrhenius plot with a transition temperature of 1755 °C. No discontinuity was found in the diffusion coefficients so that the change at 1755 °C is thought not to be due to a structural phase change. The diffusion coefficients and temperatures are given in Table 1. Fig. 3 shows the Arrhenius plot of the diffusion coefficients together with the previous reported data for comparison. A least squares analysis of the data from 1000 to 1750 °C

Table 1

Diffusion coefficients and temperatures for the diffusion of Cr^{51} in pure chromium

temperature T(°C)	diffusion coefficient D(cm²/s)	temperature T(°C)	diffusion coefficient D(cm²/s)
1014 S	1.50×10^{-14}	1723	1.76×10^{-9}
1079 S	1.00×10^{-13}	1785	3.67×10^{-9}
1105	1.57×10^{-13}	1785	9.84×10^{-9}
1200 S	5.96×10^{-13}	1797	8.27×10^{-9}
1286	2.80×10^{-12}	1797	1.89×10^{-8}
1344	1.41×10^{-11}	1824 S	1.27×10^{-8}
1400	2.52×10^{-11}	1825	1.19×10^{-8}
1440	3.49×10^{-11}	1825	3.86×10^{-8}
1547	2.18×10^{-10}	1870	3.36×10^{-8}
1642	7.39×10^{-10}	1870	1.25×10^{-7}

S means single crystal

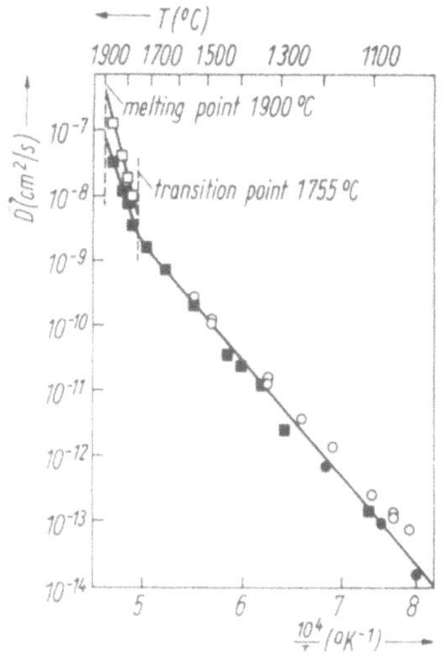

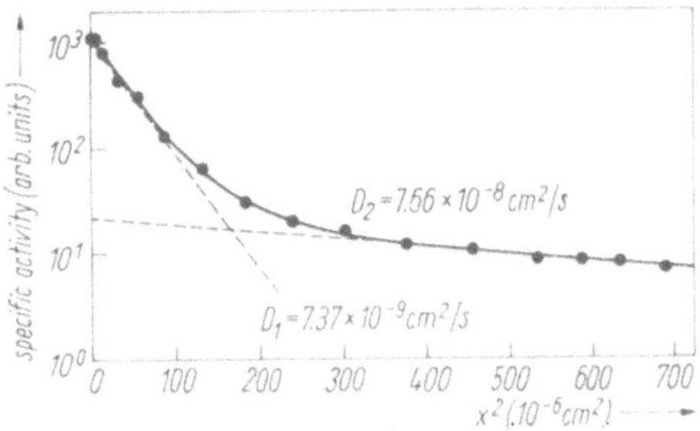

Fig. 4. Penetration plot for self diffusion of chromium in Cr + 20 wt% Ni alloy at 1454 °C

Fig. 3. Arrhenius plot for self diffusion in chromium. ○ [6]; □ polycrystals; ● single crystals

gave the relation

$$D = 1.6 \exp\left(\frac{-81.0}{RT}\right) \text{cm}^2/\text{s} .$$

The form of the Arrhenius plot for self diffusion of chromium within 150 deg of the melting point is very similar to that found for self diffusion in vanadium by Peart [11]. Since both V and Cr are similar in many respects it seems reasonable that the same explanation for the non-linearity in the Arrhenius plot may apply to both. Peart and Askill [12] have suggested that a combination of two diffusion processes, single vacancies and divacancies, could explain the observed non-linearity in the Arrhenius plot for not only Cr and V, but the other so-called "anomalous metals" Ti, Zr, Hf, and U as well.

3.1 Tracer diffusion of Cr^{51} in a 20 wt % Ni alloy

The penetration plots for the diffusion of Cr^{51} in a 20 wt% Ni alloy within 115 °C of its melting point were similar to those found for self diffusion near the melting point. The penetration plots for each of the three highest temperatures were all curved. Fig. 4 shows the penetration plot for the experiment at 1454 °C. As before, the curves were analysed into two linear regions yielding two diffusion coefficients, D_1 and D_2. The higher diffusion coefficients, D_2, is associated with the high temperature grain boundary effect.

At intermediate temperatures, from 1415 down to 1100 °C, the resultant penetration plots were all linear over several orders of magnitude in activity. The transition from a single penetration plot to a two part penetration plot occurs at 1415 °C in this particular alloys. As in the case of the self diffusion data, this is also the temperature above which there is upwards curvature in the Arrhenius plot for volume diffusion. This upwards curvature can also be explained as a divacancy contribution to the diffusion process.

At temperatures below 1100 °C, the Arrhenius plot is also non-linear. This is due to a change in phase of the alloy from the b.c.c. α-phase to the phase

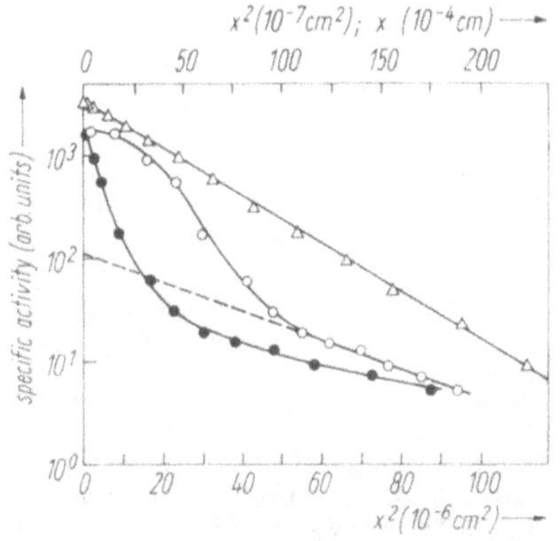

Fig. 5. Penetration plot for self diffusion of chromium in Cr + 20 wt% Ni alloy at 1091 °C. △ x^2 plot (upper abscissa) after subtraction (activity ×2); ○ x plot; ● x^2 plot (lower abscissa)

mixture α (b.c.c.) + γ (f.c.c.). Grain boundary diffusion is important below 1100 °C as can be seen from the penetration plot for the experiment carried out at 1091 °C, which is shown in Fig. 5. The figure shows an x and x^2 penetration plot. After subtraction of the grain boundary contribution, the x^2 penetration plot is linear over 2.5 orders of magnitude in activity. This phase transition temperature at 1150 °C confirms the chromium–nickel phase diagram shown in Fig. 1. This is denoted by the lowest full circle in Fig. 6.

Over the intermediate temperature range of 1100 to 1415 °C, the rate of diffusion of Cr^{51} in the 80 wt% Cr alloy is approximately the same as the rate of diffusion in pure nickel as given by Monma et al. [13]. Fig. 7 shows the Arrhenius plot of the data. The diffusion coefficients and temperatures are given in Table 2. In the course of these experiments, the melting point of the alloy was determined as 1525 ± 15 °C. This point is also shown on the phase diagram,

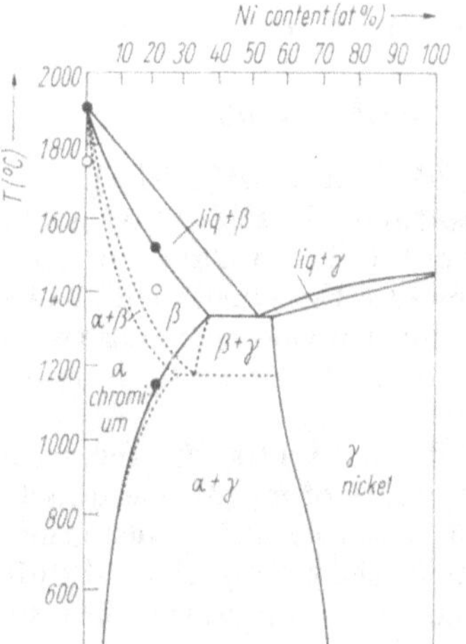

Fig. 6. The nickel–chromium phase diagram

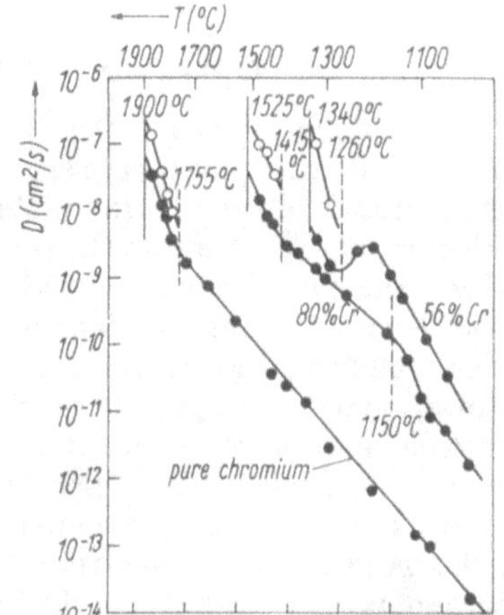

Fig. 7. Self diffusion of chromium in chromium–nickel alloys

Table 2

Diffusion coefficients and temperatures for the diffusion of Cr^{51} in Cr + 20 wt% Ni and Cr + 44 wt% Ni alloys

temperature T(°C)	diffusion coefficient D(cm²/s)	temperature T(°C)	diffusion coefficient D(cm²/s)
20 wt% Ni alloy		44 wt% Ni alloy	
1014	1.59×10^{-12}	1053	3.6×10^{-11}
1052	5.27×10^{-12}	1089	1.25×10^{-10}
1082	8.70×10^{-12}		
1095	1.71×10^{-11}	1131	4.95×10^{-10}
1122	6.03×10^{-11}	1155	1.08×10^{-9}
1160	1.57×10^{-10}		
1248	5.67×10^{-10}	1189	2.66×10^{-9}
1296	7.88×10^{-10}	1229	2.34×10^{-9}
1317	1.43×10^{-9}		
1368	2.49×10^{-9}	1289	1.45×10^{-9}
1400	2.89×10^{-9}	1289	1.20×10^{-8}
1435	6.38×10^{-9}	1321	3.67×10^{-9}
1435	3.50×10^{-8}		
1454	7.37×10^{-9}	1321	1.03×10^{-7}
1454	7.66×10^{-8}		
1477	1.49×10^{-8}		
1477	9.17×10^{-8}		

Fig. 6. The open circles in Fig. 6 represent the diffusion transition temperatures. No discontinuity was found in the diffusion rates at 1225 °C (the proposed α–($\alpha + \beta$) transition) or at 1450 °C (the proposed ($\alpha + \beta$)-β transition); furthermore, the solidus appears to be 100 °C lower (at 1525 °C) than proposed by Grant. However, it is possible that the α–($\alpha + \beta$) transition is not observable and that the transition at 1415 °C corresponds to the ($\alpha + \beta$)–β phase change.

A least squares analysis of the data from 1150 to 1415 °C gave the relation

$$D = 0.28 \exp\left(\frac{-62000}{RT}\right) \text{cm}^2/\text{s} .$$

3.2 Tracer diffusion of Cr^{51} in a 44 wt% Ni alloy

Experiments on a 44 wt% Ni alloy were carried out over the temperature range from 1050 to 1320 °C. This alloy was chosen since the phase diagram given in Fig. 6 shows a transition from the phase mixture $\alpha + \gamma$ to the phase mixture $\beta + \gamma$ at 1180 °C if, in fact, the β-Cr phase does exist.

Instead of simplifying the problem, the diffusion rates of Cr^{51} in the 44 wt% Ni alloy seem even more difficult to rationalize. At the two highest temperatures of 1289 and 1321 °C, the typical two-part penetration plot for the high temperature grain boundary effect was again observed. The penetration plots for intermediate temperatures were all linear. Fig. 7 shows the Arrhenius plot for the data. The diffusion coefficients and temperatures are given in Table 2. The transition from the two-part penetration plots to linear penetration plots occurred at 1270 °C with an apparent maximum in the diffusion coefficient occurring at about 1210 °C. It is not clear whether or not these changes could be associated with the proposed $(\alpha + \gamma)$–$(\beta + \gamma)$ phase change at 1180 °C. The melting point of this alloy was confirmed to be 1340 ± 20 °C during the course of the diffusion experiments.

The data for the diffusion of Cr^{51} in both the 20 wt% and 44 wt% Ni alloys are compared with the self diffusion data in Fig. 7.

3.3 Diffusion of Cr^{51} in a 65 wt% Ni alloy

The penetration plots for the diffusion of both Cr^{51} and Ni^{63} in the f.c.c. 65 wt% alloy were all linear over several orders of magnitude in concentration. The diffusion coefficients and temperatures are given in Table 3 and the data are shown in Fig. 8.

Table 3

Diffusion coefficients and temperatures for the diffusion of Cr^{51} and Ni^{63} in a Cr + 65 wt% Ni alloy

Ni^{63}		Cr^{51}	
temperature T (°C)	diffusion coefficient D (10^{-11} cm²/s)	temperature T (°C)	diffusion coeffcient D (10^{-11} cm²/s)
949	0.98	890.5	0.274
1024	2.00	933	0.353
1120	10.7	969	0.512
1220	28.8	989	1.275
1259	35.5	1015	3.48
1259	36.8	1052	4.34
1270	58.6	1100	6.60
1293	56.2	1140	16.9
1317	101	1219	49.1
1321	115	1240	69.3
1360	125	1279	77.5
1360	160	1318	196
		1352	259.5

A least squares analysis of the data gave the following expressions:

$$D_{Cr} = \left(0.2\begin{matrix}+0.2\\-0.1\end{matrix}\right) \exp\left(\frac{-58600 \pm 2500}{RT}\right) \text{cm}^2/\text{s}\,,$$

$$D_{Ni} = \left(4\begin{matrix}+4\\-2\end{matrix}\right) \times 10^{-3} \exp\left(\frac{-49000 \pm 2500}{RT}\right) \text{cm}^2/\text{s}\,.$$

The above values of Q and D_0 for the diffusion of Cr^{51} in the f.c.c. 65 wt% Ni alloy agree very well with those calculated by the writer from the data of Grimes et al. [14] for the diffusion of Cr^{51} in a 80 wt% Ni alloy ($Q = 59.5$ kcal/mol, $D_0 = 0.21$ cm²/s). The results of Monma [13], however, for a 70 wt% Ni alloy

Fig. 8. Arrhenius plot for the diffusion of Cr^{51} (●) and Ni^{63} (○) in Cr + 65 wt% Ni alloy

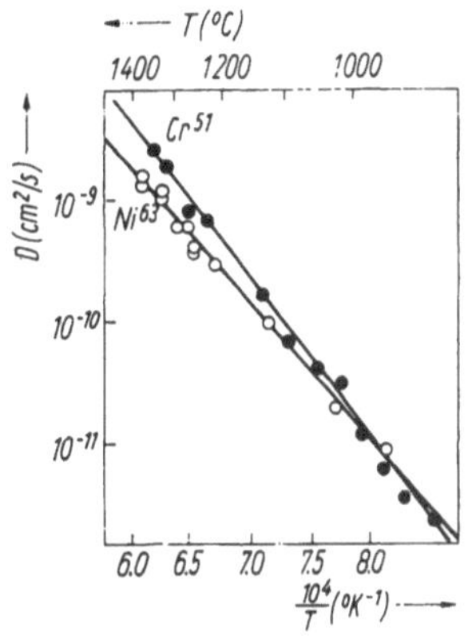

differ somewhat. This is probably due to limited temperature range of Monma's data as the actual D-values are in good agreement. Most of the available experimental data gives values of the diffusion coefficient for the diffusion of Cr^{51} in the nickel rich alloys, which vary very little with increasing Cr content up to 35 wt%. Typically at 1200 °C, the diffusion coefficient varies from about 1.5×10^{-10} to 4.5×10^{-10} cm²/s.

Using the residual activity method, Gruzin and Federov [15] found the following data for an 80 wt% Ni alloy. $D_{Cr} = 0.1 \exp(-58000/RT)$ and $D_{Ni} = 4 \times 10^{-3} \exp(-48000/RT)$ cm²/s in excellent agreement with the present data. However, the addition of 2 wt% Ti increased the activation energy to 81 kcal/mol for the diffusion of Cr in the alloy.

Diffusion of Cr^{51} in several other Ni rich alloys was studied by Monma et al. [13] in 1964. Their data is given together with all other available data on the diffusion of Cr^{51} in chromium–nickel alloys in Table 4.

Table 4

Summary of data on self diffusion of Cr^{51} in chromium–nickel alloys

Cr (wt%)	melting point (°C)	activation energy Q (kcal/mol)	Q/T_m (cal/mol deg)	frequency factor D_0 (cm²/s)	reference No., year
0	1455	65.1	37.7	1.1	[13] 1964
6.35	1440	50.5	29.6	0.01	[18] 1963
10.0	1430	66.5	39.0	1.4	[13] 1964
11.69	1426	54.7	32.2	0.037	[18] 1963
19.8	1405	58.0	34.5	0.1	[15] 1955
19.8	1405	59.5	35.5	0.21	[16] 1961
19.9	1405	67.7	40.3	1.9	[13] 1964
20.3	1403	66.0	39.3	2.0	[15] 1955
20.4	1403	64.0	38.2	0.5	[17] 1957
29.7	1380	69.4	41.9	3.2	[13] 1964
35.4	1365	58.6	35.8	0.2	this work
80.0	1525	62.0	34.5	0.28	this work
100	1890	81.6	37.8	1.6	this work

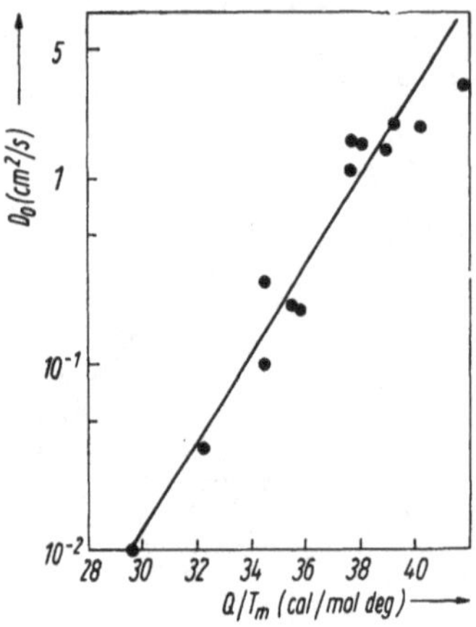

Fig. 9. Variation in lg D_0 with Q/T_m for the self diffusion of chromium in chromium–nickel alloys

It is instructive to plot the variation of log D_0 with the reduced activation energy Q/T_m where T_m denotes the melting point of the alloy. The diffusion coefficient D at any temperature T is given by the relation $D = D_0 \exp(-Q/RT)$. If D_{T_m} denotes the diffusion coefficient at the melting point, T_m, then $\log D_{T_m} = \log D_0 - Q/RT$ or $\log D_0 = \log D_{T_m} - Q/RT_m$.

If the diffusion coefficient at the melting point for the various alloys is approximately constant, we may write

$$\log D_0 = \text{constant} + \frac{1}{R}\frac{Q}{T_m}.$$

A plot of lg D_0 versus Q/T_m is shown in Fig. 9. The graph is a good linear fit with slope (theoretically $= 1/R = 0.51$) equal to 0.54. The value of the constant is 8.3×10^{-8}.

References

[1] E. P. Abrahamson and N. J. Grant, Trans. AIME **206**, 975 (1956).
[2] A. T. Grigorev, L. N. Guseva, E. M. Sokolovskaya, and M. V. Maksimova, Zh. neorg. Khim. **4**, 984 (1959).
[3] A. T. Grigorev, E. M. Sokolovskaya, Y. P. Simanov, and I. G. Sokolova, Zh. neorg. Khim. **5**, 1036 (1960).
[4] B. M. Vasyutinskii, G. N. Kartmazov, and V. A. Finkel, Fiz. Metallov i Metallovedenie **12**, 771 (1961).
[5] D. S. Bloom and N. J. Grant, Trans. AIME **191**, 1009 (1951).
[6] J. Askill and D. H. Tomlin, Phil. Mag. **11**, 467 (1965).
[7] D. Graham and D. H. Tomlin, Phil. Mag. **8**, 1581 (1963).
[8] J. Askill, Appl. Phys. Letters **9**, 82 (1966).
[9] J. C. Fisher, J. appl. Phys. **22**, 74 (1951).
[10] J. Kucera, Czech. J. Phys. **B12**, 862 (1962).
[11] R. F. Peart, J. Phys. Chem. Solids **26**, 1853 (1965).
[12] R. F. Peart and J. Askill, phys. stat. sol. **23**, 263 (1967).
[13] K. Monma, H. Suto, and H. Oikawa, J. Japan Inst. Metals **28**, 188 (1964).
[14] W. R. Grimes, G. M. Watson, J. H. DeVan, and R. B. Evans, Report U.S.A.E.C. T.I.D.-7610, 1961 (p. 63).
[15] P. L. Gruzin and G. B. Federov, Dokl. Akad. Nauk SSSR **105**, 264 (1955).
[16] D. F. Kalinovich, I. I. Kovenskii, and M. D. Smolin, Fiz. Metallov i Metallovedenie **11**, 307 (1961).
[17] A. D. Tyutyunnik and G. V. Estulin, Fiz. Metallov i Metallovedenie **4**, 558 (1957).
[18] G. B. Fedorov, E. A. Smirnov, and F. I. Zhomov, Metall. i Metalloved. Chistykh. Metall Sbornik Nauk Rabot. **4**, 110 (1963).

(Received October 18, 1971)

phys. stat. sol. (a) 8, 597 (1971)

Subject classification: 13.3 and 14.3.2; 22.1.2

Electronics Research Laboratory, Affiliated with SINTEF, Technical University of Norway, Trondheim

Cole-Cole Plotting of Surface State Admittance in MIS Capacitors

By

P. CHR. MALMIN

Information about surface states on insulator-semiconductor interfaces can be obtained from Cole-Cole plots of the surface state admittance. With this method, the surface state density can be determined with good accuracy regardless of the time constant dispersion. The other parameters of the surface states may also be obtained with this method, as shown by measurements on different dielectrics.

Informationen über Oberflächenzustände in Isolator-Halbleiter-Grenzflächen können von Cole-Cole-Kurven der Admittanz der Oberflächenzustände erhalten werden. Mit dieser Methode kann die Oberflächenzustandsdichte genau bestimmt werden, unabhängig von der Streuung der Zeitkonstante. Die anderen Parameter der Oberflächenzustände lassen sich ebenfalls mit dieser Methode erhalten, wie Messungen an verschiedenen Isolatoren zeigen.

1. Introduction

In this paper a new method to determine the interface properties of MIS structures will be described. This new method may be considered as an alternative to the MOS-conductance method [1], or to an analysis of interface properties by high frequency C–U measurements [2]. With the last method, densities of surface states greater than some 10^{11} cm^{-2} eV^{-1} may be determined by comparing the measured high frequency C–U curve with the theoretical curve for the same capacitor, without surface states. The time constant of the surface states cannot be obtained in this way [3], but may be determined both with the conductance method and the method proposed in this paper.

By use of the conductance method, the range of surface state densities that can be accurately measured, has been extended to below 10^{11} cm^{-2} eV^{-1} [1]. With the conductance method, however, the extraction of the surface state parameters from the measured data is cumbersome and time consuming. The present paper shows how the surface state density, their time constant and the surface potential may be obtained more directly from a Cole-Cole plot [4] of the surface state admittance.

In a recent work [5], it was shown that a Cole-Cole plot of the complex MOS capacitance could yield qualitative data on surface state parameters. However, this plot could not be interpreted in the same general way that is possible with this new method. The reason is that the oxide and semiconductor capacitances were not separated before the presentation of the data.

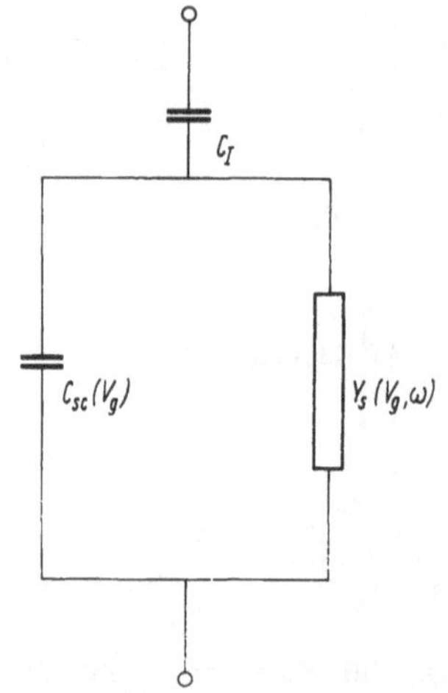

Fig. 1. Equivalent circuit of the MIS capacitor

2. Calculation of Surface State Admittance

An MIS capacitor, with the semiconductor surface biased in the depletion region, can be described by the equivalent circuit shown in Fig. 1. C_I is the insulator capacitance, C_{sc} the depletion layer capacitance, and Y_s the surface state admittance. This admittance can be considered as consisting of a capacitance C_p with a parallel conductance G_p:

$$Y_s = G_p + j\,\omega\,C_p\,. \tag{1}$$

For the case that the surface states are situated on a single energy level at the semiconductor–insulator interface, C_p and G_p are given as functions of frequency by the Debye equations [1].

In the case that the surface states are distributed in energy, these equations must be integrated over all the surface state energies to yield the admittance Y_s. Assuming the surface state density to be a slowly varying function of energy, the following result is obtained [1]:

$$Y_s = \frac{q\,N_{ss}}{2\,\tau_m} \ln\,[1 + (\omega\,\tau_m)^2] + j\,\frac{q\,N_{ss}}{\tau_m} \arctan\,(\omega\,\tau_m)\,, \tag{2}$$

where $\tau_m = \dfrac{e^{-u_s}}{c_n\,n_b}$, u_s is normalized surface potential in units of kT/q, c_n capture probability of the surface states (cm^3/s), n_b bulk majority carrier concentration (cm^{-3}), q electronic charge, k Boltzmanns constant, T temperature (°K), N_{ss} surface state density ($cm^{-2}\,eV^{-1}$), ω angular frequency (rad/s).

Measurements on the Si–SiO_2 system [1] have shown that G_p/ω varies more slowly with frequency than given by these two descriptions. Two explanations of this experimental result have been introduced. The calculated Cole-Cole plots for these two models will now be presented.

According to Nicollian and Goetzberger [1], one may assume that there are statistical fluctuations in the semiconductor surface potential, characterized by a Gaussian distribution with a standard deviation σ. In Fig. 2a are shown the Cole-Cole plots obtained for $\sigma = 1$, 2, and 3 units of kT/q. The curve labelled $\sigma = 0$ is obtained from equation (2). It is seen that for all values of σ, the distance between the crossover points with the real axis is equal to $q\,N_{ss}$.

The ratio α, defined as

$$\alpha = \frac{(G_p/\omega)_{\max.}}{q\,N_{ss}}, \tag{3}$$

can be used to determine both the standard deviation σ and the product $\omega\,\tau_m$ at the maximum value of G_p/ω. This is shown in Fig. 2b and c, respectively. The surface state time constant is here defined as follows:

$$\tau_m = \frac{e^{-\bar{u}_s}}{c_n\,n_b}, \tag{4}$$

where $\bar{u}_s$ is the mean value of the surface potential.

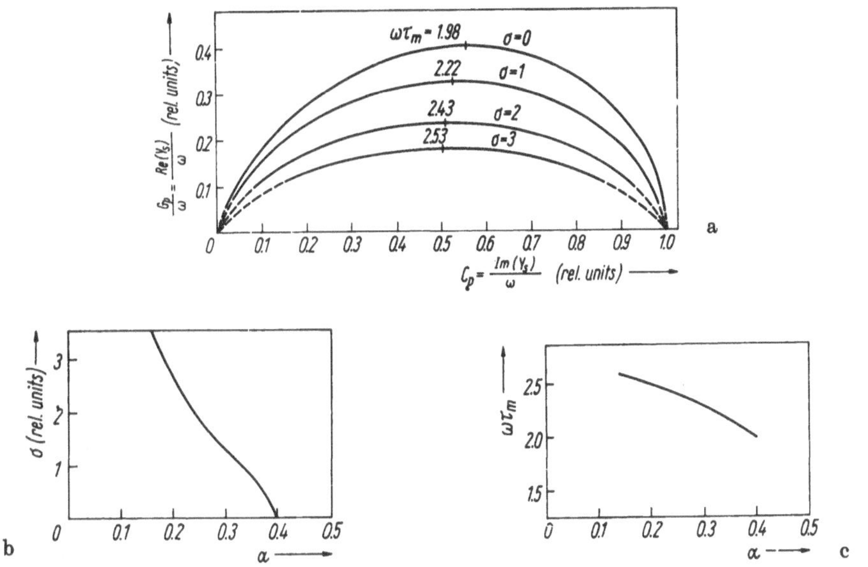

Fig. 2. a) Cole-Cole plots of the surface state admittance in units of $q\,N_{ss}$, calculated for different values of the standard deviation of the surface potential (σ in units of kT/q). The part of each curve that may be scanned within a frequency range of 3 decades is indicated by the solid lines; b) standard deviation of the surface potential, σ, in units of kT/q, as a function of the geometrical ratio α (equation (3)); c) the value of the product $\omega\,\tau_m$ at $(G_p/\omega)_{max}$. as a function of the ratio α

The other model is introduced by Preier [6], who assumed that the surface states are distributed within a certain distance, d, from the interface. Any statistical fluctuations are not considered in his description. In Fig. 3a are shown the Cole-Cole plots obtained for 4 different depths of the distribution of the surface states. The distances d are given relative to $(2\,k_0)^{-1}$, where k_0 is the attenuation constant for the electron wave function. For a barrier height of a few eV, $(2\,k_0)^{-1}$ is ≈ 1 Å [6]. Here, the curve labelled $d = 0$ is obtained from equation (2). It can be seen, that also in this case, the maximum value of C_p is equal to $q\,N_{ss}$. The value of d, and the product $\omega\,\tau_m$ at the maximum value of G_p/ω, are determined from the ratio α only, as shown in Fig. 3b and c. Here the surface state time constant is defined as

$$\tau_m = \frac{e^{-u_s}}{c_{no}\,n_b}, \tag{5}$$

where c_{no} is the capture probability for states located just at the interface.

3. Measurement Procedure

A measurement procedure based on the calculations above, will now be presented. With this procedure, the surface state parameters N_{ss}, τ_m, σ (or d), and c_n may be determined.

At a constant gate voltage, biasing the semiconductor surface into the depletion region, the small signal capacitance and conductance are measured at different frequencies. The insulator capacitance, measured in strong accumulation, is subtracted from these measurements. What is then obtained, is

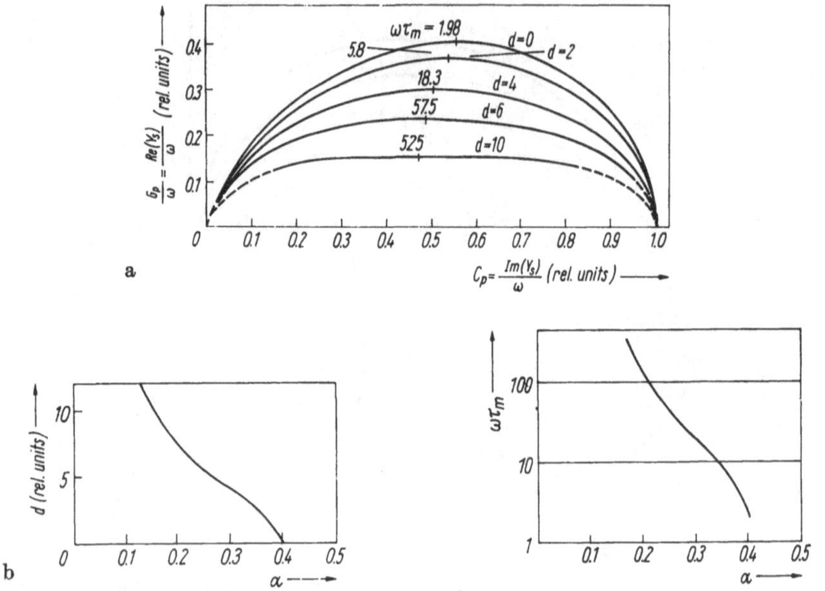

Fig. 3. a) Cole-Cole plots of the surface state admittance in units of $q\,N_{ss}$, calculated for different values of the depth of the distribution of states, d (in units of $(2\,k_0)^{-1}$); b) depth of the distribution of states, d, in units of $(2\,k_0)^{-1}$ as a function of the geometrical ratio α; c) the value of the product $\omega\,\tau_m$ at $(G_p/\omega)_{max.}$ as a function of the ratio α

the series admittance Y, where

$$Y = j\,\omega\,C_{sc} + Y_s\,, \tag{6}$$

$$Y = j\,\omega\,C_{sc} + j\,\omega\,C_p + G_p\,. \tag{7}$$

Provided the necessary frequency range is available, the following information can be obtained from a plot of $\mathrm{Re}(Y)/\omega$ versus $\mathrm{Im}(Y)/\omega$ in the complex plane:

1. The distance between the end points of the obtained Cole-Cole plot is equal to $q\,N_{ss}$, giving directly the surface state density.
2. The distance from origin to the high frequency limit of the arc is equal to C_{sc}. If the bulk majority carrier concentration is known, the corresponding value of $\bar{u}_s$ is determined from C_{sc}.
3. In order to distinguish between the two models described in connection with Fig. 2 and 3, the general shape of the Cole-Cole plot must be compared with the calculated curves, or the measurements made at different temperatures. It is not the intention here to decide which model is generally valid.
4. The ratio α (equation (3)) gives directly the standard deviation of the surface potential (Fig. 2b) or the depth of the distribution of states within the insulator (Fig. 3b), depending on which model is found to give the best description of the system.
5. The ratio α, together with the measured frequency at $(G_p/\omega)_{max.}$, gives the surface state time constant (Fig. 2c or 3c).
6. When $\bar{u}_s$ and n_b are determined, the capture probability c_n (or c_{no}) can be found from the time constant τ_m.

This measurement procedure must be repeated for different gate voltages to obtain the density of states and their capture probability for different energy levels.

4. Discussion

A condition for the Cole-Cole method to give this information on the surface states, is that the frequency range of the measurement equipment must be sufficient to determine the end points of the arc with a satisfactory degree of accuracy. In Fig. 2a and 3a are indicated the part of the curves that may be scanned within a frequency range of 3 decades. For σ less than 2 (Fig. 2a) or d less than 6 (Fig. 3a), the necessary extrapolation should give N_{ss} with an accuracy of a few per cent. A broader frequency range is needed when the time constant dispersion is larger, or a greater part of the energy gap is to be scanned.

This is true when no presuppositions are made on which model is valid, or how σ or d varies with the surface potential. In this general case, a similar frequency range is necessary with the conductance method [1]. MOS conductance measurements over a frequency range of 6 decades have been reported [7], and indicate that available measurement equipment does not seriously limit the frequency range. Only by assuming the statistical model to be valid, with the standard deviation independent of the energy level [8], is it possible to reduce the necessary frequency range. However, this assumption is not generally valid [1, 7].

When only the conductance is used for the analysis, both u_s, c_n, and σ (or d) must be determined before N_{ss} can be calculated [1, 8]. By plotting the measurements in the complex plane as shown in this paper, N_{ss} is obtained directly without the need to know the other parameters.

These may be determined from Fig. 2b, c and 3b, c without any computer analysis. The accuracy by which this last information is obtainable, is again dependent on the frequency range. However, failure to obtain capture probability and time constant dispersion due to experimental limitations, does not necessarily exclude an estimation of the surface state density, as would be the case with the conductance method.

To summarize, the Cole-Cole method gives a more direct determination of the surface state density, and a simpler, but possibly not as accurate, estimation of the other surface state parameters, compared to the conductance method. However, a somewhat larger frequency range may be necessary to obtain more accurate results, partly due to the extrapolation necessary and partly due to the fact that no presuppositions on the properties of the surface states are implied in the interpretation of the data. On the other hand, no separate determination of the relationship between gate voltage and surface potential is necessary.

5. Measurements

To demonstrate the method, a few measurements will be shown. Fig. 4 shows surface state admittance measurements on a MIS capacitor with RF-sputtered Al_2O_3. The frequency range was 100 Hz to 100 kHz, and the measurements were done at 4 different gate voltages. Within this voltage range, no change in depletion layer capacitance could be detected, due to the high density of surface states, approximately 9×10^{12} cm^{-2} eV^{-1}. These measurements were not accurate enough to yield the other parameters of the surface states.

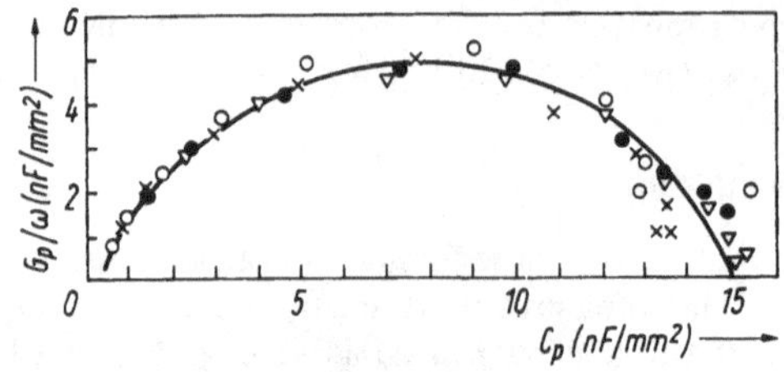

Fig. 4. Measurements on an Al–Al_2O_3–Si capacitor. $C_I = 600$ pF/mm², $N_{ss} = 9 \times 10^{12}$ cm⁻² eV⁻¹. Little dependence of applied gate voltage on C_{sc}. Applied voltage: ○ 0.5; × 1.0; ● 1.5; ▽ 2.0 V

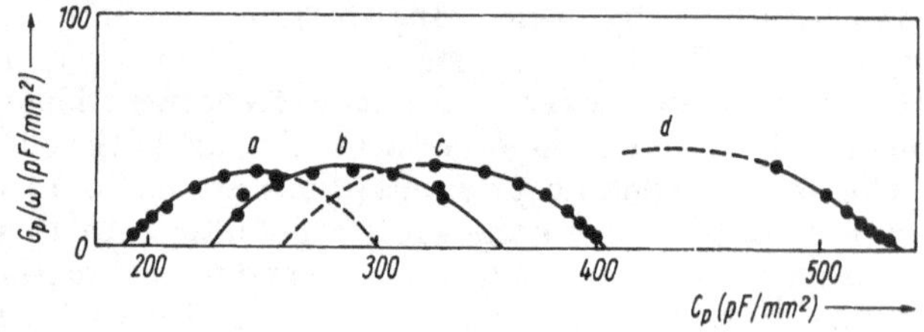

Fig. 5. Measurements on an Al–SiO_2–Si capacitor. $C_I = 300$ pF/mm², n-type Si, orientation ⟨111⟩, $n_b = 1.3 \times 10^{15}$ cm⁻³. Gate voltage range −2.5 V (a) to −2.0 V (d), giving a variation of the depletion layer capacitance between 190 pF/mm² (a) and somewhere around 350 pF/mm² (d)

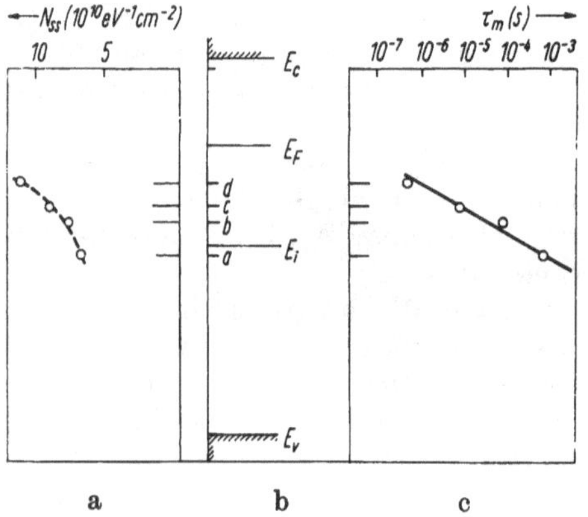

Fig. 6. Results of measurements on MOS capacitor; a) The density of states as a function of energy level; b) energy diagram of the semiconductor surface. The energy levels a, b, c, and d correspond to the four gate voltages at which the measurements were made; c) the time constant of the surface states as a function of energy level

Due to the limited frequency range, only parts of the Cole-Cole plots of the surface state admittance in a MOS capacitor with thermally grown SiO_2 could be obtained, as shown in Fig. 5. In spite of this limitation, an estimation of the surface state density and the time constant was made, as shown in Fig. 6. Assuming the statistical model to apply to this sample, a capture cross section of 4.5×10^{-15} cm² was obtained, in good agreement with other publications [1, 7].

6. Conclusions

In this paper, a new method to determine the surfacc state parameters in MIS capacitors is presented. It is based on a plot of the admittance Y_s of the surface states in the complex plane. It is shown how the various parameters of the surface states may be determined independent of each other, and more directly than in other methods. The requirements on the frequency range may be more demanding with this method than with the conductance method, but on the other hand, no presuppositions on the properties of the system are implied in the data analysis.

Acknowledgements

The author wishes to acknowledge the encouraging guidance of this work by I. Johansen, the valuable discussions with S. Bergmann and K. Kristiansen, and the technical assistance from T. Rolfsen, E. Myhre, and T. Andresen. The Central Institute for Industrial Research in Oslo is acknowledged for providing MOS capacitors with thermally grown SiO_2.

References

[1] E. H. Nicollian and A. Goetzberger, Bell Syst. tech. J. **46**, 1055 (1967).
[2] L. M. Terman, Solid State Electronics **5**, 285 (1962).
[3] K. H. Zaininger and G. Warfield, IEEE Trans. Electron Devices **12**, 179 (1965).
[4] K. S. Cole and R. H. Cole, J. chem. Phys. **9**, 341 (1941).
[5] R. Nannoni, phys. stat. sol. (a) **3**, 235 (1970).
[6] H. Preier, Appl. Phys. Letters **10**, 361 (1967).
[7] W. Fahrner and A. Goetzberger, Appl. Phys. Letters **17**, 16 (1970).
[8] E. H. Nicollian, A. Goetzberger, and A. D. Lopez, Solid State Electronics **12**, 937 (1969).

(Received September 23, 1971)

phys. stat. sol. (a) 8, 605 (1971)

Subject classification: 18.2.1; 21.1; 21.1.1

Laboratory of Ferromagnetics, Institute of Physics, Polish Academy of Sciences, Poznań

Measurement of H_k and H_c in Thin Ferromagnetic Films by the M_s Vibration Method in Low Frequency Field

By

Z. SCZANIECKI, R. GONTARZ, and H. RATAJCZAK

A method is presented of measuring the uniaxial anisotropy field H_k and coercive force H_c by forced vibrations of the magnetization vector M_s in two mutually perpendicular fields lying in the film plane, one of which, of low intensity, varies sinusoidally. The vibration amplitude of M_s exhibits maxima as a function of the dc field from which positions H_k or H_c can be determined according to the case whether the easy axis is directed perpendicular or parallel to the dc field. Theoretical predictions are confirmed experimentally using the planar galvanomagnetic effect for measuring the M_s vector amplitude. The influence of the amplitude of an ac field on the positions of the maxima is discussed.

Es wird eine Methode für die Messung des uniaxialen Anisotropiefeldes H_k und der Koerzitivkraft H_c durch erzwungene Schwingungen des Magnetisierungsvektors M_s in zwei zueinander senkrechten in der Schichtebene liegenden Feldern, von denen eins sich mit niedriger Intensität sinusförmig ändert, angegeben. Die Schwingungsamplitude von M_s zeigt in Abhängigkeit vom Gleichfeld Maxima, aus deren Lage H_k oder H_c bestimmt werden kann, in Übereinstimmung damit, ob die leichte Achse senkrecht oder parallel zum Gleichfeld gerichtet ist. Theoretische Vorhersagen werden experimentell durch Messung der Amplitude des M_s-Vektors mittels des ebenen galvanomagnetischen Effekts bestätigt. Der Einfluß der Amplitude des Wechselfeldes auf die Lage der Maxima wird diskutiert.

1. Introduction

The anisotropy field H_k and coercive force H_c are parameters of thin ferromagnetic films frequently studied. A review of some methods of H_k measurement was given by Malek and Schüppel [1]. A method of H_k and H_c determination using the planar galvanomagnetic effect is described by Kuritsyna and Vu Din Ky [2]. An extension of this method for H_k measurement was proposed by us previously [3]. The magnetoresistance effect can also be used for the determination of both the anisotropy field and coercive force [4, 11]. Ferromagnetic resonance in the frequency range from 100 to 1500 MHz has been used since long for measuring i.a. the anisotropy field H_k [5]. Somewhat later much lower frequencies, ranging from 1 to 20 MHz, were applied for this purpose [6 to 8]. Using this method, Andrä et al. obtained lower values of H_k than those obtained by the torsional method [8]. This discrepancy inexplicable in terms of ferromagnetic resonance remains unsolved so far. Frumkin [9] altogether denies the possibility of ferromagnetic resonance in real films at radio frequencies. Andrä and Klingholz [12] have shown that, in accordance with the Stoner-Wohlfarth theory [13], an increase of energy absorption in the film take place if the vector M_s of the resulting magnetic field (superposition of a dc field perpendicular to the easy axis and an ac field parallel to it) intersects the asteroid.

In this paper the authors propose a procedure of H_k and H_c determination by the method of forced vibrations of the magnetization vector $\boldsymbol{M}_s$. The system of magnetic fields applied here is similar to that used in the method of ferromagnetic resonance. Major differences are: a low frequency ω of the ac field such that magnetization reversal can be considered as quasi-static, and an amplitude of this field sufficiently large to generate forced vibrations of the magnetization vector $\boldsymbol{M}_s$ at the ac field frequency. If the ac field $H_{||}^{\sim}$ is applied along the easy axis of magnetization, the vibration amplitude of $\boldsymbol{M}_s$ as a function of the field $H_\perp$ directed perpendicularly to the easy axis has maxima from the positions of which H_k can be determined. If however the ac field is directed perpendicularly to the easy axis ($H_\perp^{\sim}$), the vibration amplitude of $\boldsymbol{M}_s$ in function of H varies stepwise at $|H_{||}| = H_c$.

2. Theoretical

The free magnetic energy density of a thin single-domain ferromagnetic film with uniaxial anisotropy in a dc magnetic field $H_\perp$ perpendicular to the easy axis and an ac field with amplitude $H_{||}^{\sim}$ parallel to this axis, is given by the expression

$$E = -H_\perp M_s \sin\varphi - H_{||}^{\sim} M_s \cos\varphi \sin\omega t + K_u \sin^2\varphi\,, \tag{1}$$

where ω is the ac field frequency, M_s the spontaneous magnetization, φ the angle between $\boldsymbol{M}_s$ and the easy axis, K_u the uniaxial anisotropy constant. At frequencies ω so low that magnetization reversal can be considered as quasi-static [10], under the condition for the position of momentary equilibrium of $\boldsymbol{M}_s$, we have

$$h_\perp = h_{||}^{\sim} \operatorname{tg}\varphi \sin\omega t + \sin\varphi\,, \tag{2}$$

where $h_\perp = H_\perp/H_k$, $h_{||}^{\sim} = H_{||}^{\sim}/H_k$, and $H_k = 2\,K_u/M_s$. Since $-1 \leq \sin\omega t \leq$ $\leq +1$, the vector $\boldsymbol{M}_s$ oscillates between limiting positions, which for a given $h_\perp$ occur at $+h_{||}^{\sim}$ and $-h_{||}^{\sim}$. The angle φ determines these limiting positions. Denoting by φ_+ the angle corresponding to the value $\sin\omega t = +1$ and by φ_- that corresponding to $\sin\omega t = -1$, we can find the dependence of φ_+ and φ_- on $h_\perp$. Fig. 1 shows this dependence for $h_{||}^{\sim} = \pm 0.01$. The graph of $\varphi_+(h_\perp)$ is continuous, whereas that of $\varphi_-(h_\perp)$ exhibits a discontinuity. The value $h_\perp = h_{cr.}$ for which the stepwise change of φ_- takes place, determined from the asteroid, is

$$h_{cr.} = (1 - h_{||}^{\sim 2/3})^{3/2}\,. \tag{3}$$

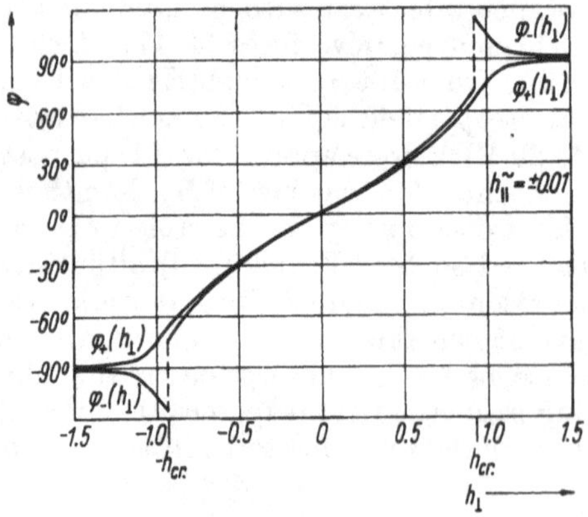

Fig. 1. Extremal positions of the magnetization vector $\boldsymbol{M}_s$ defined by the angle φ vs. the magnetic field $h_\perp$ perpendicular to the easy axis. Angles φ_+ and φ_- refer to amplitudes $+h_{||}^{\sim}$ and $-h_{||}^{\sim}$ of the ac field $h_{||}^{\sim} \sin\omega t$

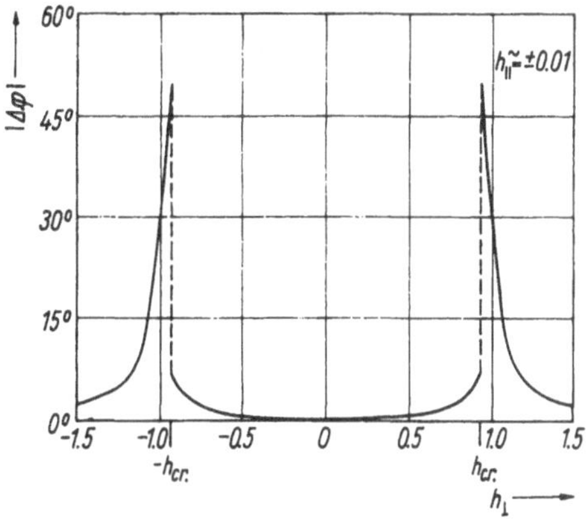

Fig. 2. Amplitude of forced vibrations of the vector M_s vs. the perpendicular dc field $h_{\perp}$

Fig. 2 shows the dependence of $\Delta\varphi = |\varphi_- - \varphi_+|$ on $h_{\perp}$ determined graphically from Fig. 1. It is readily seen that an ac field $H_{\parallel}^{\sim}$ amounting to as little as 0.01 H_k can set the magnetization vector in vibration, and that $\Delta\varphi$ at maxima, i.e. for $h_{\perp} = \pm h_{cr.}$, attains 50°. The shape of $\Delta\varphi$ vs. $h_{\perp}$ resembles the absorption curve at radio frequencies [6, 8].

The anisotropy field H_k calculated from (3) amounts to

$$H_k = (H_{cr.}^{2/3} + H_{\parallel}^{\sim 2/3})^{3/2} . \tag{4}$$

The value of $H_{cr.}$ at which the maximum of M_s vibration amplitude occurs, tends to H_k as the ac field amplitude $H_{\parallel}^{\sim}$ tends to zero. This is illustrated in Fig. 3, where the vibrations of φ vs. $h_{\perp}$ are compared for $h_{\parallel}^{\sim} = \pm 0.01$ and $h_{\parallel}^{\sim} = \pm 0.001$.

Now we assume the ac field $H_{\perp}^{\sim}$ perpendicular and the dc field $H_{\parallel}$ parallel to the easy axis. Equation (1) becomes

$$E = -H_{\perp}^{\sim} M_s \sin \varphi \sin \omega t \pm H_{\parallel} M_s \cos \varphi + K_u \sin^2 \varphi \tag{5}$$

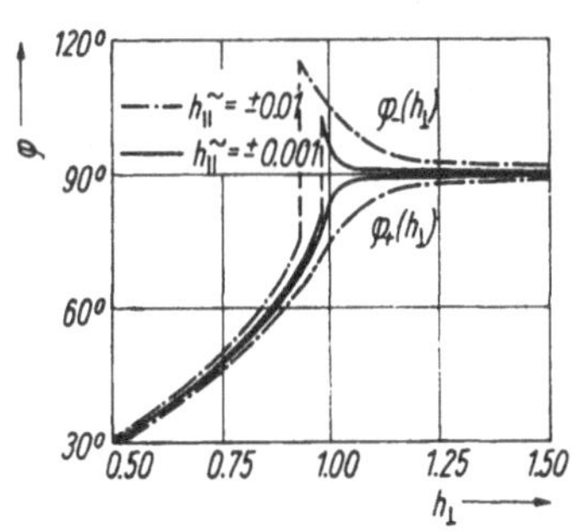

Fig. 3. A part of the dependence of the extremal positions of M_s on $h_{\perp}$, for two amplitude values of the ac field $h_{\parallel}^{\sim}$

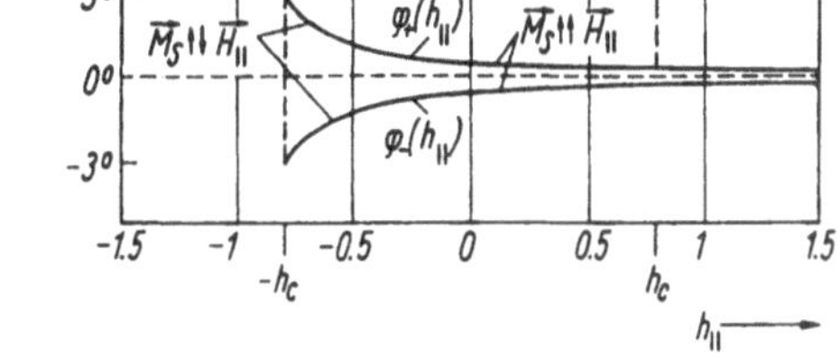

Fig. 4. Extremal positions of M_s vs. the magnetic field $h_{\parallel}$ parallel to the easy axis. Angles φ_+ and φ_- refer to amplitudes $+h_{\perp}^{\sim}$ and $-h_{\perp}^{\sim}$ of the ac field $h_{\perp}^{\sim} \sin \omega t$

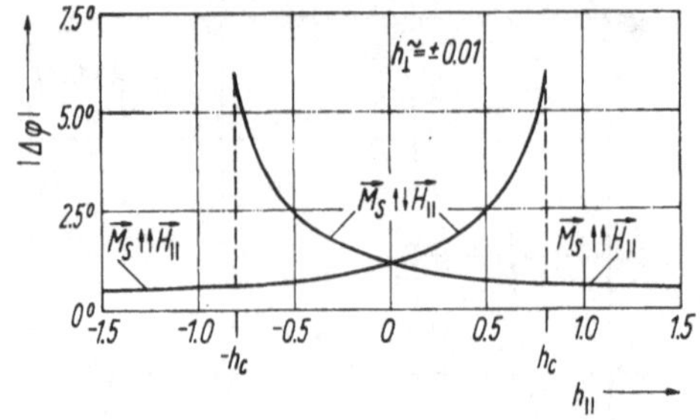

Fig. 5. Amplitude of forced vibrations of M_s vs. parallel dc field $h_{||}$

and the equilibrium condition is given by

$$\pm h = h_\perp^\sim \operatorname{ctg} \varphi \sin \omega t - \cos \varphi \,. \tag{6}$$

The sign "$-$" at $h_{||}$ corresponds to the case when $H_{||}$ and M_s are directed parallelly, whereas "$+$" indicates antiparallel directions of $H_{||}$ and M_s.

Fig. 4 shows the variations of φ_+ and φ_- vs. $h_{||}$ calculated from (6) for $h_\perp^\sim = \pm 0.01$. Only regions of interest are shown, since in the case of an ac field applied perpendicularly to the easy axis the vibration amplitudes of the vector M_s are small. As seen from the diagrams, bottom right and upper left, the vibration amplitude of M_s decreases with increasing $|h_{||}|$ when the vectors $h_{||}$ and M_s are parallel. For antiparallel directions of these vectors, the vibration amplitude of M_s grows with $|h_{||}|$, until $|h_{||}|$ becomes equal to h_c ($h_c = H_c/H_k$), when the film undergoes switching by domain wall motion. Henceforth $h_{||}$ and M_s are parallel, and $\Delta\varphi = |\varphi_- - \varphi_+|$ decreases stepwise. The dependence of $\Delta\varphi$ on $h_{||}$ plotted from Fig. 4, is shown in Fig. 5. The function $\Delta\varphi(h_{||})$ is not unambiguous in the whole range of variability of $h_{||}$. Thus, for a given value of the field $h_{||} < h_c$ we have two values of $\Delta\varphi$, one corresponding to the case $H_{||} \uparrow\uparrow M_s$, and the other to $H_{||} \uparrow\downarrow M_s$. The curve of the vibration amplitude of M_s vs. $h_{||}$ is similar to that obtained by Hasty and Boudreaux [6] at radio frequencies for analogous orientation of the magnetic fields. These considerations no longer hold when the amplitude of $h_\perp^\sim$ is relatively high. Stepwise changes in the vibration amplitude of M_s occur at those values of $|h_{||}| < h_c$ for which the resulting field (ac $h_\perp^\sim$ and dc $h_{||}$) intersects the asteroid; the condition $\partial^2 E/\partial\varphi^2 > 0$ is then no longer fulfilled.

Consider now the case when in determining H_k the ac field $h_{||}^\sim$ is not applied in the direction of the easy axis of magnetization. Denote by ϑ the angle between this field and the easy axis (Fig. 6). We can decompose the field $h_\perp$ into components $h'_\perp$ and $h''_\perp$ perpendicular and parallel to the easy axis, respectively and, similarly, the peak values $+h_{||}^\sim$ and $-h_{||}^\sim$ of the ac field. Hence, along the easy axis there act the components $h''_\perp$ and either $+h_{||}^{\sim\prime\prime}$ or $-h_{||}^{\sim\prime\prime}$. If

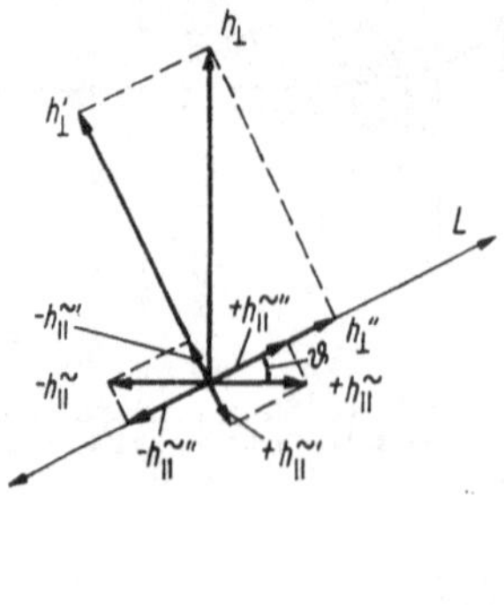

Fig. 6. Distribution of vectors of the magnetic fields $h_{||}^\sim$ and $h_\perp$ when the easy axis makes the angle ϑ with the ac field

$h''_\perp = h_\perp \sin\vartheta \geqq |\pm h_\parallel^{\sim\prime\prime}| = |h_\parallel^\sim \cos\vartheta|$, the vibration amplitude of the vector M_s has no maxima in the entire variation range of the field $h_\perp$, only a single jump of M_s can occur as $h_\perp$ increases. The limiting value of the angle ϑ_m at which maxima disappear depends on the value of $h_\parallel^\sim$. E.g. for $h_\parallel^\sim = \pm 0.01$, when the maxima occur at $h_\perp \approx 0.9$, the critical angle $\vartheta_m \approx 40'$! For $\vartheta < \vartheta_m$, the position of the maxima is a function of ϑ and, as $|h_\parallel^\sim| \sin^2\vartheta \ll 1$, (4) takes the more general form

$$H_k = [H_{cr.}^{2/3} + (H_\parallel^\sim \cos\vartheta - H_{cr.} \sin\vartheta)^{2/3}]^{3/2} . \tag{7}$$

The vibration of the magnetization vector M_s can be detected by various methods. As an indicator for the amplitude of these vibrations, one can use the change of Q of the resonance circuit coil of the generator, the voltage induced in the pick-up coil, the change of the rotation of the polarization plane of light in Kerr or Faraday effect, etc., in this study the planar galvanomagnetic effect is used for this purpose.

3. Experimental

Fig. 7 shows the circuit for the determination of the vibration amplitude of the vector M_s by the planar galvanomagnetic effect. A thin ferromagnetic film 1, equipped with electrodes 2, is placed in the magnetic field of two crossed pairs of Helmholtz coils. Coils 3 are supplied by a source 5 permitting continuous variation of the dc magnetic field; coils 4 are supplied from the sine generator 7, and provide the ac magnetic field. The measurements were performed in a 1000 Hz field whereas tests were carried out over the entire acoustic frequency range. The supply 6 is the source of dc current I_s flowing through the film 1. This leads to a drop in voltage across the film; the part dependent on the position of the vector M_s is expressed by

$$U_x(\varphi) = \Delta\varrho\, l\, j \cos^2\varphi \qquad \text{when } j \parallel L , \tag{8a}$$

or

$$U_x(\varphi) = \Delta\varrho\, l\, j \sin^2\varphi \qquad \text{when } j \perp L , \tag{8b}$$

where L is the easy direction of magnetization, $\Delta\varrho = \varrho_\parallel - \varrho_\perp$ the difference in resistivity when $M_s \parallel j$ and $M_s \perp j$, respectively, j the current density in the film, l the film length, and φ the angle between M_s and j [3].

The voltage $U_x(\varphi)$ after amplification in 8 is applied to the detector 9 and hence by way of the integrator 10 to the Y deflection system of the X–Y recorder 11. The X deflection signal, proportional to the dc field, is collected from the in series resistance R of the Helmholtz coils 3.

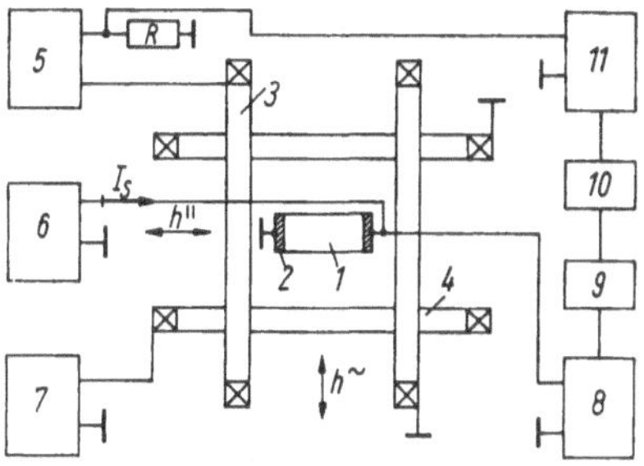

Fig. 7. Block diagram of measuring device. 1 — ferromagnetic film, 2 — electrodes, 3 — dc field Helmholtz coils, 4 — ac field Helmholtz coils, 5 and 6 — dc current supplies, 7 — ac current generator, 8 — amplifier, 9 — detector, 10 — integrator, 11 — X–Y recorder, R — resistor

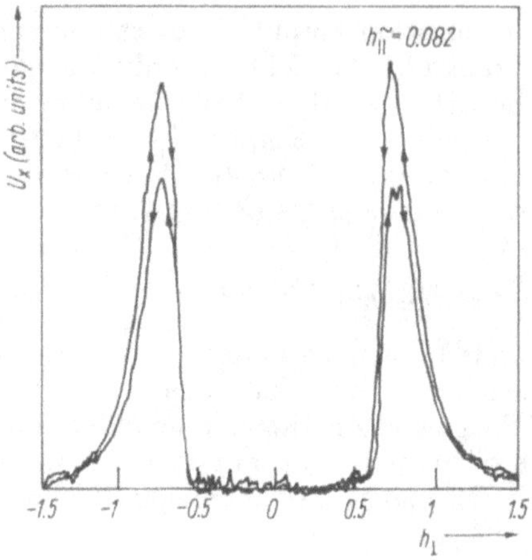

Fig. 8. Experimental curve of the amplitude of M_s vector forced vibrations vs. the dc field $h_{\perp}$, obtained by the galvanomagnetic method

The device used in our experiments enabled the orientation of the film under any angle in its plane with respect to the magnetic fields. When determining H_k the easy axis of the film was oriented perpendicularly to the dc field, and parallel when determining H_c.

Permalloy films containing 50 to 82% Ni were prepared by evaporation in vacuum of about 10^{-5} Torr onto optically flat glass substrates heated to about 300 °C during deposition and maintained in a constant magnetic field of about 120 Oe.

Typical curves obtained when determining the anisotropy field H_k are shown in Fig. 8 (the abscissa represents relative values of the magnetic field $h_{\perp}$; the anisotropy field H_k of the film was determined from equation (4)). The position of the peaks on the field axis depends on the amplitude of the ac field as predicted by (4). Fig. 8 shows that with increasing field strength the height of the peaks is smaller than when the field decreases. This result is in agreement with references [5, 6, 8]. The phenomenon can be explained as due to the hysteresis of angular dispersion of the anisotropy. Namely, when the field decreases from a value larger than H_k, the angular dispersion is lower than for a field increasing from zero. It is moreover seen (Fig. 8) that the experimental curves diverge slightly from the theoretical ones (Fig. 2); they fail to exhibit steeply rising branches at $h_{\perp} = h_{cr.}$. This discrepancy can be explained as due to the presence

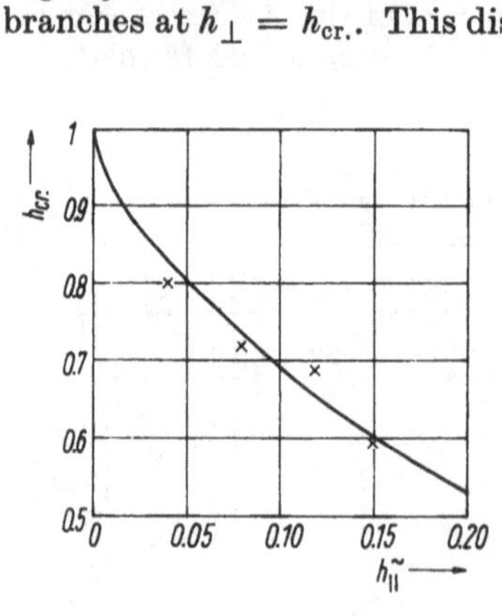

Fig. 9. Critical field $h_{cr.}$ vs. the amplitude of the ac field $h_{\parallel}^{\sim}$. Continuous line — theoretical curve, points — experimental results

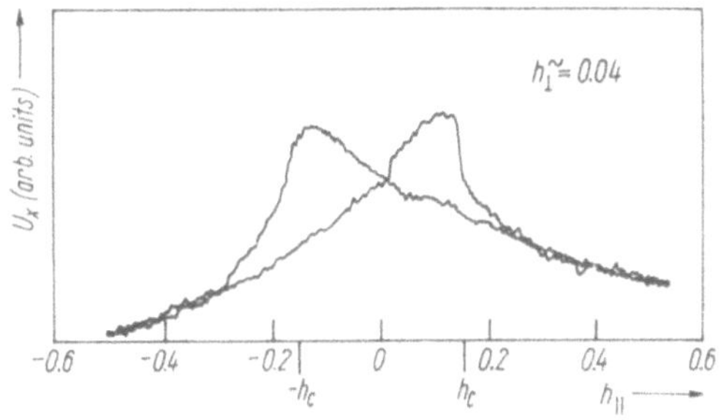

Fig. 10. Experimental curve of the amplitude of $\boldsymbol{M}_s$ vector vibrations vs. the dc field $h_{||}$ obtained by the galvanomagnetic method

of anisotropy dispersion in real films, such dispersion having been omitted in the calculations.

The dependence of the critical field $h_{cr.}$ on the amplitude of the ac field determined from (4) is shown in Fig. 9 together with the experimental points. Agreement between experiment and theory can be said to be satisfactory. The values of H_k found in various films by this method agree with the values obtained by others methods (from the reversible hysteresis loop in hard direction, and from the planar galvanomagnetic effect [3]).

Maxima of the M_s vibration amplitude were not observed, if e.g. as a result of internal stresses in the film the uniaxial anisotropy was not well-defined.

Fig. 10 shows one of the curves obtained when determining H_c by the method described above (the abscissa gives relative values of the magnetic field $h_{||}$, with H_k previously determined for this film by the method described above). In this case a signal amplification had to be used considerably larger than when determining H_k, in agreement with the predicted smaller vibration amplitude. The H_c values obtained were in agreement with those derived from the hysteresis loop determined by magnetooptical Kerr effect. As seen from Fig. 10, the experimental curves differ slightly from those theoretically predicted having no steep portions for $h_{||} = h_c$. Here, too, the discrepancy is attributable to the anisotropy dispersion.

4. Conclusions

The following conclusions can be drawn from the preceding study in which theoretical results based on the model of simple coherent magnetization rotation were compared with the experimental results:

1. The magnetization vector $\boldsymbol{M}_s$ of a thin ferromagnetic film with uniaxial magnetic anisotropy placed in crossed magnetic fields, one of which is an ac field, permforms forced vibrations of the ac field frequency. The amplitude of these vibrations as a function of a dc field applied perpendicularly to the ac field exhibits maxima in the neighbourhood of the anisotropy field H_k, if the ac field is applied along the easy axis of the film, and in the neighbourhood of the coercive force H_c, if the ac field is applied perpendicularly to its easy axis.

2. The curve of the M_s vibration amplitude versus the magnetizing field perpendicular to the ac field resembles the absorption curves at radiofrequencies.

3. H_k or H_c can be determined from the position of the maxima of the M_s vibration amplitude, according to the orientation of the easy axis with respect to the magnetic fields.

4. The experimental demonstration of the existence of ferromagnetic resonance at radio frequencies will be rendered difficult as a result of the forced vibrations of the magnetization vector $\boldsymbol{M}_s$.

Acknowledgements

The authors wish to thank Prof. Dr. S. Szczeniowski and Doc. Dr. J. Morkowski of the Institute of Physics of the Polish Academy of Sciences for their valuable and thorough discussion as well as Prof. Dr. W. Andrä, Dr. H. Gengnagel, and K. Stolzenberger of the Zentralinstitut für Festkörperphysik und Werkstoffforschung der DAW zu Berlin, Institutsteil für Magnetische Werkstoffe, Jena, for their fruitful discussion and for making available their unpublished results.

References

[1] Z. MÁLEK and W. SCHÜPPEL, phys. stat. sol. **2**, 136 (1962).
[2] E. F. KURITSYNA and VU DIN KY, Izv. AN SSSR, ser. fiz. **29**, 580 (1965).
[3] Z. SCZANIECKI, R. GONTARZ, W. KWIATKOWSKI, and H. RATAJCZAK, phys. stat. sol. (a) **7**, 597 (1971).
[4] P. ŠUDA, phys. stat. sol. **2**, 1227 (1962).
[5] D. O. SMITH, J. appl. Phys. **29**, 264 (1958).
[6] T. F. HASTY and L. J. BOUDREAUX, J. appl. Phys. **32**, 1807 (1961).
[7] A. G. LESNIK and G. I. LEVIN, Izv. AN SSSR, ser. fiz. **29**, 560 (1965).
[8] W. ANDRÄ, H. GENGNAGEL, W. SCHÜPPEL, and K. STOLZENBERGER, Fizika Magnitnykh Plenok, Akad. Nauk SSSR, Irkutsk 1968 (p. 414).
[9] A. L. FRUMKIN, Fiz. Met. Metalloved. **23**, 234 (1967).
[10] W. ANDRÄ, phys. stat. sol. **2**, 941 (1962).
[11] M. GOTO, T. HIRAOKA, and K. KUWAHARA, J. Sci. Hiroshima Univ. Soc. A-II, **26**, 117 (1963).
[12] W. ANDRÄ and F. KLINGHOLZ, phys. stat. sol. **34**, K163 (1969).
[13] E. C. STONER and E. P. WOHLFARTH, Phil. Trans. Roy. Soc. (London) **A240**, 599 (1948).

(Received October 14, 1971)

phys. stat. sol. (a) 8, 613 (1971)

Subject classification: 14.3; 6; 13.4; 22.4.1

Research Institute for Technical Physics of the Hungarian Academy of Sciences, Budapest

Electron Concentration and Mobility in CdS Single Crystals

By

B. PÖDÖR, J. BALÁZS and M. HÁRSY

n-type CdS crystals were grown by a sublimation method. Hall effect and conductivity in the crystals were measured in the temperature range from 80 to 400 °K. Analysis of the Hall coefficient data yielded donor activation energies which change with donor concentration as $E_d = (0.032 - 3.2\times10^{-8}\,N_d^{1/3})$ eV. It is supposed that the relevant donors in our undoped crystals are connected with S vacancies. The analysis of mobility showed that in the purest samples the electron mobility is limited by polar optical phonon, deformation potential, and piezoelectric scattering. Quantitative agreement between the theoretical and experimental curves was found for $m^*/m = 0.18$ and $E_1 = 13$ eV. The behaviour of mobility as a function of impurity concentration could be well described taking into account the simultaneous scattering on ionized and neutral impurities.

Методом сублимации были выращены кристаллы CdS n-типа. На кристаллах были проведены измерения эффекта Холла и удельной проводимости в области температур от 80 до 400 °K. Из анализа коэффициентов Холла была получена энергия активации донорных уровней, зависящая от концентрации доноров: $E_d = (0{,}032 - 3{,}2\times10^{-8}N_d^{1/3})$ eV. Предполагается, что в наших кристаллах доминирующие доноры связаны с S-вакансиями. На основании анализа подвижности делается вывод, что в наиболее чистых образцах подвижность определяется полярным, акустическим и пьезоэлектрическим рассеянием. Теоретические и экспериментальные кривые хорошо согласуются между собой при $m^*/m = 0{,}18$ и $E_1 = 13$ eV. Зависимость подвижности от концентрации примесей может быть хорошо объяснена, принимая во внимание рассеяние на ионизированных и нейтральных примесях.

1. Introduction

In recent years much attention has been devoted to the study of the electrical properties of CdS crystals. Several papers were published on the electrical conductivity, electron concentration, and mobility in CdS [1 to 11]. The recent improvement in the quality of CdS single crystals and the more intimate knowledge of relevant material parameters necessitated a detailed investigation and analysis of the variation of electric conductivity, electron concentration, and mobility in CdS crystals as a function of temperature and impurity content.

In this paper we describe and analyse resistivity- and Hall effect measurements performed on CdS crystals having room temperature electron concentrations between 4×10^{15} and 3×10^{18} cm^{-3}. The measurements were carried out over the temperature range of 80 to 400 °K.

Analysis of experimental results was concentrated on two particular problems. First of them was the determination of the thermal ionization energy of the relevant donors in our samples and their concentration dependence. In the literature two different shallow donors were observed generally. One is at about 0.03 eV below the conduction band edge [2, 5, 9], the other is a trap level with an energy of about 0.05 eV [11 to 13]. These values correspond to an infinitely dilute donor concentration. The concentration dependence of the ionization energy of the deeper level is known too [11, 12] but that of the 0.03 eV level has not yet been investigated systematically. The nature of these centres is still uncertain. The other point of our analysis was the investigation of the relevant scattering mechanisms, i.e. lattice scattering and the contribution to the charge carrier scattering by charged and neutral centres.

2. Experimental Technique and Results

The majority of the examined crystals are grown by a sublimation method in closed system [14]. Some additional samples were selected from a boule of ultra-high-purity grade crystal obtained from the Eagle-Picher Co.

About 4 g of luminescent grade CdS powder was charged into a quartz ampoule having a conic tip to which a quartz rod of 3 mm dia. was attached in order to conduct the heat of crystallization away. The ampoule sealed under 10^{-5} Torr pressure was placed into the homogeneous temperature zone of 1180 °C of a vertical electrical furnace, then it was pulled through a zone having a temperature gradient of 5 to 10 °C/cm with a velocity of 1 mm/h. After 50 h, the tip of the ampoule having reached the point of the furnace at 1130 °C, the pulling was stopped and the system was cooled down slowly with a cooling rate of 50 °C/h. Among the growth parameters only the pretreatment of the CdS powder was changed systematically. It was fired in dry H_2S atmosphere at various temperatures between 600 and 1100 °C, and in certain cases it was heated at 200 to 300 °C under vacuum before closing the ampoule.

Depending on the pretreatments CdS single crystals with different degrees of non-stoichiometry were grown, resulting in different concentrations of electrically active centres. The vacuum heat treatment caused a strong evaporation of sulphur from the starting material. During the growth process the deviations from the stoichiometry increased [14]. The non-stoichiometry was caused by the incorporation of excess Cd into the crystal.

Correlations between the parameters of the growth process (pretreatments, transport rate) and the physical properties (electrical, optical, and mechanical) of the crystals are described elsewhere [14]. The properties of samples are shown in Table 1, (samples 1 to 10 were grown by the sublimation method, sample EP 1 was cut from an Eagle-Picher crystal).

From the single crystalline boules platelets with approximate dimensions of $4 \times 3 \times 0.3$ mm^3 were cut out. After etching ohmic electrical contacts were made by evaporation of indium in vacuum (10^{-5} Torr), and annealing them at about 500 °C. The measurements were carried out by using the van der Pauw technique. The conventional dc compensation method was used, the magnetic field amounted to 4000 G. The usual care was taken to avoid the effects of spurious thermal gradients and contact misalignments. The cryostat used permitted to cover the temperature range between 80 and 400 °K. The estimated error of the measured values is about 10%, mostly due to the small dimensions of the samples.

Fig. 1, 2, and 3 summarize the results of electrical conductivity, Hall effect, and Hall mobility measurements. All of the samples were of n-type; their room temperature resistivities ranged from 10^{-2} to 10 Ωcm.

Table 1

sample No.	n_{300} (cm^{-3})	μ_{H300} (cm^2/Vs)	N_d (cm^{-3})	N_a (cm^{-3})	$K_n = N_a/N_d$	E_d (meV)	m^*/m
1 (15/2)	2.9×10^{18}	165	$(6 \text{ to } 8) \times 10^{18}$ *)	$(3 \text{ to } 5) \times 10^{18}$ *)	0.5 to 0.6 *)	0	0.2 **)
2 (20/1)	4.1×10^{17}	270	7.2×10^{17}	1.5×10^{17}	0.21	4.3	0.27
3 (20/3)	5.2×10^{17}	270	9.0×10^{17}	1.5×10^{17}	0.17	7.1	0.31
4 (59/1)	1.75×10^{17}	120	3.3×10^{17}	1.0×10^{17}	0.30	13.2	0.25
5 (53/3)	1.5×10^{17}	170	3.1×10^{17}	8.0×10^{16}	0.26	7.5	0.13
6 (51/3)	9.3×10^{16}	200	1.6×10^{17}	5.0×10^{16}	0.32	15.8	0.24
7 (53/1)	7.5×10^{16}	245	1.4×10^{17}	5.0×10^{16}	0.36	15.5	0.25
8 (47/1)	4.9×10^{16}	125	1.1×10^{17}	5.0×10^{16}	0.46	13.6	0.20
9 (70/1)	3.1×10^{16}	170	4.6×10^{16}	1.0×10^{16}	0.22	26.5	0.15
10 (32/1)	7.0×10^{15}	315	8.8×10^{16}	8.0×10^{16}	0.91	10.4	0.19
EP 1	3.5×10^{15}	310	4.5×10^{15} ***)	1.0×10^{15} ***)	0.22	—	—

*) Estimated from the low-temperature mobility values.
**) Assigned value.
***) Approximate values.

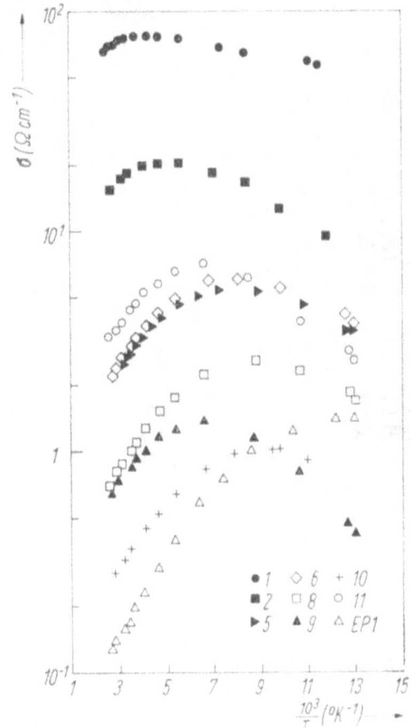

Fig. 1. Electrical conductivity vs. reciprocal temperature of CdS crystals. Notation as in Table 1

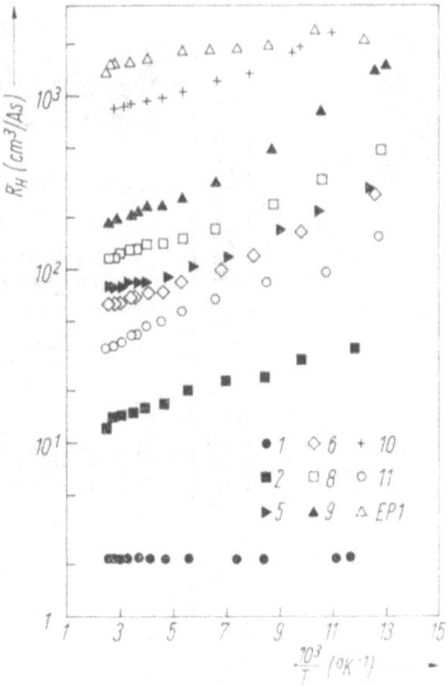

Fig. 2. Hall coefficient vs. reciprocal temperature of CdS crystals. Notation as in Table 1

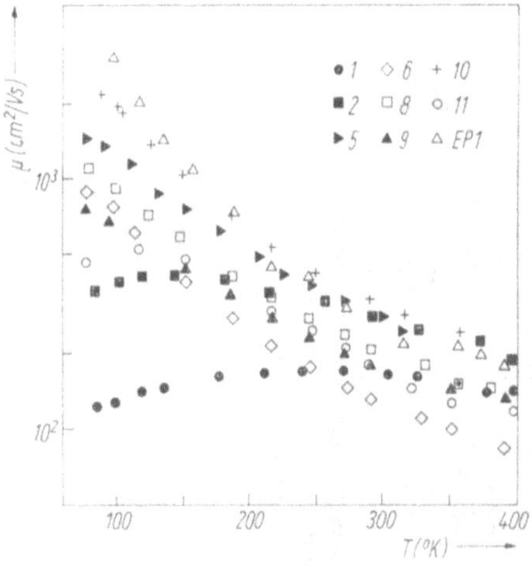

Fig. 3. Hall mobility vs. temperature of CdS crystals. Notation as in Table 1

It can be seen from Fig. 1 and 2 that all samples, except that with highest conductivity are non-degenerate. A simple estimation shows that for sample 1 at 100 °K the Fermi level is $\xi_n \approx +4$.

All of the samples except No. 1 show the presence of a shallow donor with an energy level different from zero causing the temperature dependence of the Hall coefficient. While the Hall coefficient measured on the Eagle-Picher Co.

samples exhibit saturation at room temperature, for the other samples the saturation region is missing in the given temperature range, due to the compensation.

The behaviour of the mobility as a function of temperature is more complicated. Samples 10 and EP 1 have the highest mobility in the whole temperature range. Their mobilities near room temperature are limited mainly by lattice scattering. Samples 7, 5, 2, and 1 show an increasing role of impurity scattering. The other group consisting of samples 4, 6, 8, and 9 exhibits entirely different characteristics. Their mobility vs. temperature curves run parallel to each other and their room temperature mobility is anomalously low.

3. Analysis of the Results

3.1 Carrier concentration

Converting the measured Hall coefficients to electron concentrations using $n = r/(eR_H)$ the mobility ratio was taken to be $r = 1$. This choice has been made because of the difficulties of assignment a proper unique scattering mechanism in the whole temperature range. The error caused by this approximation is usually negligible.

In analysing the carrier concentration curves obtained from Hall coefficients the ordinary single-donor–single-acceptor model was used described by the following equation:

$$\frac{n\,(n + N_a)}{(N_d - N_a - n)\,N_c'} = g\left(\frac{m^*}{m}\right)^{3/2} \exp\left(-\frac{E_d}{kT}\right), \qquad (1)$$

where $N_c' = 2\,(2\,\pi\,m\,kT/h^2)^{3/2}$, m is the free electron mass, g is the degeneracy factor of the donor level and n, N_d, N_a are the densities of conduction electrons, donors and acceptors, respectively, and E_d is the ionization energy of the donor level.

To calculate the thermal ionization energy of the relevant donors and to estimate the total concentration of donors and compensating acceptors the usual curve fitting procedure was used. According to this method N_a and $N_d - N_a$ are varied and the logarithm of the left side of (1) is plotted against the reciprocal temperature $1/T$. This plot will result in a straight line only for the correct values of N_a and $N_d - N_a$. The slope of the resulting straight line permits the determination of E_d and the intercept at $1/T = 0$ gives $g\,(m^*/m)^{3/2}$. Assuming $g = 1/2$ for the donor level degeneracy as usually, the effective mass ratio m^*/m can be determined.

Plots of equation (1) for representative samples with the final values of N_a and $N_d - N_a$ are shown in Fig. 4. The calculated values of N_d, N_a, E_d, m^*/m, and $K_n = N_a/N_d$ are given in Table 1. The parameters for sample 1 are estimated from the analysis of mobility (see Section 3.2).

It can be seen that the points fall very well on straight lines. Equation (1) is very sensitive to the choice of $N_d - N_a$ in obtaining a straight line, mainly at the high-temperature region. But the activation energy does not change significantly for a wide choice of $N_d - N_a$. The estimated values of the compensating acceptor concentrations have a relatively great uncertainity amounting to 30 to 40% except the strongly compensated sample 10 because of the limited temperature range of the measurements. Because of this fact the inter-

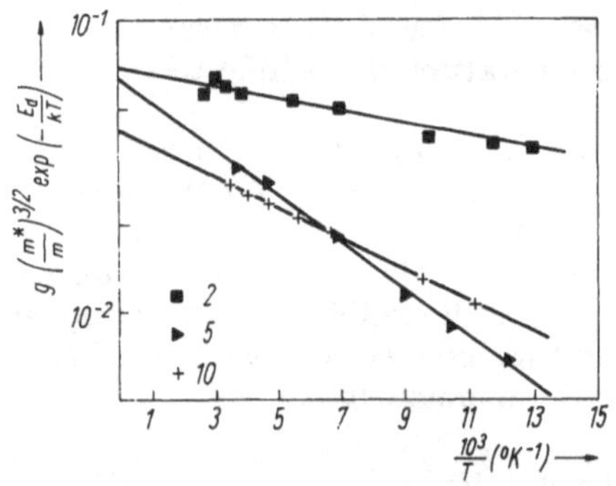

Fig. 4. Variation of the right side of equation (1) with reciprocal temperature. The parameters N_d and N_a are given in Table 1

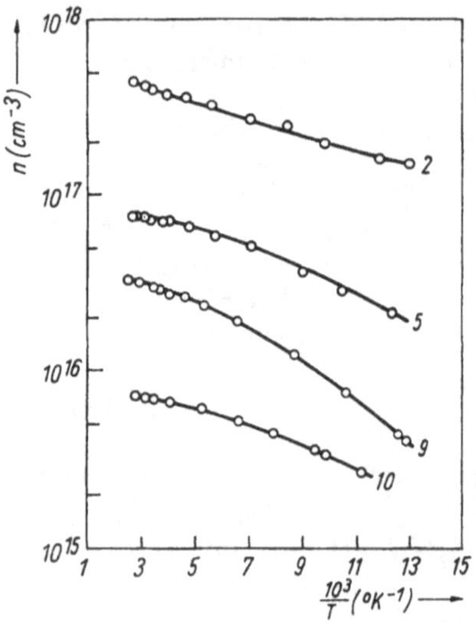

Fig. 5. Electron concentration as a function of the reciprocal temperature in CdS crystals. The points shown are the measured values, the curves are calculated with the set of parameters given in Table 1

cepts at $1/T = 0$ vary from sample to sample giving a considerable scatter in the values of the effective mass ratio m^*/m calculated from the intercepts. In spite of this inherent error the values of m^*/m for all of the samples with two exceptions are grouped around $m^*/m \approx 1/4$, giving an average value of $m^*/m = 0.23 \pm 0.05$. This value is in a good agreement with the value of 0.19 ± 0.01 obtained recently from infrared cyclotron resonance at 77 °K [15]. It is remarkable that for the two most strongly compensated samples 10 and 8 this analysis gave 0.19 and 0.20, respectively, for m^*/m.

Fig. 5 shows that the experimental electron concentrations and that ones calculated from the values given in Table 1 are in good agreement. For sample EP 1 the calculated electron concentration remains constant above 250 °K while the experimental values are increasing. This shows the presence of a second deeper donor level in this sample.

3.2 Carrier mobility

The systematic variation of mobility with carrier concentration, mainly at low temperatures permits a detailed analysis of the scattering mechanisms operating in our samples. Several scattering processes must be taken into account for the analysis of electron mobility in CdS: deformation potential scattering by acoustic phonons, scattering by polar optical phonons, piezoelectric scattering, ionized impurity and neutral impurity scattering, and scattering by space charge regions. If correlations between these scattering mechanisms are neglected, the resulting mobility can be calculated by adding the reciprocal values of the mobilities determined by the different scattering mechanisms alone.

The highest mobilities were observed in samples 10 and EP 1. Their values are close to each other in spite of the fact that their impurity concentrations deduced from the Hall curves differ by an order of magnitude. The somewhat lower mobilities of samples 10 found at the lowest temperatures are caused by the much greater impurity content and compensation degree of this sample as compared to sample EP 1. The mobilities observed in these samples at least at medium and higher temperatures are representative for the lattice-scattering-limited mobility in CdS. The highest mobilities measured at 300 and 100 °K are 315 and 3000 cm^2/Vs, respectively, in good agreement with the similar data found in the literature [5, 7, 10, 16, 17].

According to previous investigations [2, 5] the electron mobility in pure CdS is limited by polar optical mode scattering and piezoelectric scattering, of which the former predominates at higher temperatures and the latter at lower ones. Recent works have shown that in the temperature range concerned in this work acoustical mode scattering via the deformation potential coupling is not negligible, [10, 18 to 20].

The mobility, when limited by polar optical phonon scattering, piezoelectric scattering, and acoustic phonon scattering is expressed as

$$\mu_{PO} = \frac{8\,\hbar^2}{3(2\pi k)^{1/2}\, e\,\theta_D\, m^{*3/2}\, T^{1/2}} \left(\frac{1}{\varkappa_\infty} - \frac{1}{\varkappa_s}\right)^{-1} \left(\exp\frac{\theta_D}{T} - 1\right) \chi\left(\frac{\theta_D}{T}\right), \tag{2}$$

$$\mu_{PIEZO} = \frac{8\,\hbar^2}{3(2\pi)^{1/2}\, e\, m^*(kT)^{1/2}} \left\{\sum_{\text{modes}} \left[\frac{(K^2)_{av}}{\varkappa}\right]\right\}^{-1}, \tag{3}$$

$$\mu_{AC} = \frac{(8\pi)^{1/2}\, e\,\hbar^4\, c_{11}}{3\,E_1^2\, m^{*5/2}(kT)^{3/2}} \tag{4}$$

(see, e.g. [5, 21, 22]).

In equation (2) $k\theta_D = \hbar\omega_l = 0.038$ eV is the longitudinal optical phonon energy [5, 23] corresponding to $\theta_D = 441$ °K. The optical dielectric constant $\varkappa_\infty$ is 5.24 [24], and the static one is $\varkappa_s = 9.19$ resulting from a spherical average of $\varkappa_1 = 9.02$ and $\varkappa_3 = 9.53$ corresponding to transversal and longitudinal coupling [4, 17, 26] with $\varkappa_s = (1/3)\,(2\,\varkappa_1 + \varkappa_3)$.

$\chi(\theta_D/T)$ is a slowly varying function of θ_D/T [27]. We have used the more exact values of this function calculated recently by Fortini et al. [28]. This choice results in a slightly increased value of m^*/m required to fit a given mobility, as compared to older calculations.

Due to the different approximations used to average the anisotropy inherent in the electron–piezoelectric phonon coupling mechanism and due to the considerable scatter of the elastic and piezoelectric constants used to evaluate the electromechanical coupling constants, $(K^2)_{av}$ in equation (3), it is hardly possible to get an estimation of the piezoelectric mobility with an accuracy better than 15 to 20% [5, 17, 26, 29]. We have taken $(K_t^2)_{av} = 0.031$ and $(K_l^2)_{av} = 0.042$ using the isotropic model due to Hutson [22]. Equation (3) was evaluated with the only refinement of using the values for $\varkappa_1 = \varkappa_t$ and $\varkappa_3 = \varkappa_l$ already quoted [4, 17, 26], resulting in a decrease of the piezoelectric mobility by about 5% compared to the estimations of Hutson.

In equation (4) c_{11} is the properly averaged elastic constant for which an average value of 8×10^{11} dyn/cm was taken [22, 26, 29]. The crucial parameter is E_1, the deformation potential, the value of which was not properly determined till yet. According to the measurements of Gutsche [29] $E_1 = 2.5$ eV for the whole forbidden band. From the pressure dependence of the forbidden band one finds $E_1 = 2.9$ eV. But these values are too low to explain the experimental low-temperature mobilities, [10, 20]. In the recent literature values of (16 ± 2) eV [18], 15 eV [19], and 14.5 eV [20] are quoted for E_1, resulting from theoretical fits to the experimental mobilities, mainly below 77 °K.

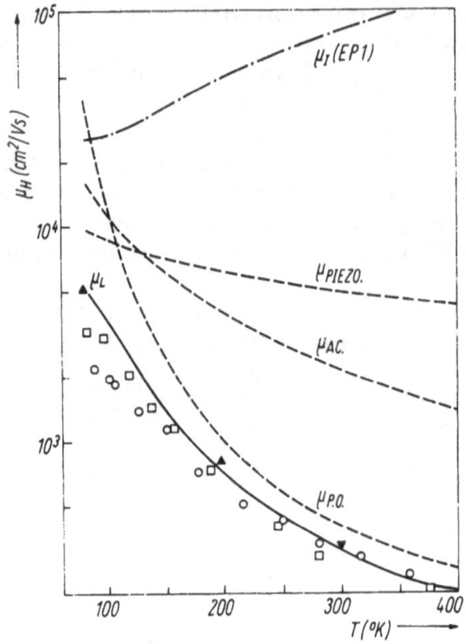

Fig. 6. Analysis of the lattice-limited mobility in CdS. ▲ data from [30], ▼ data from [16], ○ sample 10, □ sample EP 1. The curves are calculated for $m^*/m = 0.18$ and $E_1 = 13$ eV. The contribution to the mobility due to the ionized impurities for sample EP 1 is shown too

We have tried to fit the theoretical lattice mobility as given by $\mu_L^{-1} = \mu_{PO}^{-1} + \mu_{PIEZO}^{-1} + \mu_{AC}^{-1}$ to the experimental mobilities of samples 10 and EP 1. Because of the ambiguity concerning the role of deformation potential scattering the fit was tried in two different ways. In the first case $m^*/m = 0.18$ was chosen[5, 15, 20, 31] and E_1 was varied to achieve the best fit to the experimental data, allowing for a slight contribution of impurity scattering at the lowest temperatures. The best fit gave $E_1 = 13$ eV. The results are shown on Fig. 6. Beside our measurements, some data from the literature are shown too [16, 30]. This figure shows the theoretical mobilities limited by polar optical phonon, piezoelectric, and acoustic phonon scattering, respectively, and the theoretical mobility from the combination of these three mechanisms for $m^*/m = 0.20$ and $E_1 = 13$ eV. In this figure the ionized-impurity-limited mobility for sample EP 1 is shown too. In the second case we have used the older value of $E_1 = 2.5$ eV, resulting in a negligible contribution of deformation potential scattering above 77 °K. Fit was achieved changing m^*/m, the best fit gave $m^*/m = 0.20$. Above 200 °K this gave practically the same theoretical curve as in the first case, but below 200 °K the theoretical curve for the second case run higher than before, the difference was about 20% at 100 °K. We can conclude that the first model is more suitable to interprete the experimental data, therefore its results will be used later.

For all of the other samples the measured mobilities are considerably lower than the theoretical curve, the difference being the greatest at low temperatures. For the two highest mobility samples, 10 and EP 1 too, the experimental mobilities below 150 °K do not coincide with the curve calculated for lattice-limited mobility. This decrease in the mobility should be accounted for by the ionized and neutral impurity scatterings. These two mechanisms have their greatest influence at low temperatures. For scattering by ionized and neutral impurities the mobility is given by the Brooks-Herring and Erginsoy formula, respectively [5, 21]:

$$\mu_I = \frac{8 \cdot 2^{1/2} \varkappa_s^2 (kT)^{3/2}}{\pi^{3/2} e^3 N_I m^{*1/2} [\ln(1+b) - b/(1+b)]}, \tag{5a}$$

$$b = \frac{6 \varkappa_s m^* (kT)^2}{\pi n' \hbar^2 e^2}, \tag{5b}$$

$$n' = n + (n + N_a)\,[1 - (n + N_a)/N_d]; \tag{5c}$$

$$\mu_N = \frac{m^* e^3}{20\,\hbar^3\,\varkappa_s\,N_N}, \tag{6}$$

where $N_I = 2\,N_a + n$ is the ionized impurity concentration. Both N_I and n' are evaluated using the values of donor and acceptor concentrations calculated from the analysis of carrier concentration measurements. The neutral impurity concentration $N_N = N_d - N_a - n$ is evaluated using the results of the analysis of Hall curves. μ_N depends on the temperature through the temperature dependence of N_N. For m^*/m we used the value of 0.18 in (5) and (6).

It is to be emphasized that no free parameters are left in (5) and (6), therefore comparing the theoretical mobility $\mu^{-1} = \mu_L^{-1} + \mu_I^{-1} + \mu_N^{-1}$ with the measured ones, the adequacy of this model can be decided.

Fig. 7 shows the comparison of the experimental mobilities with the calculated ones for samples 7, 2, and 1 belonging to the first group as discussed in Section 3. It can be seen that the fit is satisfactory, mainly at low temperatures. We can conclude that the mobility curves of a whole series of samples can be accounted for using the model of combined lattice and ionized and neutral impurity scatterings, the densities of scattering centres corresponding to the values deduced from the analysis of carrier concentration curves. For sample 1 where the donor and acceptor concentrations could not be evaluated from the Hall curves, the concentration of scattering centres was estimated from the mobility data measured at low temperatures. The contribution to the mobility due to the ionized impurities for sample EP 1 is shown in Fig. 6. For this sample the neutral impurity scattering is negligible.

Applying this method to the strongly compensated sample 10 would result in too low values of mobility as compared to the theoretical ones. At such a high compensation the donors and acceptors are mutually screening each other, resulting in a much weaker scattering than that one given by the Brooks-Herring formula. It is reasonable to suppose that most of the donors and acceptors are forming pairs and the electrons are scattered by the dipole potential of these pairs which results in a substantially lowered scattering cross section.

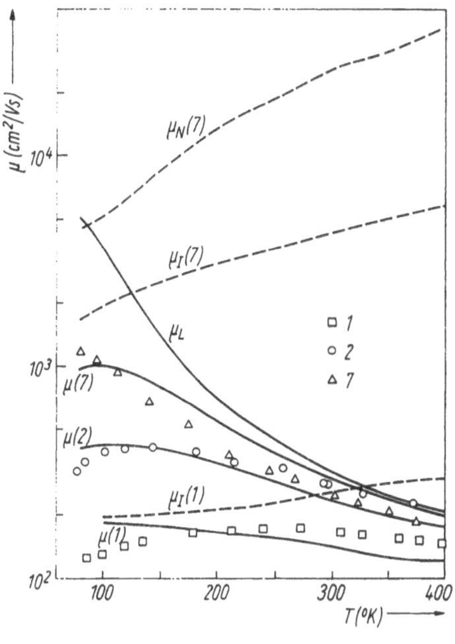

Fig. 7. Analysis of mobility in CdS. μ_L lattice-limited mobility, μ_I mobility due to the ionized impurity scattering, μ_N mobility due to the neutral impurity scattering

The fit of the combined lattice and impurity mobility for samples having anomalously low room temperature mobilities is much worse, mainly at intermediate and high temperatures. In such cases it is usually supposed that scattering due to the formation of large space charge regions around local inhomogeneities plays an important role [32]. Mobility limited by space charge scattering is calculated by Weisberg [32] to be

$$\mu_s = \frac{e}{N_s A (2\, m^* kT)^{1/2}}, \tag{7}$$

where N_s is the density of the space charge regions and A is the effective scattering area of the space charge region. Even if we take into account this mechanism, the fit between the theoretical curve and the experimental data is not improved significantly because the combined impurity and lattice scattering is sufficient to interprete the mobilities measured near 100 °K. Choosing N_s so that a fit would be achieved at 300 °K, the theoretical mobility at 100 °K would be too low.

4. Discussion and Conclusions

In this section we should like to discuss our results concerning the nature of relevant donors in our samples and we should like to compare the results of mobility analysis with other similar analyses.

The ionization energies of the relevant donors in our samples vary from zero to about 25 meV. Because of the circumstances of the crystal growth (see [14]) we can suppose that the same type of donor determines the behaviour of the whole series of samples. Therefore the variation of the observed activation energy with the carrier concentration should reflect the well-known decrease of activation energy with increasing density of centres as studied in detail by Pearson and Bardeen [33]. Fig. 8 shows the dependence of the activation energy on the cubic root of donor concentrations. In this figure some data from the literature for unspecified donors in CdS are shown too. It can be seen that a linear relationship of the form $E_d = E_{d0} - C\, N_d^{1/3}$ exists between E_d and $N_d^{1/3}$ where E_{d0} is the activation energy at infinite dilution. From our data we get $E_{d0} = (32 \pm 1)$ meV, which agrees with the value of (32 ± 2) meV obtained by Piper and Halsted [2] directly from the slopes of Hall curves measured on

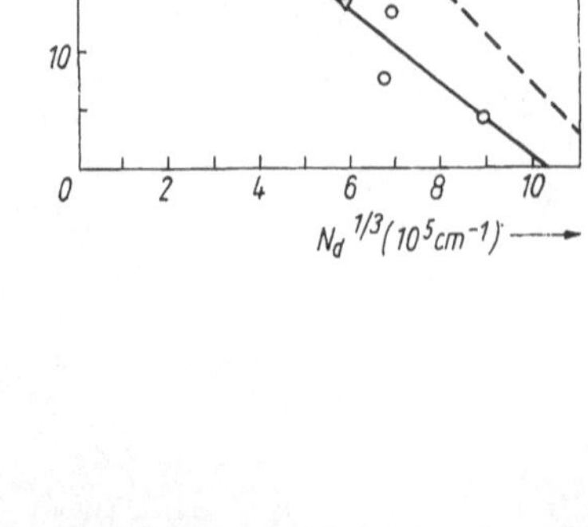

Fig. 8. Donor activation energy as a function of donor concentration. ○ our results, △ data from [2], ▽ data from [6]. Full line: our result, dashed line: curve for the 0.05 eV level [11]

reasonably pure crystals. A similar value (29 meV) is predicted for hydrogen-like impurities using $\varkappa = 9.19$ and $m^*/m = 0.18$. For C we get 3.2×10^{-8} cm in reasonable agreement with that value calculated from the theoretical model given by Pearson and Bardeen [33].

In Fig. 8 the E_d vs. N_d curve for the 0.05 eV level is shown according to Spear and Mort [11]. A similar curve is given by Chester [12] too. Our figure shows that these two levels are genuinely different from each other. Shallow donor levels of about 30 meV (or lower at higher impurity concentrations) are generally observed in undoped CdS or in crystals doped with In, Ga, Cl, etc. [2, 5, 9, 34]. Very recent optical measurements of the ionization energies of the F, In, Ga, Cl, Br, and I donors [35] gave values very close to each other between 35.1 and 32.1 meV. Our crystals were not intentionally doped, moreover during the crystal growth only the pretreatment of the CdS powder, i.e. the temperature of heating in H_2S atmosphere was varied, resulting in different donor concentrations, corresponding to the degree of non-stoichiometry (for details see [14]). Therefore we can suppose that a native defect connected with excess Cd is responsible for the observed donors in our samples.

The first ionization energy of the Cd interstitial is usually identified with the 0.05 eV level [12, 13]. On the other hand according to Woods and Wright [36] a sulphur vacancy with two electrons forms a shallow donor level with an energy level of about 0.02 eV. Clark and Woods [25] have shown, that in low-resistance non-stoichiometric crystals, grown under similar circumstances as our samples, the excess cadmium is incorporated in the form of sulphur vacancies, which in turn provide the donor centres. Therefore we believe, that the relevant donors in our samples are either S vacancies or some other defect complex connected with the S vacancies. In this case the observed ionization energy of 0.032 eV could be identified with the first ionization energy of the sulphur vacancy.

Furthermore from our results it follows that the ionization energies of the S vacancies and that of the common doping impurities are very nearly equal to each other. This explains the fact that in most cases independently of doping, etc. the same shallow donor levels are observed in the range of 0 to 30 meV.

The interesting point in the analysis of mobility is the role of deformation potential scattering. At around 100 °K the lattice mobility is equally affected by piezoelectric, deformation potential, and polar scattering, for $E_1 = 13$ eV. This value is smaller than that deduced by other authors [18 to 20], probably because of the allowance made for impurity scattering at low temperature. The results show that at room temperature the contribution of deformation potential scattering exceed that one of piezoelectric scattering.

Finally our analysis shows for CdS and probably more generally for II–VI semiconductors that the dependence of mobility on temperature and on impurity concentration can be interpreted in a consistent way using the values of impurity concentrations deduced from Hall measurements.

Acknowledgements

We should like to express our acknowledgements to Dr. M. Somogyi for her help in making the electrical contacts and to Mr. E. Lendvay for his interest in this work and for the discussions concerning the nature of the electrically active centres in our crystals.

References

[1] F. A. KRÖGER, H. J. VINK, and J. VOLGER, Philips Res. Rep. **10**, 39 (1955).
[2] W. W. PIPER and R. E. HALSTED, Proc. Internat. Conf. Semicond. Phys., Prague 1960 (p. 1046).
[3] M. ITAKURA and H. TOYODA, J. Phys. Soc. Japan **18**, 150 (1963).
[4] J. D. ZOOK and R. N. DEXTER, Phys. Rev. **129**, 1980 (1963).
[5] S. S. DEVLIN, in: Physics and Chemistry of II–VI Compounds, Ed. M. AVEN and J. S. PRENER, North-Holland Publ. Co., Amsterdam 1967.
[6] M. ITAKURA and H. TOYODA, Rev. Electr. Commun., Labor Tokyo **14**, 1 (1966).
[7] S. TOYOTOMI and K. MORIGAKI, J. Phys. Soc. Japan **25**, 807 (1968); Tech. Rep. Inst. Solid State Phys. Univ. Tokyo, No. A310 (1968).
[8] R. S. CRANDALL, Phys. Rev. **169**, 577 (1968).
[9] S. S. LI and C. I. HUANG, Phys. Rev. **B3**, 3266 (1971).
[10] K. KOBAYASHI, Internat. Conf. II–VI Semicond. Compounds, Providence, Rhode Island 1967 (p. 755).
[11] W. E. SPEAR and J. MORT, Proc. Phys. Soc. **81**, 130 (1963).
[12] R. O. CHESTER, J. appl. Phys. **38**, 1745 (1967).
[13] R. BOYN, phys. stat. sol. **29**, 307 (1968).
[14] M. HÁRSY, J. BALÁZS, P. SVISZT, B. PÖDÖR, and E. LENDVAY, J. Cryst. Growth **9**, 209 (1971).
[15] K. NAGASAKA, G. KIDO, and S. NARITA, J. Phys. Soc. Japan **28**, 1376 (1970).
[16] M. IKEDA, K. ITOH, and H. SATO, J. Phys. Soc. Japan **25**, 455 (1968).
[17] V. E. HENRICH and G. WEINREICH, Phys. Rev. **178**, 1204 (1969).
[18] M. SAITOH, J. Phys. Soc. Japan **21**, 2540 (1966).
[19] M. ONUKI and K. SHIGA, Proc. Internat. Conf. Semicond. Phys., Kyoto 1966 (p. 427).
[20] D. L. RODE, Phys. Rev. **B2**, 4036 (1970).
[21] E. H. PUTLEY, The Hall Effect and Related Phenomena, Butterworths, London 1960.
[22] A. R. HUTSON, J. appl. Phys. Suppl. **32**, 2287 (1961).
[23] R. J. COLLINS, J. appl. Phys. **30**, 1135 (1959).
[24] S. J. CZYZAK, W. M. BAKER, R. C. CRANE, and J. B. HOWE, J. Opt. Soc. Amer. **47**, 240 (1957).
[25] L. CLARK and J. WOODS, Brit. J. appl. Phys. **17**, 319 (1966).
[26] D. BERLINCOURT, H. JAFFE, and L. R. SHIOZAWA, Phys. Rev. **129**, 1009 (1963).
[27] D. J. HOWARTH and E. H. SONDHEIMER, Proc. Roy. Soc. **A219**, 53 (1953).
[28] A. FORTINI, D. DIGUET, and J. LUGANO, J. appl. Phys. **41**, 3121 (1970).
[29] E. GUTSCHE, phys. stat. sol. **1**, 716 (1961).
[30] H. FUJITA, K. KOBAYASHI, T. KAWAI, and K. SHIGA, J. Phys. Soc. Japan **20**, 109 (1965).
[31] W. S. BAER and R. N. DEXTER, Phys. Rev. **135**, A1388 (1964).
[32] L. R. WEISBERG, J. appl. Phys. **33**, 1817 (1962).
[33] G. L. PEARSON and J. BARDEEN, Phys. Rev. **75**, 865 (1949).
[34] R. H. BUBE, Photoconductivity of Solids, Wiley & Sons, 1960 (p. 160).
[35] K. NASSAU, C. H. HENRY, and J. W. SHIEVER, Proc. X. Internat. Conf. Semicond. Phys., Cambridge (Mass.) 1970 (p. 629).
[36] J. WOODS and D. A. WRIGHT, Conf. Solid State Phys. in Electronics and Telecommunications, Brussels 1958, Academic Press, New York 1960 (p. 880).

(Received October 20, 1971)

Erratum

To the paper by K. Haneda and H. Kojima

Intrinsic Coercivity of $BaFe_{12-x}Cr_xO_{19}$

phys. stat. sol. (a) **6**, 259 (1971)

the authors request the following rectification:

On page 263, line 4 from the top, read "Fe^{3+} ions which contribute to" instead of "Fe^{3+} ions with contribute to"; on page 263 in Fig. 7a read correctly ○ $x = 0$, △ $x = 0.4$, □ $x = 1.0$ instead of ○ $x = 0$, □ $x = 0.4$, △ $x = 1.0$.

See Short Note by F. FUJIMOTO et al.

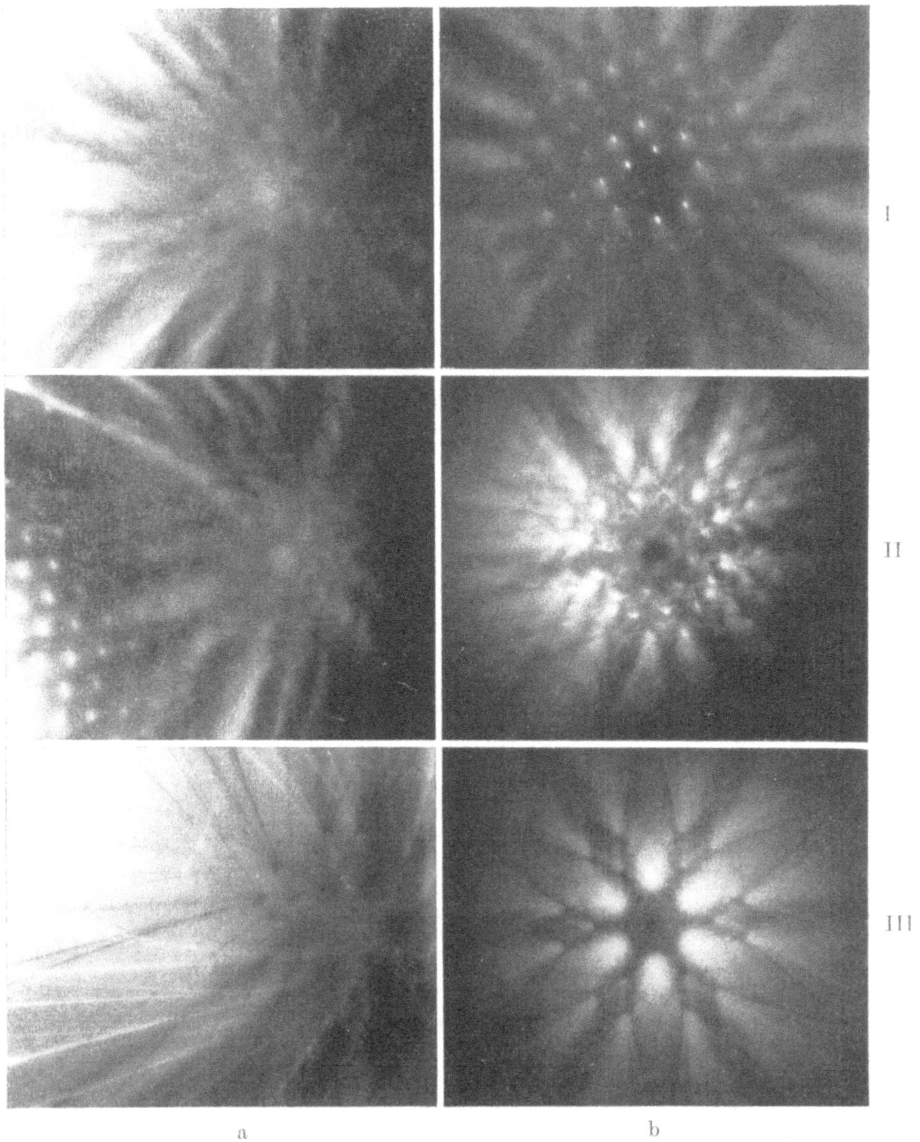

Fig. 1. Diffraction patterns from thin a) and thick b) films of gold (I), copper (II), and silicon (III) crystals. The incident beam in the case a) is tilted a few degrees towards ⟨100⟩ in gold, ⟨110⟩ in copper, and ⟨111⟩ in silicon; in the case b) the beam is almost parallel to these crystal axes

See Short Note by A. W. CZANDERNA et al.

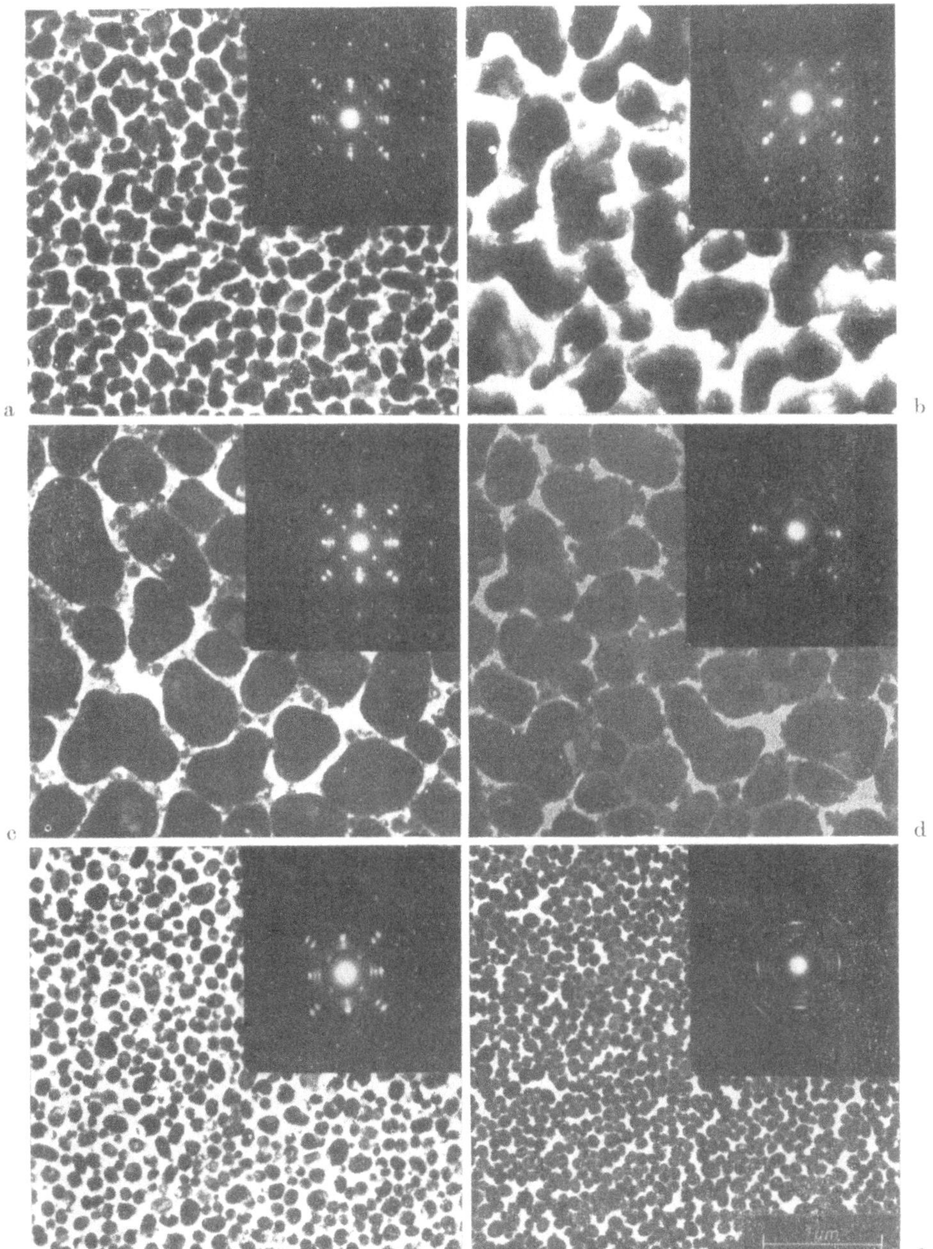

Fig. 1. Electron diffraction and transmission micrographs of copper deposited on (100) sodium chloride after exposure to water vapor: a) 432 Torr h at 25 °C; b) 4.7 Torr h at 325 °C; c) 8.6×10^{-2} Torr h at 325 °C; d) 2.5×10^{-4} Torr h at 325 °C; e) 2.5×10^{-5} Torr h at 325 °C; f) 0 Torr h at 325 °C

Short Notes

phys. stat. sol. (a) 8, K71 (1971)

Subject classification: 5; 21.1; 21.6; 22.1.2

College of General Education, University of Tokyo (a)
and Japan Electron Optics Laboratory Co. Ltd., Tokyo (b)

Axial Channeling of Electrons Observed in Kikuchi Patterns

By

F. FUJIMOTO (a), K. KOMAKI (a), and Y. UCHIDA (b)

According to the reciprocity theorem of scattering problems, the channeling experiment, in which a specimen is rotated and incident beam and detector are fixed, is equivalent to the experiment of the convergent beam techniques in electron diffraction and this identity was manifested experimentally by the present authors (1). In the present paper, we study the electron channeling by observing the Kikuchi patterns from the base of the reciprocity and the relation between the classical and wave mechanical theories in the electron channeling.

We observed the diffraction patterns from silicon, copper, and gold films by using a 1 MV electron microscope in the axial case. In Fig. 1 (Ia), the $\langle 100 \rangle$ diffraction pattern from a thin film of gold is shown, where the $\langle 100 \rangle$ axis was tilted a few degrees towards the incident beam. The Kikuchi pattern corresponding to the $\langle 100 \rangle$ axis is like a flower of chrisanthemum and a peak with strong intensity exists at the center. Such a pattern was also observed in $\langle 110 \rangle$ and $\langle 310 \rangle$ axes of gold. Fig. 1 (Ib) shows the $\langle 100 \rangle$ diffraction pattern from a thick film, where the incident beam is almost parallel to the axis. In this case, the bright region in Fig. 1 (Ia) turns into a dark one and, especially, the central peak changes into a deep dip.

The $\langle 110 \rangle$ diffraction patterns from thin and thick films of copper are shown in Fig. 1 (IIa) and (IIb), respectively. The situations in these patterns are almost the same as those in the cases of Fig. 1 (Ia) and (Ib).

Fig. 1 (IIIa) and (IIIb) show the $\langle 111 \rangle$ diffraction patterns from thin and thick films of silicon, respectively. These patterns have no distinct characteristics

Table 1

	axis	energy (keV)		α_a	ψ_1 (mrad)	ψ_c (mrad)	ψ_c/ψ_1
Si	⟨110⟩	1000	o	3.43	12.5	8.3	0.64
		500	Δ	2.29	16.7	9.9	0.59
		100	X	1.39	33.8	40	1.17
	⟨111⟩	1000	X	2.28	11.3	3.9	0.34
Cu	⟨110⟩	1000	o	9.43	22.1	14	0.63
		500	o	6.31	29.5	19	0.64
	⟨100⟩	1000	Δ	4.71	18.6	11	0.60
		500	X	3.15	24.8	15	0.61
	⟨211⟩	1000	Δ	3.14	16.8	9.6	0.57
		500	X	2.10	22.4	15	0.65
	⟨111⟩	1000	X	1.57	14.1	6.9	0.49
Au	⟨110⟩	1000	o	29.0	34.4	23	0.67
	⟨100⟩	1000	o	14.5	28.9	19	0.65
	⟨310⟩	1000	o	5.80	23.0	14	0.62

o flower pattern, Δ intermediate, X geometrical pattern.

when compared with those in a low energy case. The axial region is determined by the three {220} bands. From the foregoing results, we can classify the Kikuchi patterns for crystal axes into two groups, that is, the flower and geometrical patterns.

Lindhard (2) has shown from the standpoint of classical theory that the critical angle of axial channeling of electrons is given by

$$\psi_1 = \sqrt{\frac{4 Z_1 Z_2 e^2}{p v d}} ,$$

where $Z_1 = 1$ and Z_2 is the atomic number of crystal atoms; d is the atomic spacing along the crystal axis. p and v are the relativistic momentum and velocity of the incident electrons, respectively. In Table 1, the critical angle, ψ_c, obtained from the radius of the bright region in the axial Kikuchi pattern and the ratio ψ_c/ψ_1 are listed. Circle and cross indicate the flower and geometrical patterns, respectively; a triangle indicates an intermediate pattern. If we introduce the quantity

$$\alpha_a = \frac{4 m Z_1 Z_2 e^2}{\pi \hbar^2 N d^2} ,$$

where N is the number density of the crystal atoms and m is the electron relativistic mass the flower or geometrical pattern appears when α_a is larger or smaller than about three, respectively, though this circumstance depends somewhat on the kind of axis, too. The ratio ψ_c/ψ_1 for the flower patterns is about 0.65 which agrees with the result of Uggerhøj and Frandsen (3). From the result that the intensity distributions of positron and proton axial channeling in gold film coincide with each other (4), we can conclude that the classical model is applicable in the case of the flower pattern disregarding any fine structure features. It must be noted that the abnormal transmissions observed in Fig. 1 (IIIb), are not obvious in Fig. 1 (Ib) and (IIb). This means that the abnormal transmission is strongly related to the diffraction effects and is quite different from the channeling phenomenon described by the classical theory.

We are much indebted to Prof. S. Takagi for fruitful discussions.

References

(1) F. FUJIMOTO, K. KOMAKI, S. TAKAGI, H. KOIKE, and S. TAKASHIMA, phys. stat. sol. (a) 8, K1 (1971).

(2) J. LINDHARD, Atomic Collision Phenomena in Solids, North-Holland Publ. Co., Amsterdam 1970 (p. 1).

(3) E. UGGERHØJ and F. FRANDSEN, Phys. Rev. B2, 582 (1970).

(4) J.U. ANDERSEN, W.M. AUGSTYNIAK, and E. UGGERHØJ, Phys. Rev. B3, 705 (1971).

(Received September 23, 1971)

phys. stat. sol. (a) 8, K75 (1971)
Subject classification: 1.4 and 3.1; 21.1

Department of Physics and Institute of Colloid and Surface Science,
Clarkson College of Technology, Potsdam, New York

Effect of Water Vapor Exposure on the Epitaxy of Copper Films on Sodium Chloride

By

A.W. CZANDERNA[1)], B.W. BRENNAN, and E.G. CLARKE, Jr.

It is well known that the substrate temperature, residual vacuum pressure, and evaporation rate have an important effect on the structure of vapor deposited thin film materials (1 to 7). Numerous studies have been carried out where these three parameters have been varied during the deposition of various f.c.c. metals on NaCl cleaved in air and in vacuum. However, only a few studies have been concerned with the effect of residual gases on epitaxy (8 to 14). There is mounting evidence (14) that to achieve epitaxy, exposure of the substrate to water vapor is extremely important prior to deposition of the film, but apparently not essential (15). In the studies of the effect of residual gases on epitaxy, the exposure to water vapor was made at room temperature (10 to 12, 14). In this study, the effect of water vapor exposure at 25 and 325 $^{\circ}$C on the epitaxial growth of copper films deposited on thermally etched NaCl has been considered. It is shown that epitaxial films can be grown after exposure of NaCl to over 2.5×10^{-4} Torr h of water vapor at 325 $^{\circ}$C.

Copper thin films were prepared by vacuum evaporation of 99.999% wire from molybdenum boats onto air cleaved (100) NaCl single crystal substrates. The source-to-substrate distance was 25 cm. The evaporations were performed in a Pyrex chamber with stainless steel flanges at pressures from 3×10^{-7} to 9×10^{-6} Torr. The temperature of the substrate, which was maintained to $\pm 3^{\circ}$ C by a double layer of heating tapes, was determined from an external thermocouple and a prior calibration of the thermocouple and substrate temperature. The deposition rate of about 1 Å/s was estimated by comparing the transmittance of the film, measured during and after deposition, with values obtained from deposition onto substrates suspended from a vacuum ultramicrobalance in a parallel experiment (16).

1) Present address: Fritz-Haber-Institut der Max-Planck-Gesellschaft, 1 Berlin 33, Faradayweg 4 to 6.

The NaCl substrates were cleaved in air to 3 mm thick from a 40 mm diameter boule and mounted in the vacuum chamber. After establishing a vacuum of 10^{-6} Torr, the substrates were etched by heating at 450 ^{o}C, as described by Shirai (17), for two hours to remove about 20 monolayers from the surface (16). The substrate was cooled to 25 or 325 ^{o}C and exposed to a constant vapor pressure of water for various periods of time. These pressures were obtained by exposing the system to a water vapor source which consisted of a container of water maintained in the constant temperature baths listed in Table 1. The triply distilled water was vacuum distilled three times. The vapor pressure given was calculated from the temperature of the water or ice in the container. The deposition was made as soon as a base pressure below 10^{-6} Torr could be obtained after pumping out the water vapor. The copper films were floated from the substrates by dissolving the NaCl in distilled water. Representative sections were mounted on 400 mesh copper grids for observation and microphotography using a Phillips EM-100 electron microscope operated at 80 kV.

The pertinent results from this study are summarized in Table 1 and Fig. 1. The sample letter in Table 1 corresponds to the diffraction and electron micrographs shown in Fig. 1. The degree of single crystallinity of the copper was estimated from the diffraction pattern published by Ino et al. (3). The "extra" spots and rings in Fig. 1 a to e, result from the oxidation of the copper to Cu_2O (18) during its exposure to air at room temperature for removal from the substrate and floating onto grids. This appearance of Cu_2O has been noted by most workers with copper films (19, 20). The conclusions of this work were obtained from the study of the copper pattern, and not that resulting from the oxidation of copper during mounting on the grids. The film thickness was deduced from the optical transmittance at 400 nm by comparison with a calibration of thickness vs. percent transmittance obtained from simultaneous microbalance and optical measurements (16, 21).

It is seen from a in Fig. 1, that after a 432 Torr h exposure at 25 ^{o}C an epitaxed film is obtained. This result was expected based on previous studies (13). The diffraction pattern shown for f shows that thermal etching of air cleaved NaCl precludes the growth of epitaxial films. This result for the copper-NaCl system

Table 1

Percent single crystallinity, and thickness of copper films deposited on NaCl after thermal etching at 450 °C and exposing to water vapor at 325 °C

sample	duration of water vapor exposure (h)	bath	water temperature (°C)	water vapor pressure (Torr)	film thickness (Å)	single crystallinity (%)
a+	18	none	25	23	500	100
b	1	ice-water	0	4.7	1300	100
c	1	$CaCl_2$-water	-42	8.6×10^{-2}	600	100
d	1	CO_2-acetone	-82	2.5×10^{-4}	950	75 to 100
e	30*	none	-	8×10^{-7}	300	75 to 100
f	0**	-	-	-	250	0 to 25

+ The water vapor exposure for this film was at 25 °C.

* Estimated from residual pressure of water vapor in the vacuum system.

** Prior to deposition this film was cooled from 450 to 327 °C over a 1.5 h period at an average pressure of 1×10^{-6} Torr.

is similar to that reported by Matthews for gold on NaCl (11, 22). The diffraction patterns, b to f, show that the water vapor exposure at 325 °C necessary to yield single crystal films deposited at 325 °C and 1 Å/s is between 2.5×10^{-4} and 8.6×10^{-2} Torr. This new result clearly demonstrates that epitaxy can be obtained by exposure to water vapor at elevated temperatures without exposure at 25 °C. Furthermore, within broad limits, the critical exposure of NaCl to water vapor at 325 °C is defined. A continuation of this work has shown that exposures to water vapor at lower substrate temperatures also leads to growth of single crystal films at 325 °C. Thus, the exposure of NaCl to room air at 25 °C may merely provide the necessary source of water vapor to yield the critical exposure on heating of the substrate to the deposition temperature. The minimum exposure required at a series of lower temperatures is being investigated.

After their experiments with water vapor exposure to cleaved NaCl, Mihama et al. (12) suggested that the ratio of the residual pressure of water vapor to deposition rate was the important parameter for growing epitaxial films. Green et al. (8) reported that the ratio of the residual gas pressure of the vacuum system to the deposition rate had to be greater than 1×10^{-4} Torr per Å/s to achieve epitaxy. Based on our results, it seems reasonable to conclude that the ratio of the exposure to water vapor, viz., $\int P_{H_2O}\, dt$, to the deposition rate is the important relationship between residual pressure and deposition rate.

Acknowledgements

The authors are pleased to thank Dr. A.W. Smith for the fruitful discussions about this problem. They are also grateful to the U.S. Atomic Energy Commission for the partial support of this work.

References

(1) D.W. PASHLEY, Adv. Phys. 5, 173 (1956); 14, 327 (1965).

(2) L.E. MURR, Brit. J. appl. Phys. 15, 1511 (1964).

(3) S. INO, D. WATANABE, and S. OGAWA, J. Phys. Soc. Japan 19, 881 (1964).

(4) J.W. MATTHEWS, Phil. Mag. 12, 1143 (1965).

(5) C.F. POWELL, J.H. OXLEY, and J.M. BLOCKER (Eds.), Vapor Deposition, Wiley, New York 1966.

(6) L.E. MURR and M.C. INMAN, Phil. Mag. 14, 135 (1966).

(7) K.L. CHOPRA, Thin Film Phenomena, McGraw-Hill Publ. Co., New York 1969.

(8) A.K. GREEN, E. BAUER, and J. DANCEY, in: Molecular Processes on Solid Surfaces, Eds. E. DRAUGLIS, R.D. GRETZ, and R.I. JAFFEE, McGraw-Hill Publ. Co., New York 1969 (p. 479).

(9) M. HARSDORFF and H. RAETHER, Z. Naturf. 19a, 1497 (1964).

(10) J.W. MATTHEWS and E. GRUNBAUM, Appl. Phys. Letters 5, 106 (1964).

(11) J.W. MATTHEWS and E. GRUNBAUM, Phil. Mag. 11, 1233 (1965).

(12) D. MIHAMA, H. MIYAHARA, and H. AOE, J. Phys. Soc. Japan 23, 785 (1967).

(13) J.N. ZEMEL, in: Solid State Surface Science, Ed. M. GREEN, Marcel-Dekker, New York 1969 (p. 335).

(14) J.S. VERMAAK and C.A.O. HENNING, Phil. Mag. 16, 269 (1970).

(15) P.W. PALMBERG, C.J. TODD, and T.N. RHODIN, J. appl. Phys. 39, 4650 (1968).

(16) E.G. CLARKE, Jr. and A.W. CZANDERNA, J. Vacuum Sci. Technol. 8, 233 (1971).

(17) S. SHIRAI, Proc. Phys. Soc. Japan 20, 855 (1938).

(18) T.N. RHODIN, J. Amer. Chem. Soc. 73, 3143 (1951).

(19) W.A. JESSER and J.W. MATTHEWS, J. Crystal Growth 5, 86 (1969).

(20) L. BRUCK, Ann. Phys. 26, 232 (1936).

(21) A.W. CZANDERNA, in: Ultra Micro Weight Determination in Controlled Environments, Eds. S.P. WOLSKY and E.J. ZDANUK, Wiley-Interscience, New York 1969 (p. 315).

(22) J.W. MATTHEWS, J. Vacuum Sci. Technol. 3, 133 (1966).

(Received October 11, 1971)

phys. stat. sol. (a) 8, K81 (1971)

Subject classification: 10.2; 22.5.2

Research Group on Crystal Physics, Hungarian Academy of Sciences, Budapest

First Stage Coloration of Pure and Ca Doped NaCl at Room Temperature

By

G. CORRADI, I.M. GOMAA, and R. VOSZKA

Introduction The coloration process of alkali halides by ionizing radiation is usually separated into two stages (1, 2) the first of which is strongly affected by impurities, contrary to the second, that is due to an intrinsic process, most probably the creation of Frankel pairs. Unfortunately most measurements have been made on nominally pure crystals, where the OH^- ions and other uncontrolled impurities may considerably change the coloration mechanism. Our purpose was to investigate the first-stage coloration of Ca^{2+} doped and highly pure OH^- free NaCl.

Experimental The NaCl crystals used in the experiments were purified by methods developed in our laboratory (3) and grown OH^- free (4, 5). The basic material was of $2x10^{-7}$ mol/mol purity, controlled for divalent cations. The irradiations were made by a THX 250 X-ray equipment operated at 200 kV, 20 mA. The dose rate on the irradiation site was 3000 R/min. Optical absorption measurements were made on a Unicam SP-700 spectrophotometer. The absorption constant (α) of the F-band maximum was measured as a function of irradiation time (t) during 256 min, using samples with different Ca concentrations (c) verified by ionic conductivity measurements. The growth curves were fitted first graphically and then using a Hewlett-Packard computer to an expression of the form (6)

$$\alpha(t) = A_1(1 - e^{-a_1 t}) + A_2(1 - e^{-a_2 t}) + bt .$$

The values of the parameters are contained in Table 1. The deviation of the experimental and calculated values of α does not exceed 2%. The last row contains F-centre fading data after storage in dark for 5 h.

Table 1

preparation	zone refined	doped	doped	doped	doped
c (mol/mol)	$2x10^{-7}$	$3x10^{-6}$	$1.1x10^{-5}$	$1.6x10^{-4}$	$4.5x10^{-4}$
a_1 (min^{-1})	0.076	0.049	0.041	0.030	0.037
A_1 (mm^{-1})	0.391	0.516	1.19	6.56	10.6
a_2 (min^{-1})	-	0.633	0.452	0.331	0.315
A_2 (mm^{-1})	0.0	0.383	0.513	0.95	1.03
b (mm^{-1} min^{-1})	0.00838	0.00346	0.00458	0.0108	0.0145
F-centre fading in 5 h (%)	0.0	1.9	3.9	3.8	4.4

These data enable us to make the following statements:

1) a_1 and a_2 show little concentration dependence, a_1 can practically be considered as a constant.

2) A_1 and A_2 increase with increasing concentration.

3) At extremely small concentrations ($2x10^{-7}$ mol/mol zone refined samples) the coloration can be described by a linear and only one exponential term.

Discussion The results do not give the possibility for detailed conclusions concerning the coloration mechanism, but some comments can be made. The first-stage coloration consist of two substages having rate constants differing by about one order of magnitude. The fact that the fast process (a_2) is refound in our doped crystals, shows that the fast electronic process suggested for this substage is distinct from the photochemical reaction U→F (6). The process responsible for this substage must be a fast filling of easily available anion vacancies by electrons. The slow increase of A_2 with concentration may be accounted for by the increase of the number of hole traps introduced by doping: the cation vacancies. The simplest cation vacancy-hole association, the V_F-centre, has been found to be stable in Ca doped NaCl at room temperature (7) in contrast to the V_K-centre. The slow process (a_1) seems to be still more governed by the cation vacancies: the increase of A_1 is proportional to the square root of the Ca^{2+} concentration, as this is expected for the number of isolated cation vacancies, and the deviations find an explanation in a paper of Mecseki (8), where the square root dependence of the number of cation vacancies upon Ca^{2+} concentration was shown to be a rough approximation. Thus the process is due either to the Crawford or the interstitial stabilization mechanism (9). The apparent concentration dependence of b, that is opposite to that expected by Pooley (10) does not enable us to far-reaching conclusions because of the short measuring time. Recently Lengweiler et al. (11) found an excellent fit for pure KCl coloration at 85 ^{o}K by one linear and only one exponential term, this was now confirmed for pure NaCl. The small deviations cannot be reduced by introducing a second exponential term.

The high stability of coloration in the pure crystal as opposed to that of the doped samples leave the possibility that a mechanism related to the high dislocation density is active in the pure crystal, explaining the somewhat outstanding value of a_1 for this sample.

References

(1) I. H. SCHULMAN and W. D. COMPTON, Colour Centers in Solids, Pergamon 1963.

(2) J. H. CRAWFORD, Adv. Phys. 17, 93 (1968).

(3) K. RAKSÁNYI and R. VOSZKA, Kristall und Technik 4, 227 (1969).

(4) R. VOSZKA, I. TARJAN, L. BERKES, and J. KRAJSOVSZKY, Kristall und Technik 1, 423 (1966).

(5) R. VOSZKA, K. RAKSÁNYI, and L. BERKES, Kristall und Technik 5, 409 (1970).

(6) J. L. ALVAREZ RIVAS and P. W. LEVY, Phys. Rev. 162, 816 (1967).

(7) A. WATTERICH, M. GÉCS, and R. VOSZKA, phys. stat. sol. 31, 571 (1969).

(8) A. MECSEKI, Acta phys. Acad. Sci. Hungar., to be published.

(9) M. IKEYA, N. ITOH, T. OKADA, and T. SUITA, J. Phys. Soc. Japan 21, 1304 (1966).

(10) D. POOLEY, Proc. Phys. Soc. 89, 723 (1966).

(11) K. LENGWEILER, P. L. MATTERN, and P. W. LEVY, Phys. Rev. Letters 26, 1375 (1971).

(Received October 20, 1971)

phys. stat. sol. (a) 8, K85 (1971)

Subject classification: 8 and 18; 21.1; 21.1.1; 22; 22.6

Bhabha Atomic Research Centre, Chemistry Division, Bombay

Results of a New Method for Determination of the Temperature of Magnetic Transitions

By

M.D. KARKHANAVALA and U.R.K. RAO

Magnetic transitions, e.g., ferro(F) or antiferro(AF) to para(P) are normally studied utilising the variation in magnetic or electrical properties either directly or indirectly by methods such as neutron diffraction or Mössbauer spectroscopy. Since such transitions are of the second order no heat change is involved ($\Delta H = 0$) at the transition temperature (T_t), but the specific heat goes through the well known lambda type transition.

A general survey of the known changes in specific heat as a function of temperature in magnetically ordered systems through the T_t shows that these changes are too small and gradual over a wide temperature interval around T_t to be detected by the conventional DTA apparatus. Hence it was considered worthwhile to investigate such systems using a differential scanning calorimeter (DSC) which is designed to compensate for the variation in temperature of the sample due to either variation in specific heat or to liberation or absorption of heat of reaction. This compensation is achieved by variation in the heater power supplied to the sample or to the reference cups.

The Perkin-Elmer model DSC-1B was used. The manufacturers recommend use of about 5 mg sample to detect first order transition such as crystallisation. However, when magnetic transitions were studied using 5 to 10 mg samples, the peaks obtained were exceedingly small and hence much larger samples (70 to 130 mg) were tried. These gave peaks which were well characterized. The magnetic transitions can be located with extreme ease and precision. The minor variation in T_t of nickel and FeGe may be due to impurities in either the samples used by us or by earlier workers. Besides, variations in heat treatment could possibly also contribute to this difference.

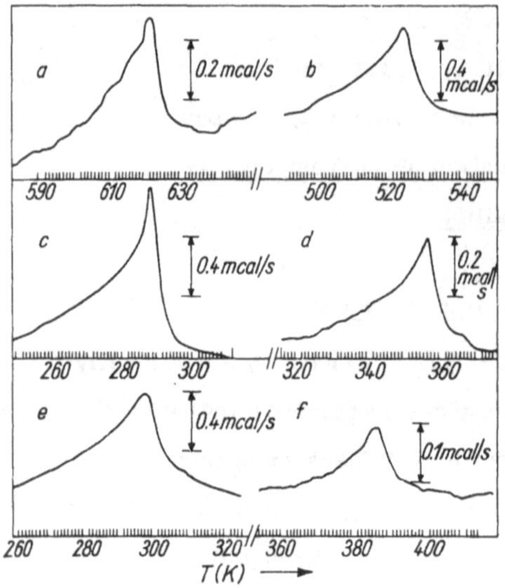

Fig. 1. DSC curves for different magnetically ordered materials through ordering temperatures (heating rate 16 deg/min). a) Ni metal (83 mg); b) NiO (80 mg); c) CoO (77 mg); d) Fe_2As (92 mg); e) Mn_5Ge_3 (129 mg); f) FeGe (119 mg)

The advantages of this new way of determining such transition temperature are:

1. Although the specific heat is on an arbitrary scale when using DSC, yet with a sample as small as 100 mg the transition temperature can be determined within 10 min. Conventional methods of determining specific heat require much larger sample and are time consuming as the specific heat has to be measured at each temperature. Same is true with techniques involving measurement of magnetic or electric properties or those involving magnetic interaction. We could not make the study quantitative since we do not as yet have the specific heat kit which is available as an accessory for the DSC.

2. The DSC permits traverse through the transition temperature at rates of heating and cooling varying from 0. 5 to 64 deg/min in both up and down directions. We have found that 16 and 32 deg/min yielded the best results.

3. Usually the peaks are sharp and the temperature can be read to within one degree, thus making it superior to the magnetic susceptibility studies where it is often difficult to precisely locate the transition temperature.

4. Inspite of several thermal cyclings through the transition temperature, the T_t as observed on DSC is reproducible. This is not always true in the conventional magnetization and electrical resistivity measurements.

5. In some exceptional cases such as $FeGe_2$ where magnetic susceptibility failed (5) to detect the AF ⟶ P transition, two typical lambda transitions (6) at 371 and 389 K were observed using DSC.

6. It is interesting that in case of B35 type FeGe, several neutron diffraction studies (7, 8) failed to observe ordering while susceptibility (9) and Mössbauer (10)

Table 1

sample No.	substance	type of magnetic order	transition temperature (K)	
			reported	observed
1	nickel metal	AF	631* (1)	623
2	nickel oxide	AF	523* (1)	523
3	cobalt oxide	AF	290* (1)	289
4	Fe_2As	AF	350+ (2)	357
5	Mn_5Ge_3	F	300+ (3)	298
6	FeGe	AF	412+ (4)	389

*From specific heat data.

+Specific heat data not available. These values are from other data such as magnetization, neutron diffraction.

studies indicated magnetic ordering. In this compound we have obtained a weak maximum around 389 K using the DSC at its maximum sensitivity (1 mcal/s full scale), which clearly indicates a second order transition.

We thank our colleagues in Bhabha Atomic Research Centre, Dr. N.S. Satyamurty of Nuclear Physics Division for a sample of FeGe and Dr. R. Krishnan of Metallurgy Division for advice and assistance in making some alloys.

References

(1) A. GOLDSMITH, T.E. WATERMAN, and H.J. HIRSCHORN, Handbook of Thermophysical Properties of Solid Materials, Revised Edition, Vol. 1 and 3, Macmillan Co., New York 1961.

(2) H. KATSURAKI and N. ACHIWA, J. Phys. Soc. Japan 21, 2238 (1966).

(3) R. FONTAINE and R. PAUTHENET, C.R. Acad. Sci. (France) 254, 650 (1962).

(4) K.A. BLOM, O. BECKMAN, and M. RICHARDSON, Solid State Commun. 5, 977 (1967).

(5) G. AIROLDI and R. PAUTHENET, C.R. Acad. Sci. (France) 258, 3994 (1964).

(6) J.V. YAKHMI, M.Sc. Thesis, Bombay University, under preparation.

(7) N.S. SATYAMURTY, R.J. BEGUM, C.S. SOMANATHAN, and M.R.L.N. MURTY, Solid State Commun. 3, 113 (1965).

(8) E. ADELSON and A.E. AUSTIN, J. Phys. Chem. Solids 26, 1795 (1965).

(9) T. OHOYAMA, K. KANEMATSU, and K. YASUKOCHI, J. Phys. Soc. Japan 18, 589 (1963).

(10) S. TOMIYOSHI, H. YAMAMOTO, and H. WATANABE, J. Phys. Soc. Japan 21, 709 (1966).

(Received October 27, 1971)

phys. stat. sol. (a) 8, K89 (1971)

Subject classification: 13.4 and 14.3; 22.2.1

Institut für Physikalische Elektronik, Technische Hochschule Wien
and Ludwig-Boltzmann-Institut für Festkörperphysik, Wien

Ionized Impurity Density and Mobility in n-GaAs[1)]

By

D. KRANZER and G. EBERHARTER

Recently Wolfe et al. (1) havengiven empirical curves by which the total impurity concentration can be estimated if the mobility μ and carrier concentration n is known at 77 K only. However, lattice scattering is neglected completely in their treatment. Using the relaxation time approximation Wolfe et al. (2) calculated the temperature dependence of μ in high purity epitaxial n-GaAs. The relaxation time approximation, however, is not a priori strictly valid because of scattering by polar optical modes. The purpose of this paper is: (i) to reanalyse Wolfe's et al. (2) results (Fig. 3 in reference (2)) avoiding any serious approximations in the calculation, (ii) to compare our theoretical with the empirical results obtained by Wolfe et al. (1), and (iii) to show that in a wide lattice temperature range a good agreement between the experimental and theoretical temperature dependence of μ is obtained if combined lattice and ionized impurity scattering is taken into account only. A similar agreement was found by Rode and Knight (3) using a different method to calculate μ. Measurements at very low temperatures, however, should not be included because the mobility is affected by hopping, impurity band, or degenerate conduction if the impurity concentration is high. Therefore we restrict ourselves to lattice temperatures greater than 40 K (our temperature range being 40 to 300 K).

The samples used in this study were n-type epitaxial layers of GaAs. The resistivity ϱ and the Hall constant were measured by the van der Pauw method on 5x5x0.02 mm^3 samples with a magnetic field of 4.5 kG. Four point contacts were arranged in each corner of the sample. In this case there are two possibilities to measure the resistivity and two possibilities to measure the Hall constant. It is important to note that these properties of the van der Pauw method enables one to check the uniformity of the samples. In this investigation only samples were used

1) This work was sponsored by the Fonds zur Förderung der wissenschaftlichen Forschung.

for which both values of the resistivity and of the Hall constant were in good agreement.

The temperature was measured with a Au-Fe thermocouple wire (5). In the temperature range considered the dominant scattering mechanism are (2): polar optical phonon, acoustic deformation potential, electric acoustic phonon, ionized impurity. Since the relaxation time approximation fails for scattering by polar optical modes we calculated the mobility by means of the method reported in (4, 6, 7). Further assumptions are homogeneous material and an energy-momentum relation according to Kane (8): $p^2 = 2m^* \varepsilon (1 + \frac{\varepsilon}{\varepsilon_g})$, where m^* is the effective mass at the edge of the conduction band and ε_g the energy gap.

For small electric fields E the solution of the Boltzmann equation may be written as

$$f(\mathfrak{p}) = f_o(\varepsilon) + \frac{\mathfrak{p}\mathfrak{E}}{p E} f_1(\varepsilon) . \qquad (1)$$

$f_o(\varepsilon)$ is the Maxwell-Boltzmann distribution and $f_1(\varepsilon)$ is the unknown function. To calculate $f_1(\varepsilon)$ we insert $f(\mathfrak{p})$ into the Boltzmann equation. We obtain the same difference equation as Grigorev and Dykman (6) which was solved numerically. In this difference equation there are a number of parameters for which we chose values according to Wolfe et al. (2). We want to emphasize that this calculation yields the mobility μ exactly to first order in the electric field E.

Results for three samples listed in Table 1 are shown in Fig. 1. At 40 K μ is fitted by means of the total ionized impurity concentration. We obtain a good agreement in the whole lattice temperature range considered (Fig. 1). To obtain the total ionized impurity concentration it is essential to know the effect of anomalous scattering mechanisms (9, 10) on the mobility. A large temperature range must be considered to study this anomaly. Within our experimental accuracy our samples did not show any effect due to "mobility killers" in the temperature range considered. The agreement between the theory and the experiment can be regarded as an additional check of the uniformity of our samples. In Fig. 2 we compare Wolfe's et al. (1, 2) and our results. Concerning the relaxation time approximation a deviation is found for low as well as high impurity densities. Because of the strong inelasticity of the polar optical scattering the discrepancy must be attributed to the approximate

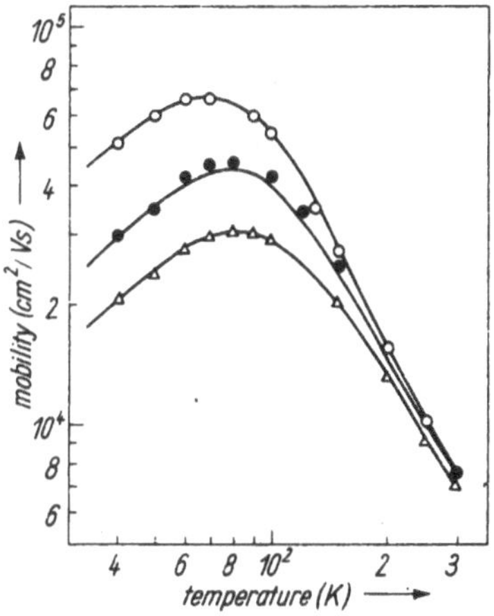

Fig. 1

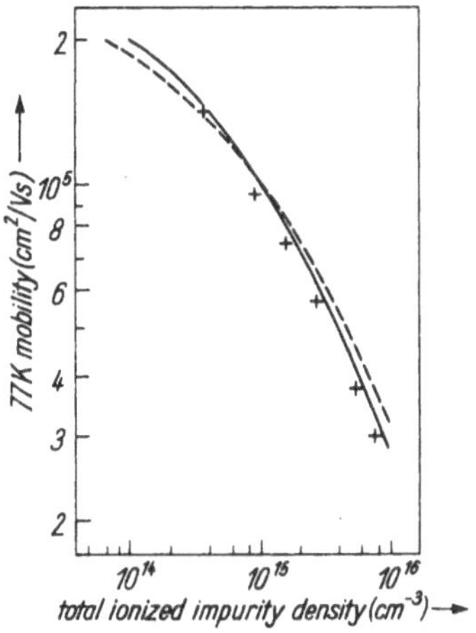

Fig. 2

Fig. 1. Temperature dependence of the mobility for different samples listed in Table 1. △ sample 1, ● sample 2, ○ sample 3, — calculated

Fig. 2. Mobility vs. total ionized impurity concentration (N_A is the total compensating acceptor concentration, N_D is the shallow donator concentration) —— present work, -- Wolfe et al. (2), + Wolfe et al. (1)

Table 1

sample No.	μ (T = 77 K) (cm^2/Vs)	n (cm^{-3})	$N_A + N_D$ (cm^{-3})
1	3.0×10^4	1.15×10^{15}	5.4×10^{15}
2	4.6×10^4	4.3×10^{14}	2.95×10^{15}
3	6.6×10^4	9.85×10^{14}	1.88×10^{15}

consideration of the polar optical scattering. The agreement between the empirical and our calculated total ionized impurity concentration is improved as compared with the relaxation time approximation. We want to emphasize, however, that the experimental error is usually comparable with the deviations found between the different curves in Fig. 2. These results justify the application of the simple relaxation time approximation to analyse mobility experiments at 77 K in n-GaAs.

Strictly speaking the final value derived for the compensation ratio remains an adjustable parameter. However, Wolfe et al. (1) have estimated the compensation ratio by Hall constant analysis and mobility analysis in n-GaAs. They found that the compensation ratio obtained by different methods agrees very well. On the other hand, our results (Fig. 2) are in better agreement with the empirical curve (1) than the mobility analysis by Wolfe et al. (2). These results suggest that the compensation ratios derived from our mobility analysis are reliable.

Acknowledgements

We wish to thank Prof. Dr. H.W. Pötzl for his interest in our work and many valuable discussions. We are very thankful to the Institut für Numerische Mathematik, Technische Hochschule Wien, for being allowed to use the computer.

References

(1) C.M. WOLFE, G.E. STILLMAN, and J.O. DIMMROCK, J. appl. Phys. 41, 504 (1970).

(2) C.M. WOLFE, G.E. STILLMAN, and W.T. LINDLEY, J. appl. Phys. 41, 3088 (1970).

(3) D.L. RODE and S. KNIGHT, Phys. Rev. B3, 2534 (1971).

(4) A. FORTINI, D. DIGUET, and J. LUGAND, J. appl. Phys. 41, 3121 (1970).

(5) R.L. ROSENBAUM, Rev. sci. Instrum. 39, 890 (1968).

(6) N.N. GRIGOREV and I.N. DYKMAN, Soviet Phys.-Solid State 10, 837 (1968).

(7) D. KRANZER, phys. stat. sol. 46, 591 (1971).

(8) E.O. KANE, J. Phys. Chem. Solids 1, 249 (1957).

(9) L.R. WEISBERG, J. appl. Phys. 33, 1817 (1962).

(10) C.M. WOLFE, Appl. Phys. Letters 18, 205 (1971).

(Received October 27, 1971)

phys. stat. sol. (a) 8, K93 (1971)

Subject classification: 14.3; 1.2; 22

Centre d'Études d'Électronique des Solides, associé au C.N.R.S.,
Université des Sciences et Techniques du Languedoc, Montpellier

On the Electrical Properties of Cu_xS

By

J. BOUGNOT, F. GUASTAVINO, H. LUQUET, and D. SODINI

We present experimental results on the electrical properties of Cu_xS for x between 1.85 and 2. In our work, sulfides are melted and afterwards crystallised by the Stockbarger-Bridgman method. We performed two series of conductivity measurements: for temperatures above 300 $^{\circ}K$ and at low temperatures.

1. Conductivity for 300 < T < 900 $^{\circ}K$ We note two peaks: the first situated at around 90 $^{\circ}C$ corresponding to the passage of monoclinic chalcocite to hexagonal chalcocite, the second lying between about 400 and 90 $^{\circ}C$ according to the content of sulphur ($1.85 < x \leqq 2$) (Fig. 1 and 2). When we compare the conductivity and the D.T.A. plots for samples cut from the same crystal, it appears that apart from a slight shift in temperature, for each discontinuity on the conductivity curves we have a phase transformation peak (D.T.A.), in particular at high temperatures, corresponding to the passage to high digenite as Hirahara (1) suggested. After the abrupt increase to about 440 $^{\circ}C$, the samples with concentration near 2 show a negative slope as in Hirahara's curves (2) (Fig. 1). This slope becomes positive as the molar copper to sulphur ratio drops below 2.

The variation of conductivity before the second phase transition may be interpreted by the combination of the conductivities of high digenite and hexagonal chalcocite which composed the sample considered. This explains the shift of the position and the value of the slope vs. the concentration; for samples near Cu_2S this slope decreases abruptly as the influence of high digenite conductivity increases. Gassanova (3) and Abdullaev et al. (4) interpreted this abrupt increase, for Cu_2S samples, as the passage to intrinsic character and give the values 1.71 and 1.8 eV respectively, for the corresponding activation energy.

For the concentration x = 1.90 we find the same curve as Kamigachi (5), with a difference of a factor 10 in the value of σ. Our curves show no great variation for

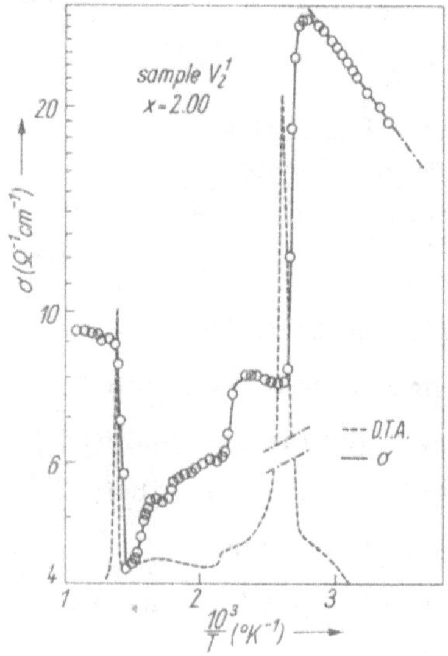

Fig. 1

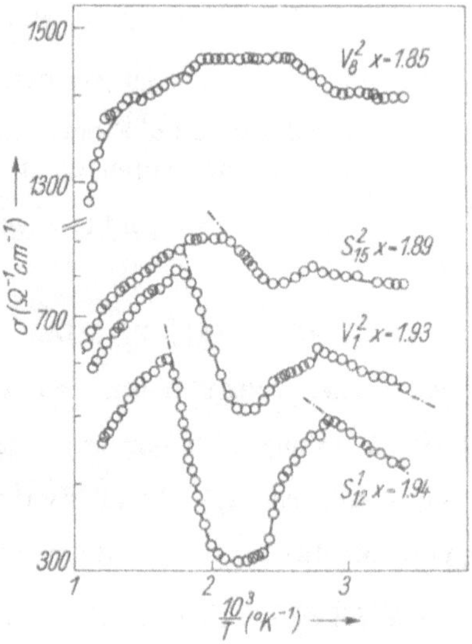

Fig. 2

samples with a higher sulphur percentage and could be situated between Kamigachi's curves for x = 1.90 and x = 1.85. In this range of concentration the inaccuracy in x determination becomes greater, and the conditions of fabrication could influence the value of the conductivity.

2. Conductivity at low temperatures $77 < T < 300$ °K On Fig. 3 we note that, for temperatures below the first phase transition, Cu_xS shows a classical semiconductor resistivity which changes to a metallic one as we move from chalcocite

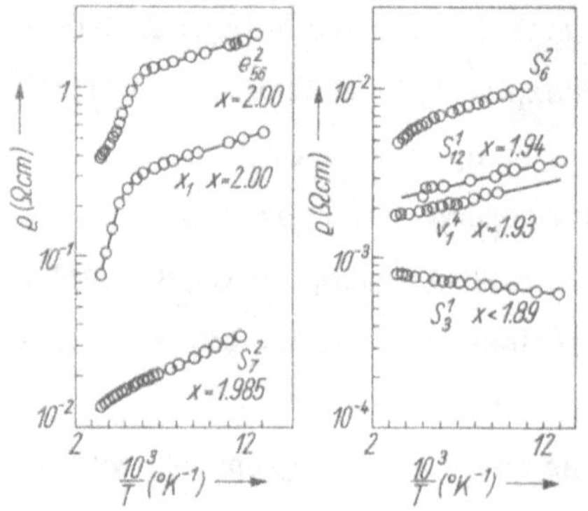

Fig. 3

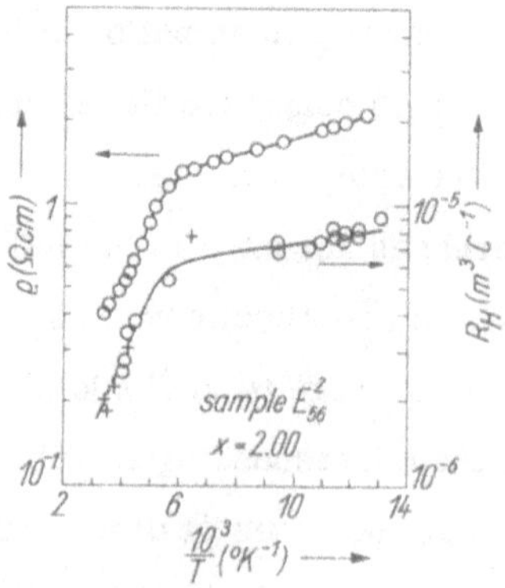

Fig. 4

Table 1

sample No.	x	σ (Ω^{-1} cm^{-1})	p (cm^{-3})	μ (cm^2 V^{-1} s^{-1})	E_1 (eV)	E_2 (eV)	E_3 (eV)
V_2^1	2.00	19				0.11	1.80
V_{27}	1.98	86				0.12	0.80
S_{12}	1.94	430	4×10^{20}	7	0.010	0.06	0.42
V_1	1.93	550	7×10^{20}	5	0.01	0.04	0.28
V_8	1.85	1400	1.8×10^{21}	5			

to digenite. For chalcocite (Cu_2S) the variation of the resistivity and the Hall constant R_H show two clearly defined slopes (Fig. 4) and when we leave this composition, the first slope, corresponding to the highest activation energy, tends to disappear and the temperature of the point of discontinuity approaches 300 ^oK.

3. Results For samples whose composition is between 2 and 1.85, we obtained, at room temperature, the results presented Table 1, where E_1 and E_2 correspond to the two slopes situated below 90 ^oC and E_3 corresponds to the steep slope near the second phase transition. (The determinations of E_2 and E_3 are made in short temperature interval.)

An analysis the results in the table brings out an increase in the number of carriers and a decrease of activation energies (E_2 and E_3) as we leave the copper rich limit; the variation of the mobility vs. x does not appear exactly because the precision of our measurements is not sufficient. From a first study on about twenty samples, we could presume a simple law for the variation of σ with x ($1.85 < x \leq 2.00$) with apparent discontinuities where normally a simple compound exists. Our future studies should allow us to establish a relation between E_3, p, and x by optical and electrical measurements.

References

(1) E. HIRAHARA, J. Phys. Soc. Japan 6, 422 (1951).

(2) E. HIRAHARA, J. Phys. Soc. Japan 6, 428 (1951).

(3) N. A. GASSANOVA, Izv. Akad. Nauk Azerbaidzhan SSR, Ser. fiz. mat. tekh. Nauk 3, 91 (1963).

(4) G. B. ABDULLAEV, Z. A. ALIYAROVA, E. A. ZAMANOVA, and G. A. ASADOV, phys. stat. sol. 26, 65 (1968).

(5) T. KAMIGACHI, J. Sci. Hiroshima Univ. A16, 325 (1952).

(Received November 1, 1971)

phys. stat. sol. (a) 8, K97 (1971)
Subject classification: 6.1; 22.8.1

Institute for Telecommunications (a), Central Research Institute for Physics (b), and Roland Eötvös University of Sciences (c), Budapest

Mössbauer Study of the Fe Ions in CaVBi Garnets

By

E. BEREGI (a), L. CSER (b), T. STERK (a), A. VÉRTES (c),
I. VINCZE (b), and B. ZSOLDOS (a)

The Mössbauer spectra of Fe ions in CVB garnets which can be described by the general formula $\left\{Bi^{3+}_{3-2x} Ca^{2+}_{2x}\right\} Fe^{3+}_{2} \left(Fe^{3+}_{3-x} V^{5+}_{x}\right) O^{2-}_{12}$ were measured on samples prepared from metallic oxides by powder metallurgy.

The Mössbauer spectra were taken at room temperature by an automatic Mössbauer spectrometer of constant acceleration. Spectra were decomposed by a computer assuming a Lorentzian curve for each spectral line. The parameters of two typical spectra are listed in Table 1.

The spectra can be divided into three sets of six lines. The set (of six lines) showing the highest internal field with the lowest intensity clearly indicates the existence of α-Fe_2O_3, which means that a part of the base material has not undergone any reaction.

The iron oxide concentration in the two samples with x = 1.45 and 1.25, sintered at 1150 and 1100 $^{\circ}$C, respectively, were 1.6 and 4.5%. A greater fraction of base material remained unchanged by sintering at lower temperatures.

The two intense sixtets specified in the table result from Fe ions located at either octahedral or tetrahedral sites. The problem is to identify the sites of the Fe ions from the corresponding lines.

The chemical composition indicates that there is a larger number of iron ions at octahedral than at tetrahedral lattice sites and consequently the higher intensity sixtet would be attributed to the former type.

However, the ratio of the area in the two cases does not show numerical agreement with the ratio of the numbers of iron atoms. One of the reasons for this discrepancy is that the fitting by Lorentzian curves is not good for the analysis of the Mössbauer spectrum. This fact may introduce systematic errors in the computed

Table 1

	x = 1.45			x = 1.25		
a_1	2900	± 170	(arb. units)	1900	± 100	(arb. units)
a_2	2000	± 180	(arb. units)	840	± 110	(arb. units)
δ_1	0.416	± 0.01	(mm/s)	0.367	± 0.02	(mm/s)
δ_2	0.356	± 0.01	(mm/s)	0.589	± 0.04	(mm/s)
H_1	403.6	± 1.4	(kOe)	397.0	± 1.4	(kOe)
H_2	373.4	± 1.3	(kOe)	437	± 3	(kOe)
Γ_1	0.79	± 0.09	(mm/s)	0.78	± 0.06	(mm/s)
Γ_2	0.57	± 0.06	(mm/s)	0.69	± 0.14	(mm/s)

a_i are the amplitudes, δ_i the chemical shifts, H_i the hyperfine fields, Γ_i the linewidths.

areas. Because this, and considering the deviations from the results of Bokov et al. (1), the identification based on the relative intensities of the lines is doubtful.

The quadrupole splitting could not be well observed and did not yield any valuable information about the microstructural environments, since the quadrupole term becomes zero for cubic symmetry.

The markedly broadened linewidth may be due to fluctuation of the composition. It is obvious that a change in x will induce a change in the environment of the iron atoms. It follows from the crystal structure that the next neighbours of the atoms in octahedral sites are those occupying tetrahedral and Bi-type sublattices. With increasing x the effective valence charge becomes larger at the tetrahedral sites and smaller at the Bi-type sites and the two changes compensate each other. The situation is quite different in the environment of tetrahedral sites, which consist of octahedral and Bi-type sites. While variation in x does not lead to a change in the population of the octahedral sites, the effective charge on the Bi-type sites decreases with increasing x. Consequently, the chemical shift on the average does not change at octahedral sites, while at tetrahedral sites it does. It is clear that the identification from the area ratio can thus be supported by the chemical shift.

Similar inferences can be made from the concentration dependence of the hyperfine magnetic splitting. The hyperfine field at iron atoms in octahedral sites does not change appreciably, while that at the tetrahedral sites exhibits a substantial variation. Since there is no contribution from the orbital momentum, the total change is attributable to a change of the 4s-electron density. As shown by Watson and Freeman (2), the contribution from the 4s-electrons has a positive sign, while the resultant hyperfine field has a negative sign. With increasing chemical shift the number of 4s-electrons, and hence their contribution to the hyperfine field, decreases and the value of the resultant negative field grows. Mössbauer data on FeF_3 (3) and α-Fe_2O_3 (4) show that the chemical shift and internal field of the more ionic FeF_3 are appreciably higher than those of α-Fe_2O_3. These results are in good agreement with our model.

References

(1) V.A. BOKOV, G.V. POPOV, and S.I. YURCHUK, Fiz. tverd. Tela 11, 1994 (1969).

(2) R.E. WATSON and A.J. FREEMAN, Phys. Rev. 123, 2027 (1961).

(3) O.C. KISTNER and A.W. SUNYAR, Phys. Rev. Letters 4, 412 (1960).

(4) G.K. WERTHEIM, Mössbauer Effect, Principles and Applications, Academic Press, New York 1964.

(Received November 4, 1971)

phys. stat. sol. (a) 8, K101 (1971)

Subject classification: 10.2 and 20.3; 22.5.3

Physics Department, Indian Institute of Technology, Kharagpur

Cathodoluminescence of CaF_2:Mn and CaF_2:Ce:Mn Crystals

By

H. D. BANERJEE and V. V. RATNAM

It is an established fact that characteristic luminescence emission and colour centre formation have the same origin in alkali halides (1 to 3). Characteristic luminescence is pronounced at low temperatures and stable colour centres are formed at temperatures above which the characteristic luminescence has ceased. These phenomena are well explained by Hersh's excitonic mechanism. According to this, the progenitors for characteristic luminescence and colour centre formation are the bimolecule halide ions in the excited states, $X_2^{=*}$, or excitons formed of an X_2^- molecule (V_k or V_H centre) and an electron. Luminescence occurs when the (e + V_k, V_H) exciton radiatively decays to the ground state according to the equation

$$X_2^{=*} \; ((e + V_k, V_H) \text{ exciton}) \longrightarrow 2X^- + h\nu'c_L . \qquad (1)$$

Difficulties are encountered when these ideas are applied to the system CaF_2. In CaF_2, characteristic luminescence occurs with high efficiency from LNT to RT, but ionising radiations as such are not very effective in forming intrinsic colour centres. With a view to understand the nature and behaviour of the luminescence centres in crystals of Mn and Ce, Mn doped CaF_2, the present measurement on the cathodoluminescence of single crystals of CaF_2:Mn(1%) and CaF_2:Ce(0.4%): :Mn(2.4%) was undertaken. The formation of colour centres in these systems will be reported in the subsequent note.

Fig. 1 shows the cathodoluminescence spectra of 1% Mn doped CaF_2, irradiated by 10 kV, 40 μA cathode rays at temperatures as indicated. For comparison, the corresponding spectra for undoped CaF_2 are shown in Fig. 2. Fig. 3 shows the spectra for CaF_2:Ce(0.4%):Mn(2.4%).

The prominent bands of the spectra in undoped and manganese doped samples are the characteristic luminescence emission (occurring according to equation (1) (reference (4)), the one occurring at 280 nm is due to (e + V_k) exciton recombination

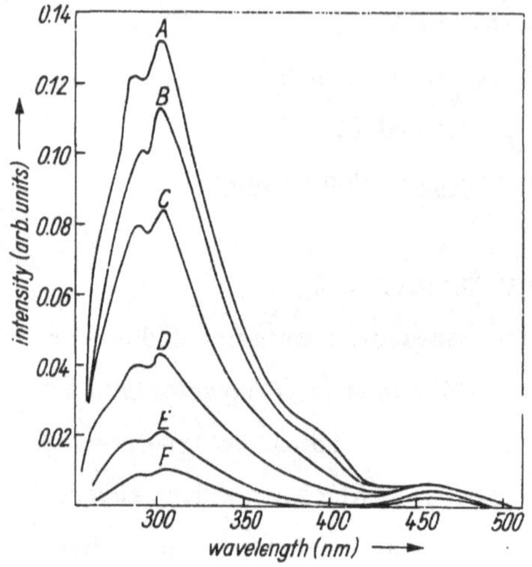

Fig. 1

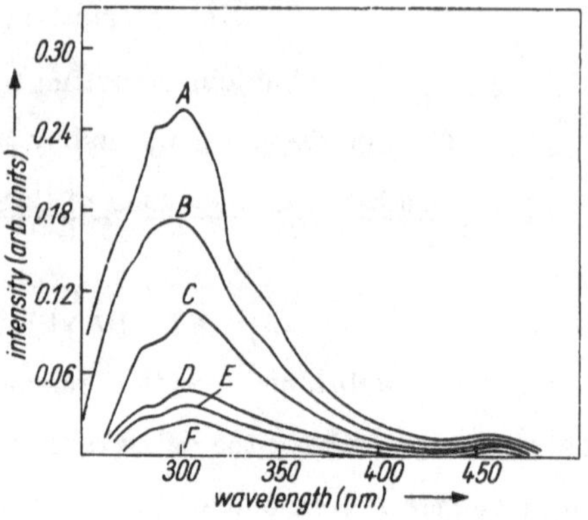

Fig. 2

Fig. 1. Cathodoluminescence spectra of single crystal of CaF_2:Mn(1%) at different temperatures as indicated. The crystal is bombarded by 10 kV, 40 μA cathode rays; (A) at 28 ^oC, (B) at 50 ^oC, (C) at 100 ^oC, (D) at 150 ^oC, (E) at 200 ^oC, (F) at 250 ^oC

Fig. 2. Cathodoluminescence spectra of single crystal of CaF_2 (undoped) at different temperatures as indicated. The crystal is bombarded by 10 kV, 40 μA cathode rays; (A) at 28 ^oC, (B) at 50 ^oC, (C) at 100 ^oC, (D) at 150 ^oC, (E) at 200 ^oC, (F) at 250 ^oC

luminescence and the other occurring at 300 nm is probably due to (e + V_H) luminescence. With rise in temperature, the band peaks shift and their intensities decrease. The variations of the total emission, measured by the areas under the curves in the 250 to 400 nm region are plotted as a function of temperature and shown in Fig. 4. For the undoped and Mn doped samples, the intensities diminish monotonically with temperature. Larger values of the intensities for the Mn doped sample occurring in the same region as for the undoped sample at corresponding temperatures, should be attributed to the formation of a large number of $F_2^{=*}$ molecules, made available by the incorporation of Mn. This is possibly due to the existence of a large number of F^- intersititials prior to irradiation. These F^- interstitials, during irradiation, help the formation of a large number of V_k and V_H centres resulting in large emission intensities. The characteristic emission of Mn is not present in the luminescence

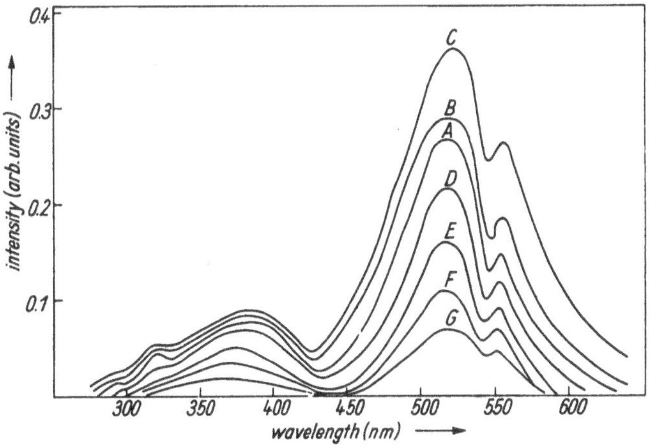

Fig. 3. Cathodoluminescence spectra of single crystal of CaF_2:Ce(0.4%):Mn(2.4%) at different temperatures as indicated. The crystal is bombarded by 10 kV, 40 μA cathode rays; (A) at 28 °C, (B) at 50 °C, (C) at 75 °C, (D) at 100 °C, (E) at 150 °C, (F) at 200 °C, (G) at 250 °C

spectra because direct excitation of Mn^{2+} to higher excited states is forbidden.

In the sample doped with Ce and Mn, the spectrum is markedly different. Along with a strong Mn emission, it consists of a weak band at 280 nm (V_k recombination luminescence), a weak band around 320 nm (Ce emission) and a fairly strong band around 380 nm. There are two interesting features of the emission spectra in the Ce and Mn doped samples. Firstly, the characteristic luminescence bands are greatly suppressed and secondly, the overall emission intensities increase with temperature upto 75 °C and decrease thereafter with temperature (Fig. 4). The suppression of the V_k and V_H luminescence emission can be attributed to the fact that this emission of the parent lattice is readily absorbed by the Ce^{3+} ions which have a characteristic absorption band at 305 nm; it may also be explained by saying that the (e + V_k, V_H) excitons are destroyed at the Ce^{3+} sites, leaving the latter in excited states. The release of this absorbed energy gives the observed emission. The increase of intensities upto 75 °C and their fall thereafter upto 250 °C can be understood in the following way: Ce^{3+} ions in cubic sites (which form only 10% of the total number) are favourable for absorbing the excitonic energy. So at room

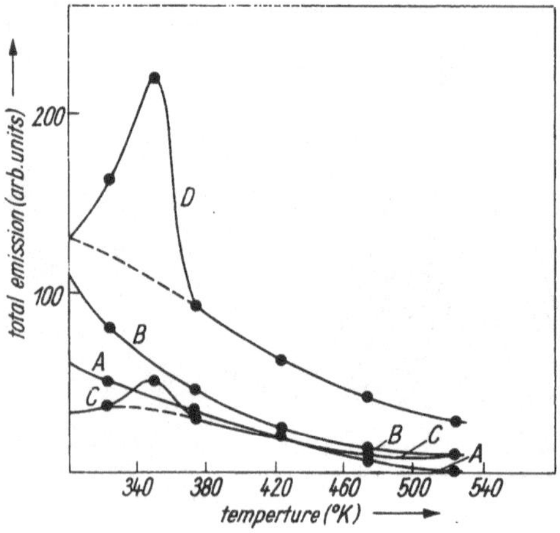

Fig. 4. Variation of cathodoluminescence intensities with temperature. (A) UV emission for CaF_2 (undoped); (B) UV emission for CaF_2:Mn(1%); (C) UV (Ce emission), and (D) visible (Mn emission) for CaF_2:Ce(0.4%):Mn(2.4%). Dashed lines in (C) and (D) indicate the trend of variation in the absence of the transient effects due to the presence of cerium

temperature it is only these ions that are excited. As the temperature is raised, more of the Ce^{3+} ions are freed from their local compensators (4, 5), making a large number of Ce^{3+} ions available in cubic sites. This increases the Ce absorption, consequently the UV emission due to Ce and the visible emission due to Mn are also increased. This seems to be happening in the temperature range 50 to 75 ^{o}C. Above 75 ^{o}C more of Ce^{3+} ions should be in cubic sites, but some of them are probably converted into Ce^{2+} ions by trapping electrons, effectively reducing the number of Ce^{3+} ions in cubic sites. As a result of this, the overall emission intensities would be less. At still higher temperatures nonradiative effects predominate and the emission intensities are further reduced. Thus as shown in Fig. 4, the initial increase and the subsequent decrease of emission intensities, observed between 28 and 100 ^{o}C, are transient effects superimposed on the otherwise regular behaviour observed in undoped and Mn doped samples.

In conclusion it may be pointed out that the (e + V_k, V_H) excitons are playing a major role in determining the luminescence characteristics of doped samples also.

The authors wish to acknowledge their gratefulness to Prof. H. N. Bose, Head of the Department of Physics for helpful discussions.

References

(1) H. N. HERSH, Phys. Rev. 148, 928 (1966).

(2) M. N. KABLER, Phys. Rev. 136, A1296 (1964).

(3) R. B. MURRAY and F. J. KELLER, Phys. Rev. 137, A942 (1965).

(4) J. H. BEAUMOUT, W. HAYES, D. L. KIRK, and G. P. SUMMERS, Proc. Roy. Soc. A315, 69 (1970).

(5) J. L. MERZ, Ph. D. Thesis, Harvard University, 1966 (available as Technical Report No. 514, Office of the Naval Research, USA).

(Received November 1, 1971)

phys. stat. sol. (a) 8, K107 (1971)

Subject classification: 10.2 and 20.1; 22.5.3

Physics Department, Indian Institute of Technology, Kharagpur

Colour Centres in CaF_2:Mn and CaF_2:Ce:Mn Crystals

By

V.V. RATNAM and H.D. BANERJEE

The system CaF_2 has proved itself to be uniquely different from other ionic crystals, particularly alkali halides, as regards the formation of colour centres by ionising radiations (1 to 4). Earlier attempts to produce intrinsic colour centres in this system by ionising radiations have failed because free negative ion vacancies do not seem to exist prior to irradiation. Even if they are produced by irradiation, they do not seem to remain isolated, since these vacancies produced by irradiation get associated (4). Zintal and Udgard (5) and Ure (6) have actually shown that they exist in an associated form. Only under certain experimental conditions, irradiation can generate free negative ion vacancies in undoped crystals, e.g., X-ray irradiation in the temperature range 180 to 250 ^{o}C (4) and cathode ray irradiation in the temperature range 150 to 300 ^{o}C (7).

In this note some new features of cathode ray colouration in Mn doped and Ce, Mn doped CaF_2 are reported. (i) In the crystal doped with Mn (1 mol%), cathode ray irradiation at 28, 50, and 100 ^{o}C forms the α-band at 375 nm. (ii) In the same crystal cathode ray irradiation at 150, 200, and 250 ^{o}C both the α- and β-bands are formed as in undoped crystals; the α-band is formed in the same position (375 nm) but the β-band shifts from 580 nm at 100 ^{o}C to 540 nm at 200 ^{o}C. (iii) Upto 100 ^{o}C, the α-band formed, exclusive of the β-band, gets saturated at a small irradiation dose (1 h of 10 kV, 40 μA cathode rays) (Fig. 1). In undoped crystals the α- and β-bands formed at temperatures above 100 ^{o}C are due to the free negative ion vacancies produced by irradiation (7); the similarity in the behaviour of these bands in both systems (undoped and doped) suggests that they are of the same origin. The α-band formed (upto 100 ^{o}C), only in Mn doped samples must then be due to negative ion vacancies of a different origin. Since room temperature (28 ^{o}C) irradiation cannot produce these free vacancies they must have been ex-

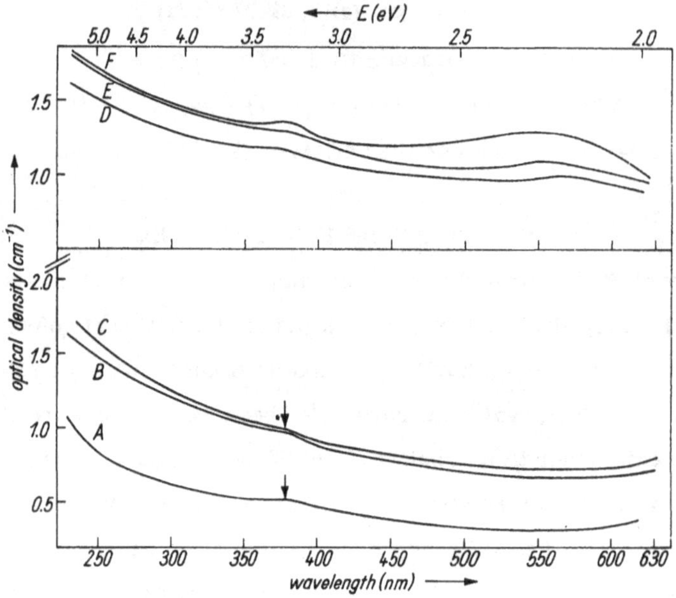

Fig. 1. Optical absorption spectrum of single crystal of CaF_2:Mn(1%) measured at 28 ^{o}C. The crystal is irradiated by 10 kV, 40 μA cathode rays at temperatures as indicated: (A) at 28 ^{o}C, (B) at 50 ^{o}C, (C) at 100 ^{o}C, (D) at 150 ^{o}C, (E) at 200 ^{o}C, (F) at 250 ^{o}C

isting prior to irradiation. Irradiation upto 150 ^{o}C is unable to produce free negative ion vacancies because these vacancies, even if formed by irradiation, must be getting associated with F ion interstitials to form A. F. pairs since the energy of formation of a free F ion vacancy (4.6 eV) is much larger than the energy of formation of a negative ion vacancy-interstitial pair (2.7 eV) (8).

The absorption spectra of a crystal containing both Ce and Mn (Ce(0.4%) and Mn(2.4%)) show an entirely different behaviour as regards the colour centre formation by cathode rays is concerned. The two main features noticed are: (i) the nonformation of the α- and β-bands at any temperature of irradiation and (ii) the absence of any perceptible change in the strength of the absorption bands (246 and 305 nm bands), which exist before irradiation (Fig. 2). These facts indicate that free negative ion vacancies are not made available at any temperature of irradiation and the centres giving rise to the above two absorption bands are little affected by

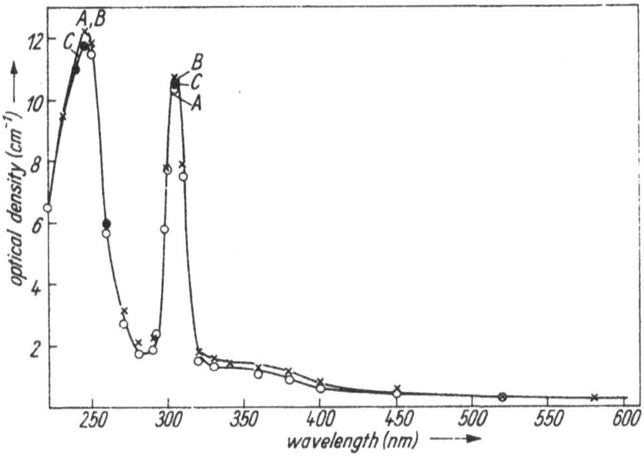

Fig. 2. Optical absorption spectrum of single crystal of CaF_2:Ce(0.4%): Mn(2.4%) measured at 28 °C. The crystal is irradiated by 10 KV, 40 μA cathode rays at temperatures as indicated: (A) at 28 °C, (B) at 50 °C, (C) at 75 °C, (D) at 100 °C, (E) at 150 °C, (F) at 200 °C; (A)○○○, (B)×××, (C), (D), (E), (F)●●●

irradiation. These two results can be understood in the following way: the rapid diffusion of the large flux of neutral fluorine atoms formed by irradiation, which would favour the formation of free negative ion vacancies, does not seem to take place in the Ce doped samples possibly because these trapped holes are partly destroyed and partly made available. The results on the cathodoluminescence spectra of this sample (previous note) have, in fact, indicated that these hole centres like V_k, V_H form the self trapped excitons (e + V_k, V_H), which are readily destroyed at the Ce^{3+} sites by parting their energy with Ce^{3+} ions. Hence free negative ion vacancies normally formed at higher temperatures of irradiation in undoped and Mn doped samples are not formed in this system. The energy received by Ce^{3+} ions is released partly as characteristic emission of the Ce^{3+} ions, partly as emission at lattice imperfection levels and mostly as Mn emission (9), and thus the Ce^{3+} ions directly affected by irradiation return to states existing prior to irradiation. This is the reason why the 305 nm band, the characteristic band of Ce in this system, is practically unaffected by irradiation. It can then be inferred that cathode ray irradiation mainly produces transient changes in the Ce centres of this site system.

In conclusion it may be pointed out that the present results confirm that cathode ray irradiation is inefficient in forming intrinsic colour centres in Mn doped and Ce:Mn doped CaF_2 too.

The authors gratefully acknowledge their indebtedness to Prof. H. N. Bose, Head of the Department of Physics for helpful discussions.

References

(1) A. SMAKULA, Phys. Rev. 77, 408 (1950); Z. Phys. 138, 276 (1954).

(2) F. LÜTY, Z. Phys. 134, 596 (1953).

(3) P. GÖRLICH, H. KARRAS, and R. LEHMANN, phys. stat. sol. 1, 389 (1961); 3, 98 (1963).
P. GÖRLICH, H. KARRAS, W. LÜDKE, H. MOTHES, and R. REIMANN, phys. stat. sol. 3, 478 (1963).

(4) V. V. RATNAM, phys. stat. sol. 16, 559 (1966).

(5) E. ZINTAL and A. UDGARD, Z. anorg. allg. Chem. 240, 150 (1939).

(6) R. W. URE, Jr., J. chem. Phys. 26, 1363 (1957).

(7) D. R. RAO and H. N. BOSE, J. Phys. Soc. Japan 28, 152 (1970).

(8) A. D. FRANKLIN, J. Phys. Chem. Solids 29, 823 (1968).

(9) S. C. SEN and H. N. BOSE, Z. Phys. 201, 368 (1967).

(Received November 1, 1971)

phys. stat. sol. (a) 8, K111 (1971)
Subject classification: 13.1; 6; 8; 22.2.1

Institute of Experimental Physics, University of Warsaw

Temperature Dependence of the Energy Gap in GaAs

By

A.M. HENNEL

It was shown by Fan (1) that the temperature dependence of the energy gap in semiconductors is caused by two different effects: the thermal expansion of the crystal and the temperature variation of the lattice vibrations. This fact can be expressed by the simple thermodynamic equation

$$\left(\frac{\partial E_g}{\partial T}\right)_P = \left(\frac{\partial E_g}{\partial T}\right)_V - \frac{\beta}{\varkappa}\left(\frac{\partial E_g}{\partial P}\right)_T ,$$

where V is the volume, T the temperature, P the pressure, $\beta = \frac{1}{V}\left(\frac{\partial V}{\partial T}\right)_P$ the volume coefficient of thermal expansion, $\varkappa = -\frac{1}{V}\left(\frac{\partial V}{\partial P}\right)_T$ the coefficient of isothermal compressibility.

In this equation the first term is connected with the electron-phonon interaction, and the second with the dilatation effect.

For gallium arsenide the following quantities have been measured in numerous experiments: $(\partial E_g/\partial T)_P$, $(\partial E_g/\partial P)_T$, β, $\varkappa$. It is therefore possible to use the above equation for determining the phonon term $(\partial E_g/\partial T)_V$.

The value and the temperature dependence of $(\partial E_g/\partial T)_P$ can be found from the Panish and Cassey equation (2)

$$E_g(T) = 1.522 - 5.8\text{x}10^{-4}\,\frac{T^2}{T + 300}\ \text{eV}$$

which is valid in the temperature range from 0 to 1000 K.

From this equation we get

$$\left(\frac{\partial E_g}{\partial T}\right)_P (T) = -5.8\text{x}10^{-4}\,\frac{T(T + 600)}{(T + 300)^2}$$

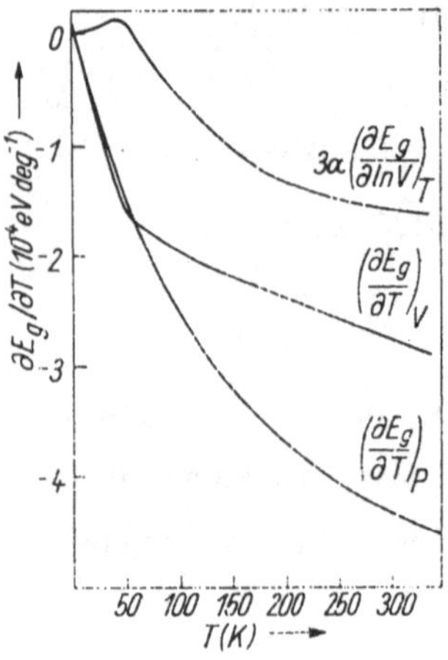

Fig. 1. The temperature dependence of $(\partial E_g/\partial T)_P$, $(\partial E_g/\partial T)_V$, and $3\alpha\ (\partial E_g/\partial \ln V)_T$

On the basis of many experiments performed by different methods in a large range of temperatures, it can be concluded that $(\partial E_g/\partial P)_T$ is independent of temperature.

The same applies to the coefficient $\varkappa$ which is equal to $3(S_{11} + 2S_{12})$, where S_{11} and S_{12} are the elastic constants. Their temperature dependence is so small (3) that between 0 and 300 K the coefficient $\varkappa$ is reduced only by about 5%.

The volume coefficient β is for cubic structures equal to 3α, where α is the linear expansion coefficient. The value of α was measured by Novikova (4) between 30 and 340 K. She has found that for temperatures $T < 55$ K α is negative and then rises to 6×10^{-6} deg^{-1} for $T = 340$ K. The minimum value of α, equal to -0.5×10^{-6} deg^{-1}, is reached at $T = 40$ K. For $T = 0$ α becomes equal to zero in accordance with the Nernst theorem.

Finally the first equation can be written in the form

$$\left(\frac{\partial E_g}{\partial T}\right)_P = \left(\frac{\partial E_g}{\partial T}\right)_V + 3\alpha\left(\frac{\partial E_g}{\partial \ln V}\right)_T ,$$

where $(\partial E_g/\partial \ln V)_T$ is the deformation potential designated by Pikus and Bir (5) with the symbol a. On the basis of the above discussion we can state that this potential is temperature independent. For gallium arsenide it is equal to about -9 eV.

The results of calculations performed by the use of this equation are shown in Fig. 1.

It can be seen that $(\partial E_g/\partial T)_V$ is a monotonic function of temperature. For temperatures $T < 55$ K the dilatation term is negligible and for $T > 150$ K it is approximately equal to one third of the total temperature coefficient of E_g. So it is twice as large as for germanium, where the dilatation term amounts to about 17% (6) of the total $(\partial E_g/\partial T)_P$.

The author wishes to express his gratitude to Prof. L. Sosnowski and Dr. J. Ginter for their helpful discussions, and also to Mgr. M. Górski for computer calculations.

References

(1) H.Y. FAN, Phys. Rev. 82, 900 (1951).

(2) M.B. PANISH and H.C.J. CASSEY, J. appl. Phys. 40, 163 (1969).

(3) C.W. GARLAND and K.C. PARK, J. appl. Phys. 33, 759 (1962).

(4) S.I. NOVIKOVA, Fiz. tverd. Tela 3, 178 (1961).

(5) G.E. PIKUS and G.L. BIR, Fiz. tverd. Tela 1, 1642 (1959).

(6) W. PAUL and H. BROOKS, Phys. Rev. 94, 1128 (1954).

(Received November 8, 1971)

phys. stat. sol. (a) 8, K115 (1971)

Subject classification: 8; 21

Institute for Low Temperature and Structure Research,
Polish Academy of Sciences, Wrocław

On the Graphical Determination of the Temperature Dependence of the Thermal Conductivity of Pure Metals in the Low-Temperature Maximum of the Conductivity for the Case of Aluminium

By

J. RAFAŁOWICZ and B. SUJAK

The commonly used relation defining the low-temperature dependence of the thermal conductivity on the temperature, has for pure metals the form

$$K = (\alpha T^2 + \beta T^{-1})^{-1} \quad (1)$$

This equation was theoretically derived, e.g., by Wilson (1). However, experiments on real metals reported in (2, 3) show the dependence (1) to be

$$K = (\alpha T^n + \beta T^{-1})^{-1} , \quad (2)$$

where $2 \leqq n \leqq 3$. α depends weakly on the purity of the samples, whereas β depends strongly, changing its value, in the case of aluminium, even by three orders of magnitude. If the theoretical relation (1) is to hold, the following inequality, according to Klemens (2), must be satisfied:

$$W_i \gg \varrho_i / L_o T , \quad (3)$$

hence

$$\frac{W_i}{W_o} \gg \frac{\varrho_i}{\varrho_o} , \quad (3a)$$

where the index i refers to the thermal resistivity caused by scattering by phonons and to the ideal term of electrical resistivity, and the index o to the residual electrical resistivity and to the thermal resistivity due to the scattering by static imperfections and chemical impurities. L_o is the Lorentz number.

In practice, inequality (3a) never holds, as even for the purest aluminium samples the ratio W_i/W_o does not exceed 10^3, whereas ϱ_i/ϱ_o for aluminium samples investigated in (4) is equal to 24000.

The values recommended in (3) for the exponent n of equation (2) for 23 metals are between 2 and 3, and only for 8 metals n = 2.0.

As it is shown in (3), drawing, for hypothetical metal samples, a family of curves K(T) given by equation (1), (where the parameter β varied from 0.1 to 2, and $\alpha = 0.00015$) we get on a logarithmic scale a linear dependence of the height of the maximum of the thermal conductivity K_m on the temperature T_m of this maximum.

The slope of this linear dependence is equal to minus two, and its absolute value is equal to the value of the exponent in the expression αT^2 for the temperature dependence of the ideal term of the thermal resistivity in equation (1).

Following the reasoning presented in (3), and referring to equation (1), we put the first derivative dK/dT of the relation (2) equal to zero, and we get the expression for the temperature T_m corresponding to the thermal conductivity maximum:

$$T_m = \left(\frac{\beta}{n\alpha}\right)^{1/n+1}, \qquad (4)$$

where n, α, β are coefficients from relation (2). Substituting (4) into (2) we get the expression for the maximum of the thermal conductivity coefficient

$$K_m = \frac{n}{n+1}\left(\frac{1}{n\alpha\,\beta^n}\right)^{1/n+1} \qquad (5)$$

Eliminating the coefficient β from (4) and (5) we obtain the relation between the maximum of the thermal conductivity coefficient and the temperature T_m corresponding to this maximum

$$K_m = \frac{1}{\alpha(n+1)}\,T_m^{-n}. \qquad (6)$$

The experimental values of K_m and T_m, compiled in (5) from available scientific literature, for investigated aluminium samples of different chemical purity, are presented on a logarithmic scale on Fig. 1. Also, the points established

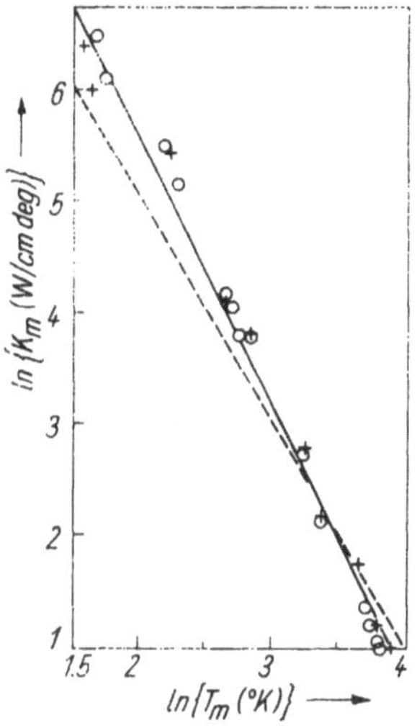

Fig. 1. The dependence of the conductivity maximum on the temperature corresponding to this maximum for aluminium samples of different chemical purity;
+ points established graphically on the basis of literature data compiled in (5);
o points determined analytically on the basis of numerical values for coefficients α and β from equation (1), as given in (4).
The dashed curve refers to the straight line with a slope of 2, as predicted by theory

analytically by using the data from (4) for different Al samples, are plotted.

The slope of the straight line on Fig. 1 is equal to 2.34 what is in a very good quantitative agreement with the value 2.30 suggested for Al in (3).

The value of the coordinate obtained by the intersection of the straight line on Fig. 1 with the axis in K_m makes it possible to determine graphically the coefficient α, which for aluminium is equal to about $1.2\text{x}10^{-5}$ and lies within the minimum values for α, as given in (4).

Drawing the relation (2) in the coordinate system T/K and T^{n+1}, we are able to determine graphically the coefficient β which is different for samples of different chemical purity.

It is the aim of this paper to emphasize the fact that based on the experimental curves K(T) for samples of different chemical purity, but the same metal, we can graphically determine the coefficients n, α, and β appearing in relation (2).

The procedure used, i.e. in (4) for elaborating the experimental results of measuring the thermal conductivity for pure aluminium, based on equation (1), was employed to extract the ideal term W_i from the total thermal resistivity. Next, W_i and T were presented graphically on a logarithmic scale, and since the slope of the straight lines was found to be equal to 2.00, it was looked on as a proof of the correctness of the formula (1) in describing the temperature dependence of the aluminium thermal conductivity, The result obtained by the authors of the paper (4) that n = 2.00 is obvious, as the equation (1) serving as a starting point for their analysis already contains the assumption of the square dependence of the ideal term of the aluminium thermal resistivity on temperature.

References

(1) A. H. WILSON, Proc. Camb. Phil. Soc. 33, 371 (1937).

(2) P. G. KLEMENS, in: Thermal Conductivity, Vol. 1, Ed. R. P. TYE, Academic Press, London, New York 1969 (p. 42).

(3) A. CEZAIRLIYAN and Y. S. TOULOUKIAN, Advances in Thermophysical Properties at Extreme Temperatures and Pressures, Amer. Soc. Mech. Engng., New York 1965 (p. 301).

(4) P. SEEBERG and T. OLSEN, Phys. Norveg. 2, 198 (1967).

(5) J. RAFALOWICZ, K. BALCEREK, B. SUJAK, E. PEGA, and L. LIPIŃSKI, phys. stat. sol. (a) 5, 785 (1971).

(Received October 15, 1971)

phys. stat. sol. (a) 8, K119 (1971)

Subject classification: 20.3; 22.2.1

Institute of Radio Physics and Electronics,
University of College of Science & Technology, Calcutta

Doping Dependence of the Energy of Peak Spontaneous Emission from n-Type GaAs Excited by a 40 kV Electron Beam at 300 °K

By

S. N. BISWAS, S. RAKSHIT, and A. N. CHAKRAVARTI

In recent years, with the advent of electron-beam-pumped GaAs lasers, considerable interest has grown in studying the spectral characteristics of spontaneous emission from specimens of GaAs excited by an electron beam. Though several workers (1 to 4) have already reported experimental results on the doping dependence of these characteristics, yet a suitable method for obtaining the same from theoretical computation is still not available. The purpose of the present communication is to theoretically determine the doping dependence of the energy of peak spontaneous emission from n-type GaAs excited by a 40 kV electron beam at 300 °K.

Assuming thermal equilibrium to exist for carriers within each band and applying the principle of detailed balance (which is quite general and is applicable to both optical and electron beam excitations), the net rate of electron-hole recombinations per unit volume within a specimen under excitation may be expressed (5) in terms of the interband absorption coefficient $\alpha(h\nu)$ and the excess carrier concentration Δp as

$$U_{eh} = \int_0^\infty B\,\Delta p\,\frac{n^2 (h\nu)^2}{\pi^2 c^2 \hbar^3}\,\alpha(h\nu)\,\exp\left(-\frac{h\nu}{kT}\right) d(h\nu), \qquad (1)$$

where n is the index of refraction, c is the velocity of light, and $B = 1/p_0$ for degenerate n-type specimens respectively, p_0 being the concentration of holes at thermal equilibrium. The excess hole concentration Δp created in an n-type specimen by the irradiation of an electron beam has an indepth distribution which may be obtained by solving the diffusion equation with proper boundary conditions. Assuming a Gaussian distribution for the carrier generation rate per unit volume, Δp can be expressed (6) in terms of the depth z below the irradiated surface in a direction normal to it by

$$\Delta p(z) = \frac{G_o L}{2 D_p} \left[1 + \mathrm{erf}\,(a\, z_o)\right]^{-1} \exp\left(\frac{1}{4a^2 L^2}\right) \left\{ \exp\left(-\frac{z}{L} + \frac{z_o}{L}\right) \mathrm{erf}\left(\frac{1}{2a\,L} + a\, z_o\right) - \right.$$

$$- \left(\frac{S-1}{S+1}\right) \exp\left(-\frac{z}{L} - \frac{z_o}{L}\right) \mathrm{erfc}\left(\frac{1}{2a\,L} - a\, z_o\right) + \exp\left(\frac{z_o}{L} - \frac{z}{L}\right) \times \qquad (2)$$

$$\left. \times\, \mathrm{erf}\left(az - \frac{1}{2a\,L} - a\, z_o\right) + \exp\left(\frac{z}{L} - \frac{z_o}{L}\right) \mathrm{erfc}\left(az + \frac{1}{2a\,L} - a\, z_o\right)\right\},$$

where a and z_o are constants for a particular beam voltage, G_o is the total carrier generation rate in $cm^{-2}\, s^{-1}$, L and D_p are the diffusion length and the diffusion coefficient of the minority carriers respectively, and $S = s\tau_n/L$ is the reduced surface recombination velocity (s and τ_n being the surface recombination velocity and the lifetime of the minority carriers respectively). Incidentally, a fraction of the photons which are emitted by radiative recombination within the specimen will be reabsorbed. Taking the reabsorption to be due mostly to interband transitions (7), we can also write

$$\int_0^\infty I(h\nu)\, d(h\nu) = H \int_0^\infty U_{eh} \exp\left[-\alpha(h\nu) z\right] dz\,, \qquad (3)$$

where $I(h\nu)$ is the intensity of the radiation emerging from the crystal at the frequency ν and H is a constant of proportionality. Hence, using the dimensionless variables $u = \varrho z/R$, $\Delta u = \varrho/Ra$, $u_o = \varrho z_o/R$, and $w = R/\varrho L$, where ϱ is the mass density and R is the maximum electron range in $g\,cm^{-2}$ and noting that u_o is to be replaced by $(u_o - \frac{d}{wL})$ to take into account the effect of a dead layer which has been assumed elsewhere (8) to exist at a depth d below the irradiated surface, we can express the intensity from equations (1 to 3) as

$$I(h\nu) = \mathrm{const}\,(h\nu)^2 \exp\left(-\frac{h\nu}{kT}\right) \alpha(h\nu) \int_0^\infty \left\{ \exp\left[-uw\left\{1 + \alpha(h\nu) L\right\}\right] \times \right.$$

$$\times \exp\left(\frac{w^2 \Delta u^2}{4} + u_o w - \frac{d}{L}\right) \mathrm{erf}\left[\frac{w\Delta u}{2} + \frac{1}{\Delta u}\left(u_o - \frac{d}{wL}\right)\right] - \left(\frac{S-1}{S+1}\right) \times$$

Fig. 1. Doping dependence of the energy of peak spontaneous emission from n-type GaAs excited by a 40 kV electron beam at 300 °K
—— with reabsorption, ---- without reabsorption, o experimental (3)

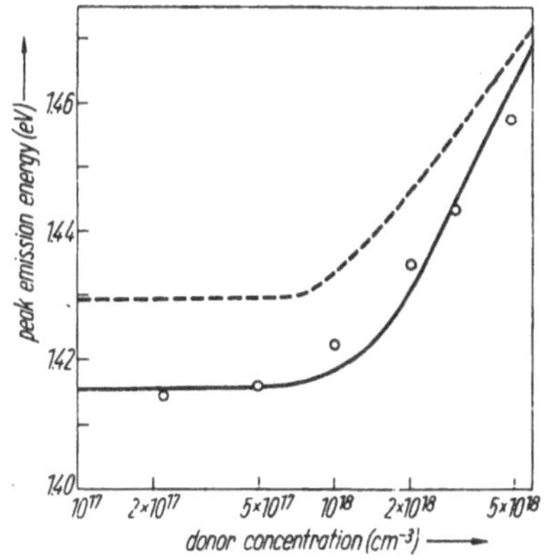

$$\times \exp\left[-uw\left\{1+\alpha(h\nu)L\right\}\right] \exp\left(\frac{w^2\Delta u^2}{4} - u_o w + \frac{d}{L}\right) \mathrm{erfc}\left[\frac{w\Delta u}{2} - \frac{1}{\Delta u}\left(u_o - \frac{d}{wL}\right)\right] +$$

$$+ \exp\left[-uw\left\{1+\alpha(h\nu)L\right\}\right] \exp\left(\frac{w^2\Delta u^2}{4} + u_o w - \frac{d}{L}\right) \mathrm{erf}\left[\frac{u}{\Delta u} - \frac{w\Delta u}{2} - \frac{1}{\Delta u} \times\right.$$

$$\left.\times \left(u_o - \frac{d}{wL}\right)\right] + \exp\left[uw\left\{1-\alpha(h\nu)L\right\}\right] \exp\left(\frac{w^2\Delta u^2}{4} - u_o w + \frac{d}{L}\right) \times$$

$$\left.\times\ \mathrm{erfc}\left[\frac{u}{\Delta u} + \frac{w\Delta u}{2} - \frac{1}{\Delta u}\left(u_o - \frac{d}{wL}\right)\right]\right\} du\ . \tag{4}$$

In the absence of reabsorption effects, $I(h\nu)$ may, therefore, simply be expressed as

$$I(h\nu) = \mathrm{const}\ (h\nu)^2 \exp\left(-\frac{h\nu}{kT}\right)\alpha(h\nu). \tag{5}$$

As reported elsewhere (8), we can take $R = 2.53\times10^{-3}$ $g\,cm^{-2}$ at 30 kV, $u_o = 0.125$, $\Delta u = 0.350$, $d/L = 0.05$, and $S = 50$, and assume that these parameters are mainly dependent on the mass density of the specimen and are more or less independent of temperature and doping (9). Thus using the absorption and diffusion data of Hwang (5) together with the other parameters as mentioned above and also noting that $R = \mathrm{const}\ (V_o)^{1.7}$, where V_o is the beam voltage in kV, the relative spectral intensity distribution curves for n-type specimens of GaAs (having donor concentrations ranging from 2.2×10^{17} to 4.9×10^{18} cm^{-3}) excited by a 40 kV electron beam at 300 °K are obtained from equations (4) and (5) with the help of a digital computer. From these emission spectra, the energy of peak spontaneous emission is obtained as a function of the doping concentration as shown in Fig. 1. In the same figure the experimentally measured (3) values of the energy of peak spontaneous emission are also shown. It appears, therefore, that the computed results are in reasonable agreement with the experimental observation only when the effect of photon reabsorption is taken into consideration. It can also be concluded that the effect of reabsorption on the energy of peak spontaneous emission gradually decreases with increasing donor concentration.

Acknowledgements

The authors are indebted to Prof. J. N. Bhar for his keen interest in the work. Two of the authors (S. N. B. and S. R.) are also grateful to the Council of Scientific and Industrial Research of India for the award of senior research fellowships.

References

(1) D. A. CUSANO, Solid State Commun. 2, 353 (1964).

(2) J. I. PANKOVE, J. Phys. Soc. Japan Suppl. 21, 298 (1966).

(3) H. C. CASEY, Jr. and R. H. KAISER, J. Electrochem. Soc. 114, 149 (1967).

(4) B. TUCK, J. Phys. Chem. Solids 29, 615 (1968).

(5) C. J. HWANG, J. appl. Phys. 40, 3731 (1969).

(6) D. F. KYSER and D. B. WITTRY, Proc. IEE 56, 733 (1967).

(7) S. RAKSHIT, S. N. BISWAS, and A. N. CHAKRAVARTI, phys. stat. sol. (a) 4, 249 (1971).

(8) D. B. WITTRY and D. F. KYSER, J. appl. Phys. 38, 375 (1967).

(9) T. S. RAO-SAHIB, private communication.

(Received November 5, 1971)

phys. stat. sol. (a) 8, K123 (1971)

Subject classification: 17; 22.3

Sektion Physik der Universität Rostock

TSEE Spectra of Si_3N_4 Layers

By

W. WILD and H. GLAEFEKE

It is possible to realize the analysis of electron traps in insulating layers by means of the TSEE (thermally stimulated electron emission) technique (1). The results for Si_3N_4 layers produced under different conditions are briefly reported in this note.

The Si_3N_4 layers were deposited on Si wafers by reactive rf sputtering in a nitrogen stream. Technical nitrogen as well as spectroscopically pure nitrogen was used for the sputtering process. The produced amorphous layers were about 3000 Å thick.

The Si wafers covered by the Si_3N_4 layers were excited by a 10 min bombardment with 1 keV electrons. The exoelectron glow curves were obtained by linear heating of the samples. A 17-stage open multiplier detected the emitted electrons in a vacuum of $3x10^{-7}$ Torr.

Using technical nitrogen for the sputtering process an entire different composition of the layers was achieved than by using spectroscopically pure nitrogen. Si_3N_4 layers produced in technical nitrogen contain essentially more impurities than layers produced in spectroscopically pure nitrogen. This could be confirmed by analysing the layers by IR spectroscopy and by micro-electron beam analysis (2).

The obtained TSEE spectra of the same layers showed the expected dependence on the sputtering process conditions. Intensity and temperature position of the peaks of the glow curve reflect the number and the energy position of the electron traps caused by the impurities in the layer. There-

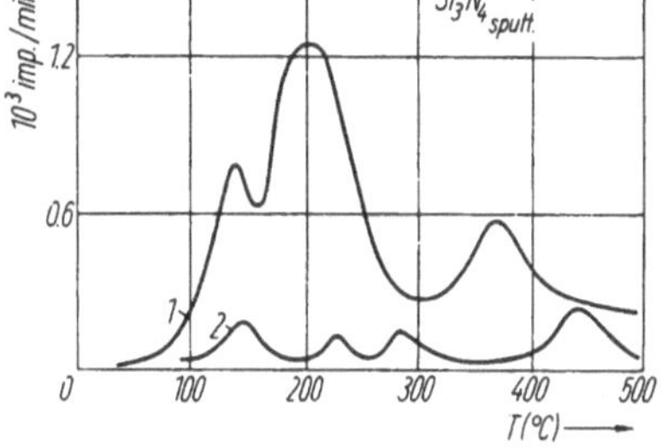

Fig. 1. TSEE spectra of Si_3N_4 layers sputtered in (1) technical nitrogen, (2) spectroscopically pure nitrogen

fore layers with a high number of impurities show an essentially higher intensity of emission and a different glow curve spectrum in comparison with layers having a small content of impurities (see Fig. 1).

At present we cannot assign the reproducible peaks to definite impurities.

We are much indebted to Dipl.-Phys. P. Ahrenholz, Sektion Technische Elektronik der Universität Rostock for producing the layers.

References

(1) J. DRENCKHAN, H. GROSS, and H. GLAEFEKE, phys. stat. sol. (a) 4, K17 (1971).

(2) P. AHRENHOLZ, to be published.

(Received November 8, 1971)

phys. stat. sol. (a) 8, K125 (1971)
Subject classification: 10.1; 21.1

Laboratoire de Métallurgie Physique, associé au C.N.R.S., Université de Poitiers

The Stacking Fault Energy in Cu-Al-Zn Alloys

By
M. F. DENANOT and J. P. VILLAIN

The composition of the alloys studied has been chosen in order to have only solid solutions; the following table shows the aluminium and zinc contents in terms of atomic fractions, the electron-atom ratio e/a, and the shear modulus G measured by an ultrasonic method (1).

alloy	1	2	3	4	5
Al (at%)	11	12.1	10	11.3	4.3
Zn (at%)	0	4.1	10.3	13.1	22.8
e/a	1.220	1.285	1.305	1.355	1.315
G (dyn/cm^2)	46.8×10^{10}	46×10^{10}	44.3×10^{10}	43.5×10^{10}	40.5×10^{10}

The intrinsic stacking fault energy γ has been deduced from an electron microscope study of extended dislocation nodes observed in samples after a tensile strain of about 1%. We have selected for each alloy some twelve isolated nodes which had a good symmetry with regard both to their geometrical configuration and to the character of the dislocations. The dimensionless ratio γ/Gb (b length of Burgers vector of partial dislocations) and the energy γ have been calculated with Brown's relationships (2, 3), choosing as principal geometrical parameter the width of the node; the determination of this parameter, which is certainly far from easy, is here more rigorous than that of the curvature radius of partial dislocations.

The found energies (points 1, 2, 3, 4, 5) and the value of γ corresponding to brass (point B) (4) are plotted in Fig. 1a and 1b as a function of total atomic fraction solute content (C_a = Al + Zn) and of electron-atom ratio (e/a); we have also plotted on these figures, using results collected by Gallagher (4), the variations of γ for

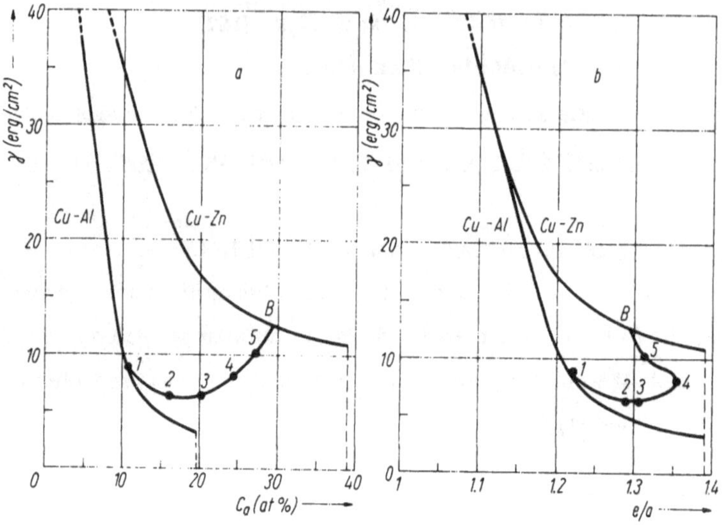

Fig. 1. Variations of the stacking fault energy with total atomic fraction solute content (C_a) and with electron-atom ratio (e/a)

Cu-Al and Cu-Zn solid solutions. The stacking fault energy of the binary alloy 1 is in very good agreement with the literature data. The values of γ relative to ternary solutions are situated between the curves corresponding to Cu-Al and to Cu-Zn. Initially the energy decreases slightly when the total solute content increases; but then, as the influence of zinc becomes preponderant, γ increases rather quickly (Fig. 1a). The representation of the variations of γ with e/a enables us to cancel the influence of the difference in valence between solute atoms; Fig. 1b shows obviously that the stacking fault energy of Cu-Al-Zn alloys does not depend only on the total number of free electrons brought by aluminium and zinc, but also on the relative proportions of these two elements. This result is in good agreement with the model of Tisone et al. (5) which, in addition to the part of free electrons, emphasizes, in the case of noble metal allovs, the likely influence on γ of the d electron band of so-

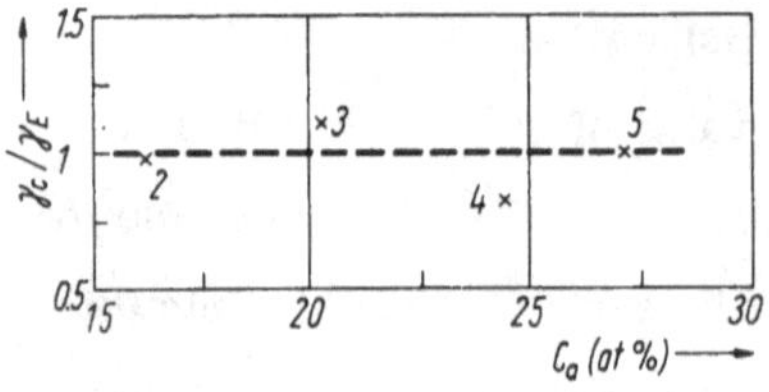

Fig. 2. Comparison of calculated and experimental values of the stacking fault energy in the alloys

lute atoms; aluminium which has an empty d band, could induce a more important decrease of γ than zinc which has a saturated d band.

In order to relate the stacking fault energy of Cu-Al-Zn ternary solid solutions to energies of Cu-Al and Cu-Zn alloys, we have been led to consider an expression of the following type:

$$\frac{\gamma - \gamma_{Cu-Al}}{\gamma_{Cu-Zn} - \gamma_{Cu-Al}} = \frac{C_a - (Al)_E}{(Zn)_E - (Al)_E} = \frac{2C_a}{e/a - 1} - 1 .$$

In this relation γ_{Cu-Al} and γ_{Cu-Zn} are the stacking fault energies of binary alloys with the same electron-atom ratio as the ternary solid solution and with the respective atomic fraction contents $(Al)_E$ and $(Zn)_E$. The ratio γ_c/γ_E of calculated energies to experimental energies in the alloys studied, plotted as a function of C_a in Fig. 2, remains close to one. Then the suggested relationship would be satisfactory, at least as a first approximation, and would enable to determine "a priori" the stacking fault energies of Cu-Al-Zn alloys.

References

(1) C. GAULT, Thèse troisième cycle, Poitiers 1971.

(2) L. M. BROWN, Phil. Mag. 10, 441 (1964).

(3) L. M. BROWN and A. R. THÖLEN, Disc. Faraday Soc. 38, 35 (1964).

(4) P.C.J. GALLAGHER, Met. Trans. 1, 2429 (1970).

(5) T.C. TISONE, R.C. SUNDAHL, and J.Y. CHIN, Met. Trans. 1, 1561 (1970).

(Received November 3, 1971)

phys. stat. sol. (a) 8, K129 (1971)

Subject classification: 18.4; 12; 21.1

Department of Solid State Physics, Institute of Advanced Studies, Australian National University, Canberra

Magnetoelastic Behaviour of Dilute Chromium Alloys

By

B. C. MUNDAY

The dynamic Young's modulus of dilute binary alloys of Cr with V, Ru, Rh, and Re has been measured in the temperature range 90 to 500 °K. The modulus is anomalously low in the AF_1 phase between T_{sf} and T_N. In alloys which exhibit a commensurate SDW phase the magnitude of the anomaly is enhanced, indicating that the magnetoelastic coupling between the spin system and the lattice is stronger for this arrangement.

The specimens were polycrystalline bars cut from arc-melted ingots homogenized by annealing at 1300 °C for 100 h in vacuo and furnace cooled. The compositions of the master alloys were estimated by X-ray fluorescence spectroscopy.

Typical results are shown in Fig. 1. The onset of magnetic ordering at T_N gives rise to an anomalous component of elastic compliance. This behaviour is basically similar to that observed in pure Cr, and has been interpreted in terms of stress-induced rotation of the SDW polarization vectors (1). Within the temperature range of our experiment it was possible to observe the spin-flip transition in only the most dilute alloys. As with Cr the anomalous behaviour vanishes below T_{sf}.

Koehler et al. (2) have shown that the addition of V reduces the size of the average magnetic moment per Cr atom, whereas the addition of elements to the right of Cr in the periodic table increases the size of the moment. Our results show that the magnitude of the Young's modulus anomaly in AF_1 is only weakly dependent on the size of the magnetic moment in the range we have studied ($0.3 < \mu < 0.75\ \mu_B$ (2)). As the concentration of Ru, Rh or Re is increased the commensurate SDW phase (AF_0) becomes stable above the characteristic temperature T_{co} and the anomalous compliance increases. The transition appears relatively broad due to the coexistence of AF_1 and AF_0, presumably due to internal strains (3).

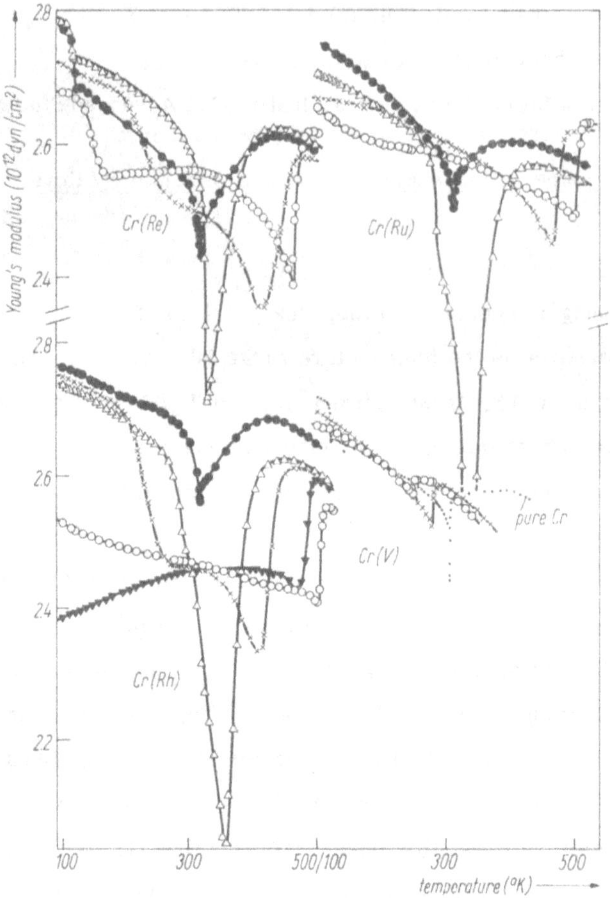

Fig. 1. Temperature dependence of Young's modulus of chromium alloys

at%	Re	Ru	Rh	V
● ● ●	0.15	0.15	0.12	-
△ △ △	0.30	0.27	0.20	-
× × ×	0.62	0.69	0.30	0.29
▼ ▼ ▼	-	-	0.71	-
○ ○ ○	0.91	1.05	1.14	0.61

Mössbauer spectra for Cr Mn alloys have shown (4) that the amplitude of the SDW is not changed by the onset of the commensurate structure. Furthermore we have been unable to observe in AF_0 any significant increase in the internal friction which might otherwise be associated with irreversible domain wall movement. We conclude therefore that the anomalous compliance in AF_o is still due to stress-induced rotation of spins, but is enhanced by an increase in the strength of the spin-lattice coupling. It follows that the magnitude of the differential thermal expansion measured along the axis of spin polarization should increase on heating through T_{co}.

The Young's modulus curve has a very deep cusp at T_N when the electron/atom ratio is such that $T_N \approx 340\ ^oK$. From the work of Jayaraman et al. (5) we infer that this is the condition where $T_N \approx T_{co}$. Jayaraman et al. have shown furthermore that hydrostatic pressure applied to Cr-Ru and Cr-Mn alloys lowers T_N. They also conclude from their results that increasing pressure raises T_{co} until a triple point is reached in the p-T phase diagram. For alloys with a triple point at ≈ 1 atm we might speculate on the effect of an alternating stress applied at temperatures near $T_N \approx T_{co}$. This stress will, by altering the free energy of the system, initiate a cycle of first order transitions between the non-magnetic, AF_o and AF_1 states. A large magnetoelastic strain will be associated with the critical spin ordering and re-ordering processes and so give rise to a cusp in the Young's modulus curve. This will be broadened by chemical inhomogeneities.

References

(1) B.C. MUNDAY and R. STREET, J. Phys. F 1, 498 (1971).

(2) W.C. KOEHLER, R.M. MOON, Al.L. TREGO, and A.R. MACKINTOSH, Phys. Rev. 151, 405 (1966).

(3) T.J. BASTOW, Proc. Phys. Soc. 88, 1 (1966).

(4) R. STREET, B.C. MUNDAY, B. WINDOW, and I.R. WILLIAMS, J. appl. Phys. 39, 1050 (1968).

(5) A. JAYARAMAN, T.M. RICE, and E. BUCHER, J. appl. Phys. 41, 869 (1970).

(Received October 25, 1971)

phys. stat. sol. (a) 8, K133 (1971)
Subject classification: 11; 22.1.2

Institute of Nuclear Physics, Academy of Sciences of the Uzbek SSR, Tashkent

Some Special Features of Subthreshold Radiation Effect in Silicon

By

M.A. ZAIKOVSKAYA, A.E. KIV, and O.R. NIYAZOVA

Previous investigations of damage production in silicon by electron irradiation below 100 keV indicated that defect production occurs even at energies which are far below the threshold for the generation of knock-on displacements (1, 2). It is assumed that the underlying process is caused by the ionization of inner electron shells and a subsequent displacement mechanism analogous to the Varley mechanism (7) in irradiated ionic crystals.

We report on experiments with boron doped p-silicon single crystals of 4 to 10 Ωcm resistivity. Slabs were cut at right angles to ⟨111⟩. Prior to contactation with palladium the crystals were treated as described in (2). The mean dislocation density in the samples ranged from 10^4 to 10^5 cm^{-2}.

The temperature greatly influences the radiation damage production rate (3 to 6). This points to a temperature dependent rate of defect annealing or defect reactions with impurities which in general decreases the overall production rate with increasing temperature. On the other hand, the subthreshold defect production rate itself is likely to increase with temperature as it is observed in the electrostatic ejection by the Varley mechanism (7). Thus, for a given radiation flux density a maximum defect production rate is likely to occur at a certain intermediate temperature.

The electron irradiation was performed between 90 and 380 ^{o}K. The temperature dependence of the defect production at 16 and 100 keV electron irradiation is shown in Fig. 1 where the quantity $\varphi = \Delta R/R_o \Phi$ is plotted against the irradiation temperature. R_o is the initial resistance of the sample, ΔR is the increase due to an integrated electron flux Φ. The maximum production rate occurs between 140 and 200 ^{o}K.

Wertheim and Buchanan (5) and also Stein (6) noted an increase of the damage rates produced in silicon by electron irradiation at 1 MeV or γ-irradiation by

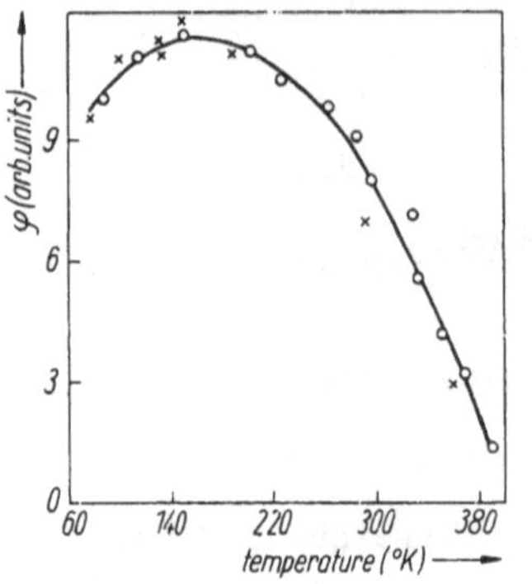

Fig. 1. Temperature dependence of damage production in p-silicon;
o irradiation with 16 keV electrons, flux density 3.4×10^{14} electrons/cm^2s,
x irradiation with 100 keV electrons, flux density 6.8×10^{14} electrons/cm^2s.
Duration of the irradiation 60 min at each point. The resistivity change was measured at 90 K

increasing the temperature from 76 to 300 °K. In the same temperature range no dependence was found with 45 MeV electrons or neutron irradiation (3). These observations are in contrast to the present finding of the maximum in defect production.

In addition we investigated the dependence of the subthreshold defect production with 16 keV electrons on silicon samples doped with various boron concentrations. The data exhibit a proportionality between the defect production rate and the hole concentration in p-type silicon.

The energy level of defects produced with electrons up to 20 keV were estimated according to Morin and Maita (9) and Vitovskii et al. (10). The result, $E_V + 0.2$ eV, is consistent with the level position found in γ-irradiated p-silicon (8).

References

(1) M.A. ZAIKOVSKAYA, A.E. KIV, and O.R. NIYAZOVA, Fiz. Tekh. Poluprov. 1, 1131 (1967).

(2) M.A. ZAIKOVSKAYA, A.E. KIV, and O.R. NIYAZOVA, phys. stat. sol. (a), 3, 99 (1970).

(3) Z.J. CHENG, and I.C. CORELLI, Phys. Rev. 140, 2130 (1965).

(4) A.I. MAKAREVICH, L.Y. RAINES, and N.F. KURILOVICH, Radiation Physics of Nonmetallic Crystals, Naukova Dumka, Kiev 1967 (p. 122).

(5) G.K. WERTHEIM and D.N.E. BUCHANAN, J. appl. Phys. 30, 1232 (1959).

(6) H.J. STEIN, J. appl. Phys. 37, 3382 (1966).

(7) H.J.O. VARLEY, J. Phys. Chem. Solids 23, 985 (1962).

(8) R.R. HASIGUTI and S. ISHINO, Symp. Radiation Damage in Semiconductors, Dunod, Paris 1965.

(9) F. E. MORIN and MAITA, Phys. Rev. 96, 28 (1954).

(10) N. A. VITOVSKII, T. V. MASHOVETS, S. M. RIVKIN, Fiz. tverd. Tela 4, 2845 (1962).

(Received November 22, 1971)

phys. stat. sol. (a) 8, K137 (1971)

Subject classification: 14.3; 13.1; 14.1; 22.4.4

Institute of Physics, Committee of Nuclear Energy, Bucharest

Transport Properties of HgS_xSe_{1-x} Solid Solutions

By

D. NICULESCU, R. IONESCU, M. POPA, and M. VANTZA

Recent studies (1) carried out on HgS_xTe_{1-x} solid solutions have evidenced the semimetal character of the energy band structure of β-HgS, with an overlap of the valence and conduction bands higher than that of HgTe. Mercury selenide (HgSe) has also semimetal character with an overlap higher than that of HgTe (2).

In this note, some transport properties of HgS_xSe_{1-x} solid solutions are reported.

Solid solutions in this system have been obtained up to temperatures of 1200 °C (depending on the HgS content) and an external pressure of 150 atm, followed by a quick cooling to 700 °C and annealing for several hours to obtain a good homogenisation. Homogeneous solid solutions with sphalerite structure have been obtained in the complete composition range for this system, the variation of the lattice parameter obeying Vegard's law.

The temperature dependence of the electrical conductivity is given in Fig. 1. It is obvious that all samples show a metallic behaviour in the electrical conductivity vs. temperature curves. At all temperatures the conductivity is lower for the HgS-rich solid solutions.

The Hall coefficient measurements performed by a conventional method using dc and a direct magnetic field of 2000 Oe as shown in Fig. 2, suggest that all solid solutions are n-type with large carrier concentrations. The carrier concentration increases as the HgS percentage increases.

The slope of the $\mu(T)$ curves is negative in the whole temperature range (Fig. 3). Therefore, the mobility is determined by lattice scattering. At low temperatures, above 78 °K, the mobility is proportional to $T^{-1/2}$, in a temperature range which is the wider the smaller x is this variation shows a piezoelectric scattering (Fig. 3). Above 180 °K the mobility is proportional to $T^{-3/2}$. This variation indicates a deformation potential scattering (Fig. 3). It is remarkable that with increasing HgS

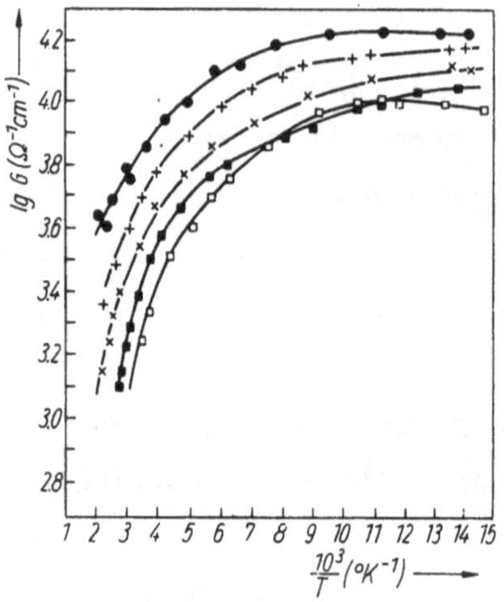

Fig. 1

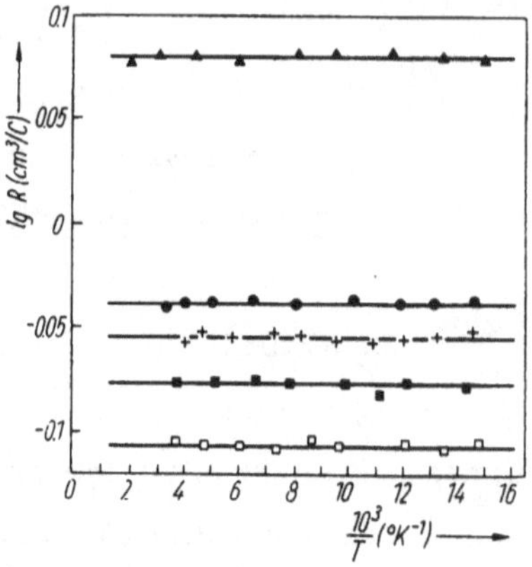

Fig. 2

Fig. 1. Temperature dependence of the electrical conductivity of the HgS_xSe_{1-x} system
● x = 0.2, + x = 0.3, × x = 0.4, ■ x = 0.6, □ x = 0.8

Fig. 2. Temperature dependence of the Hall coefficient for the HgS_xSe_{1-x} system
▲ x= 0.1, ● x = 0.2, + x = 0.3, ■ x = 0.6, □ x = 0.8

concentration the carrier mobility decreases.

The temperature dependence of the Seebeck coefficient, α , measured for all compositions in the range 77 to 450 °K showed the same variation from 14 to 90 μV/deg for all values of x.

The effective electron masses were calculated at 200 °K assuming an acoustical phonon scattering (r = -1/2) and a value α = 40 μV/deg. A nonparabolic band model was used with nonparabolicity parameter $\gamma = 0.3$ (assuming γ to be the same for HgS_xSe_{1-x} as for HgSe) using the relation (3)

$$\alpha = \frac{\pi^{2/3} k_0^2}{3 e \hbar^2} \; \frac{m^* T}{(3n)^{2/3}} \; (2 r + 3 - 6 \gamma) .$$

The results are shown in Fig. 4. It is evident that the effective carrier mass increases with increasing percentage of HgSe and the values for HgS-rich samples are in good agreement with those reported for HgS (4).

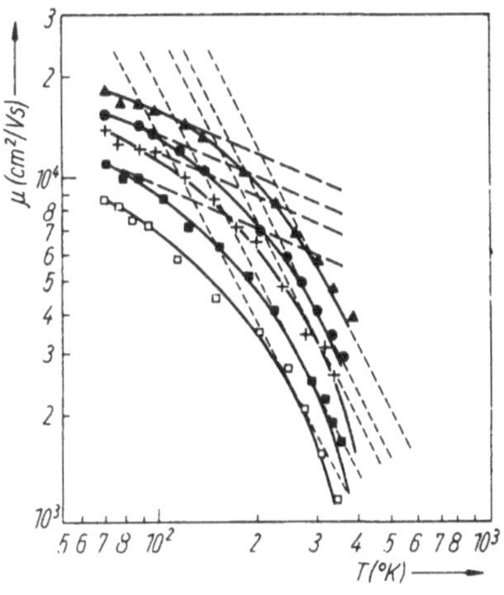

Fig. 3

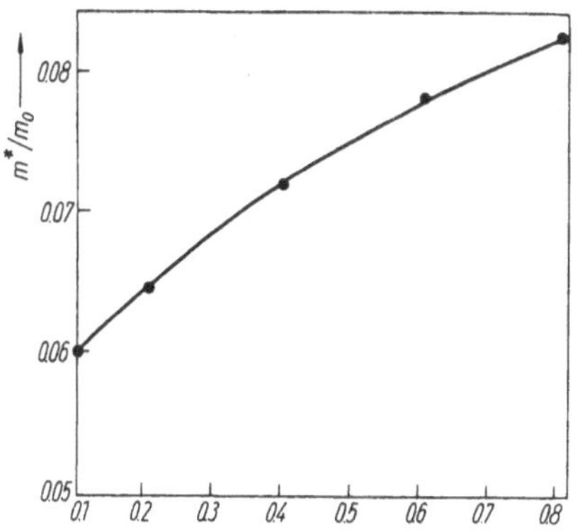

Fig. 4

Fig. 3. Comparison of the experimental curves $\mu = f(T)$ with the theoretical curves —— $\mu = kT^{-1/2}$ and ---- $\mu = kT^{-3/2}$
▲ x = 0.1, ● x = 0.2, + x = 0.3, ■ x = 0.6, □ x = 0.8

Fig. 4. Electron effective mass of HgS_xSe_{1-x} solid solutions as a function of x

The present experimental results lead to the conclusion that β-HgS is a semimetal with an overlap of the valence and conduction bands (according to the Groves-Paul model (5)) higher than of HgSe. Therefore, in the series of mercury chalcogenides, the overlap of energy bands and accordingly the metallization, increases from HgTe to HgS, unlike the series of cadmium chalcogenides, in which the metallic character increases from CdS to CdTe.

The effective masses were calculated at different Fermi levels for different compositions, but from the continuously increasing effective mass from HgSe to HgS, the presence of an inverted band structure is obvious in both cases, i.e. a negative Γ_6 - Γ_8 gap, as well as for HgTe.

References

(1) D. NICULESCU, N. NISTOR, M. POPA, M. VANTZA, and R. IONESCU, J. Mater. Sci. 5, 385 (1970).

(2) T.C. HARMAN and A.J. STRAUSS, J. appl. Phys. 32, 2265 (1961).

(3) S.S. SHALYT and S.A. ALIEV, Fiz. tverd. Tela 6, 1979 (1964).

(4) R. ZALLEN and M. SLADE, Solid State Commun. 8, 1291 (1970).

(5) S. H. GROVES and W. PAUL, Proc. Internat. Conf. Phys. Semicond., Paris 1964 (p. 41).

(Received November 24, 1971)

Pre-printed Titles

Papers to be published
in "physica status solidi (a) - applied research"
Vol. 9, No. 1

Review Article

P. P. BARASHEV, Moscow
The Multiquantum Photoemissive Effect in Condensed Media and Statistical Characteristics of the Multiquantum Photocurrent (I)

Original Papers

H. E. ALBRECHT, Rostock
Zur Berechnung der Elektronenaustrittsarbeit von mit metallischem Adsorbat bedeckten Emitteroberflächen

M. L. SWANSON and A. F. QUENNEVILLE, Chalk River, Ontario
Low Temperature Recovery of Plastically Deformed Cadmium, Indium, Tin and Lead

V. B. VYKHODETS, P. V. GELD, V. B. DEMIN, A. N. MEN, I. A. MURTAZIN, and A. YA. FISHMAN, Sverdlovsk
Isotope Effect in the Solubility of Hydrogen in F. C. C. Metals

J. ILCZUK and J. W. MOROŃ, Katowice
Further Investigations on Small Local Extrema in the Magnetic Loss versus Temperature Curves of Iron and Its Alloys

W. F. DRUYVESTEYN, R. SZYMCZAK, and R. WADAS, Eindhoven and Warsaw
Calculation on the Behaviour of a Cylindrical Magnetic Domain in a Finite Plate

F. P. HERRMANN and J. HEYDENREICH, Berlin and Halle/Saale
On the Dislocation Geometry in Antimony Cleavage Foils

I. TUREK, Žilina

The Illumination Dependence of the Velocity and Attenuation of Ultrasonic Waves in Gallium Arsenide

R. PRATAP and R.K. GUPTA, Varanasi

Integral-Breadth Analysis of X-Ray Line Broadening in Cold-Worked Natural Molybdenite (MoS_2) Crystals

K. OHORI and K. SUMINO, Sendai

Internal Friction in Deformed Germanium Single Crystals

A.V. ANTONOV, CH.V. KOPEZKII, L.S. SHVINDLERMAN, and Y.M. MUKOVSKII, Chernogolovka

The Shape of a Moving Grain Boundary

C.S. PANDE, Parks Road, Oxford

The Obstacle Radius of a Pile-up of Screw Dislocations

P. THERNQUIST, CH. KÄRRQVIST, and A. LODDING, Gothenburg

Electro-Transport and Thermo-Transport of Tracer Sodium in the Lithium Matrix

J. BRÁDLER and M. POLCAROVÁ, Prague

Study of Magnetostrictive Deformation of Fe-Si by X-Rays

H.P. GESERICH, H.D. HAUSEN, K. KROGMANN, and P. STAMPFL, Karlsruhe

Optical Investigations on the Metallic Behavior of $K_2Pt(CN)_4Cl_{0.3} \cdot 2.6\ H_2O$ and $K_2Pt(CN)_4Br_{0.3} \cdot 2.6\ H_2O$ Single Crystals

C. VITTORIA and G.C. BAILEY, Washington, D.C.

Intrinsic Magnetic Resonance Linewidth in Permalloy Films

B. BURAS, T. GIEBULTOWICZ, W. MINOR, and A. RAJCA, Copenhagen, Risö, and Warsaw

Explanation of Neutron Diffraction Phenomena Observed in Vibrating Piezoelectric Crystals

H. BIALAS, Heidelberg

Acoustic Amplification in Germanium at 9.3 GHz

B. KORNEFFEL and W. LUDWIG, Berlin

Energy-Momentum Dispersion in Al_2O_3 Tunnelling Structures

J. L. HUTCHISON and J.S. ANDERSON, Oxford

The Direct Correlation between Lattice Fringes and Atomic Positions in Defect Structures in N-Niobium Oxide

H.-G. UNRUH and H.-J. WAHL, Münster

Critical Slowing Down at Ferroelectric Transitions

A.W. SZAFRAŃSKI and B. BARANOWSKI, Warsaw

The Electrical Resistance of the Pd-Ag-H System at 25 $^{\circ}C$ in a Wide Range of Hydrogen Pressure

FL. CIORĂSCU, I. SPÎNULESCU-CARNARU, and C. STĂNESCU, Bucharest

The Dependence of the Electrophysical Properties of ZnTe Thin Films on Stoichiometry

B.N. DEY and W.R. TYSON, Don Mills, Ontario

Plastic Deformation of $CuAl_2$

G. GUIZZETTI, E. REGUZZONI, and G. SAMOGGIA, Pavia

A New Method for the Determination of ΔE_C in Heterojunctions

W. BOLLMANN and M. GERNAND, Jena

On the Disorder of $LiNbO_3$ Crystals

H. SIETHOFF, Würzburg

Versetzungen in kohärenten Zwillingsgrenzen von Indiumarsenid

J. HORÁK, L. TICHÝ, M. FRUMAR, and A. VAŠKO, Pardubice and Prague

The Influence of Iodine Impurities on the Electrical Conductivity of Sb_2Te_3 Crystals

C.W. ALLEN, P. DELAVIGNETTE, and S. AMELINCKX, Mol

Electron Microscopic Studies of the Laves Phases $TiCr_2$ and $TiCo_2$

R. PUTMAN and R. VAN OVERSTRAETEN, Heverlee

Influence of Band-Bending on the Tunnelling Probability in MIS Structures

R. A. COLLINS, I. A. EDGE, and K. O. LEGG, Lancaster

Electrical Properties of Al-SiO_x-Ag Thin Film Cathodes

K. HESS and H. KUZMANY, Vienna

Nonlinear Solution for Single-Mode Operated Acousto-Electric Oscillators

E. FABRE, Limeil-Brévannes

Copper Contamination during the Vapour Epitaxial Growth of GaAs

S. NASU, M. TAMAKI, H. TAGAWA, and T. KIKUCHI, Tokai-mura,

Magnetic Behavior of Hexagonal U_2N_3 in the Temperature Range from 77 to 300 ^oK

K. OZAWA, F. FUJIMOTO, K. KOMAKI, M. MANNAMI, and T. SAKURAI, Tokai-mura, Tokyo, and Kyoto

Effect of Channeling on the Lattice Defects Produced by Proton Irradiation in Alkali Halide Crystals

H. VAN HOVE and R. LEYSEN, Heverlee

LEED Study on the Polar Surfaces of ZnO

Č. KOŇÁK and J. STUKE, Marburg

Electron Tunnelling into Amorphous InSb and GaSb Films

O. LØPEN, Trondheim

Observation of Cd Crystal Growth in Thin CdSe Films during an Airbake

A. KEHL, Göttingen

Dielektrische Eigenschaften zweier halbleitender, einkristalliner Äthyl-phenothiazin-Jod-Komplexe bei 10 GHz

Papers to be published

in "physica status solidi (b) - basic research"

Vol. 49, No. 1

Review Article

M.R. TUBBS, Warwick, Coventry, Warwickshire

The Optical Properties and Chemical Decomposition of Halides with Layer Structures

Original Papers

E.M. GERSHENZON, N.M. PEVIN, and M.S. FOGELSON, Moscow

Spin Relaxation of Conduction Electrons in Highly Doped Semiconductors

B.P. SRIVASTAVA, H.N.K. SHARMA, and D.L. BHATTACHARYA, Varanasi

Energy Shift in the Resonant Absorption of γ-Rays due to Disordering of the Lattice

Y. WASEDA and K. SUZUKI, Sendai

Structure Factor and Atomic Distribution in Liquid Metals by X-Ray Diffraction

S.K. DUBEY, Varanasi

Localized Electrons and Their Interaction with the Local Modes in the Electrical Conduction in Metals

E.M. GERSHENZON, N.M. PEVIN, and M.S. FOGELSON, Moscow and Barnaul

Electron Spin Resonance in As-Doped Ge in a Wide Range of Impurity Concentration

R.I. DZHIOEV, V.G. FLEISHER, L.M. KANSKAYA, O.A. NINUA, and B.P. ZAKHARCHENYA, Leningrad

Optical Spin Orientation of Non-Equilibrium Electrons in GaAs and GaAs-AlAs Solid Solution

J.S. MELVIN and T. SMITH, Aberdeen, Scotland

Resolution of Apparent APW - OPW Energy Band Differences in NaCl

B. UNGER, Darmstadt

Raman Scattering Intensity of Polaritons for General Wave Vectors (II)

H. VAN CONG and G. MESNARD, Villeurbanne
Transport Properties of Heavily Doped Semiconductors at Low Temperatures in a Weak Magnetic Field

J.K. SRIVASTAVA and R.P. SHARMA, Bombay
Magnetic Dilution Effects on Morin Phase Transition in Hematite

J. ROSSAT-MIGNOD, G. QUEZEL, A. BERTON et J. CHAUSSY, Grenoble
Chaleur spécifique et propriétés magnétiques de l'oxysulfure d'yttrium

G.R. TSCHULENA and R. KEIL, Wien and Graz
Energy Loss of Hot Carriers due to Optical Phonons in Degenerate Nonpolar Semiconductors

F.D. AMBROGIO, P. BRÜESCH, and H. KALBFLEISCH, Baden
Raman and Infrared Measurements and the Study of Phase Transitions in $DyVO_4$

R. CHAMARD-BOIS, NGUYEN VAN NHUNG et J. PIERRE, Grenoble
Propriétés magnétiques des composés équiatomiques terres rares rhodium TRh
T = Tb, Dy, Ho (Er)

S.M. KLOTSMAN, I.P. POLIKARPOVA, and A.N. TIMOFEEV, Sverdlovsk
Self-Diffusion of Cesium-134 and Diffusion of Rubidium-86 in Cesium Iodide Single Crystals

V.K. JAIN and F. LÜTY, Salt Lake City, Utah
Thermal Electric Alignment and Stark Effect of Z_1 Centers in KCl

K.H. TIMMESFELD, München
Impurity Spin States in the Paramagnetic Phase of Ferromagnets

J. STANKIEWICZ and W. GIRIAT, Warsaw
Pressure and Temperature Dependence of the Energy Gap in $Cd_xHg_{1-x}Te$

A.L. EFROS, B.I. SHKOVSKII, and I.Y. YANCHEV, Leningrad
Impurity Conductivity in Low Compensated Semiconductors

T.Y. STOKOE and J.F. CORNWELL, St. Andrews, Scotland
High Field Transport in Semiconductors. The Drifted Maxwellian Approach (I)

H. ZIEGLER, Stuttgart

ENDOR Investigations of Cr^+ Centres in NaF, NaCl, and NaBr

A. P. PATHAK and M. YUSSOUFF, Kanpur

Charged Particle Energy Loss to Electron Gas

P. WEIGHTMAN, B. HENDERSON, and D. E. DUGDALE, Keele, Staffordshire

An EPR Study of an S = 3/2 Defect in $CaWO_4$

W. KRESS, München

Lattice Dynamics of Highly Polarizable Homopolar Crystals with Diamond Structure

R. A. VAUGHAN, Dundee, Scotland

Electron Spin Resonance and Relaxation of Tetrahedrally Coordinated Copper(11) in Some Complex Halides

H. J. FISCHBECK, Berlin

Theory of Disordered Systems (I)

YA. A. IOSILEVSKII, Chernogolovka

On the Spectrum of Acoustical Phonons of the Plane-Parallel Plate

G. KLEIN and H.-U. CHUN, Frankfurt am Main

Determination of Optical Interband Transitions in Crystalline Quartz from X-Ray Spectroscopical Data

R. R. DOGONADZE and A. A. KORNYSHEV, Moscow

The Magnetic Scattering of Neutral Particles in Polar Media and the Study of the Permittivity Dispersion in Space and Time

G. BOUGNOT, G. GALIBERT, and J. P. DESFOURS, Montpellier

Ultrasonic Attenuation in p-Type GaSb at Microwave Frequencies

I. GREGORA and J. PETZELT, Prague

Far Infrared Reflectivity of $CdAs_2$

V. A. RYABOV, Obninsk

Monte-Carlo Calculations of the Blocking Effect

G. S. PAWLEY, Edinburgh, Scotland

Analytic Formulation of Molecular Lattice Dynamics Based on Pair Potential Functions

S. STAMENKOVIC and S. ZEKOVIC, Beograd
On the Properties of the KDP Type Ferroelectrics with Impurities

M.R. PARKER, Salford
Magneto-Optical Anisotropy in Nickel

V.G. LEVIN, V.G. NEUDACHIN, and YU.F. SMIRNOV, Moscow
On Investigation of the Structure of Energy Bands Using Quasielastic Knock-Out of an Electron by an Electron (e, 2e)

H. LAMATSCH, J. ROSSEL, and E. SAURER, Neuchâtel
The Optical Constants of CsI in the Excitonic Region

P.A. SLOWDOWY and J.M. BARANOWSKI, Warsaw
Absorption Spectra of $Ti(d^2)$, $V(d^3)$, and $Cr(d^4)$ Ions in CdTe

A.S. DAVYDOV and G.M. PESTRYAKOV, Kiev
On the Canonical Transformation in the Theory of Exciton-Phonon Interaction

J. STUKE and G. ZIMMERER, Marburg
Optical Properties of Amorphous III-V Compounds (I)

H. BUCK and G. ALEFELD, Jülich
Hydrogen in Palladium-Silver in the Neighbourhood of the Critical Point

19. Magnetic Resonance in Solids

20. Optical Properties of Solids
    20.1 Spectra, Optical Constants (Colour Centres See 10.2; X-Ray Spectra See 20)
    20.2 Lasers
    20.3 Luminescence of Solids (Colour Centres See 10.2; X-Ray Spectra See 20)

21. Metals. Alloys
    21.1 Transition Metals (Group Sc to Zn) and Their Alloys
        21.1.1 Fe and Fe Alloys
    21.2 Alkali Metals
    21.3 Alkali-Earth Metals
    21.4 Rare-Earth Metals
    21.5 Actinides
    21.6 Noble Metals
    21.7 Semi-Metals (Semiconducting Alloys See 22.7)

22. Semiconductors and Ionic Crystals
    22.1 Elements
        22.1.1 Germanium
        22.1.2 Silicon
        22.1.3 Group VI Elements
    22.2 III–V Compounds
        22.2.1 Arsenides
        22.2.2 Phosphides
        22.2.3 Antimonides
        22.2.4 Mixed Crystals
    22.3 IV-IV, II-V, and III-VI Compounds
        22.3.1 Mixed Crystals
    22.4 II-VI Compounds
        22.4.1 Sulphides
        22.4.2 Selenides
        22.4.3 Tellurides
        22.4.4 Mixed Crystals
    22.5 Halides
        22.5.1 Silver Halides
        22.5.2 Alkali Halides
        22.5.3 Alkali-Earth Halides
        22.5.4 Mixed Crystals
    22.6 Oxides (CdO, ZnO See 22.4; Spinels See 18.3 or 14.4.2)
        22.6.1 Mixed Crystals
    22.7 Semiconducting Intermetallic Compounds
    22.8 Tertiary and Higher Compounds
        22.8.1 Salts of Oxy-Acids (Carbonates, Phosphates, Silicates Including Mica etc.)
    22.9 Organic Semiconductors

23. Solidified Gases

Attention

The preceding classification scheme is closely akin to the scheme used by Physics Abstracts to classify the various fields of solid state physics. A paper will either be classified under one ore more of the subsections or, if none of these are relevant, it will just appear under the general section heading (i. e. sections and subsections are equivalent in this sense).

The *subject classification* given at the head of each paper begins with the main subject matter of the paper, and this is followed, if necessary, by figures referring to subsidiary subject matter and the type of solids investigated.

The papers published in this issue are listed in the order of their subject classification at the end of the table of contents.

32914

Manuscripts and letters for physica status solidi (b) — basic research and physica status solidi (a) — applied research

should be sent to

DDR, 102 Berlin, Neue Schönhauser Str. 20

or

Internationaler Buchversand GmbH, 1 Berlin 19, Eschenallee 21.

The object of "physica status solidi (b)" and "physica status solidi (a)" is to publish rapidly important and topical results concerning fundamental research in the field of solid state physics and aspects of application and experimental techniques of fundamental solid state research, respectively. The issues of "physica status solidi (b)" and "physica status solidi (a)" will be published on the 1st and the 16th day of each month, respectively. Either series will contain the following categories:

1. Review articles written by arrangement with the editors. They will be remunerated.
2. Original papers selected from those which have not been submitted for publication elsewhere, after being reviewed. (The latest date for acceptance of contributions is the 8th and the 23rd day of each month for "physica status solidi (b)" and "physica status solidi (a)", respectively, 50 days before the delivery of the journal.)
3. Short notes published in the off-set part of the journal. They are, if necessary after being reviewed, selected by the editorial staff. (The latest date for acceptance of manuscripts for this part is the 12th and the 27th day of each month for "physica status solidi (b)" and "physica status solidi (a)", respectively, 16 days before the delivery of the journal.)
4. Pre-printed titles of papers to be published in the next issue of both series.

Editorial Information

Manuscripts should be submitted preferentially in English. Besides, papers in German, French, and Russian are admitted. Authors are asked to observe the following rules:

1. Manuscripts have to be written as concisely as possible. As most, 20 and 3 pages (28 lines per page; abstracts, tables, references, captions of figures included) are admitted for original papers and short notes, respectively. All manuscripts should be type-written on one side only. Two copies are required. Badly arranged manuscripts are returned.
2. The author's name, initials, and institution should be given.
3. An abstract of about ten lines in English and a second one in German, French, or Russian are required for original papers.
4. Formulae should not be written too small and with pencil. Separate lines for formulae are desirable. Indistinct writing will lead to a return of the manuscript.
5. Drawings should be enclosed on separate sheets of paper; photographs should be submitted, on white paper, only for review articles and original papers. Duplicates are required.
6. Captions for the figures should be submitted on separate sheets of paper, and special instructions with regard to the size of illustrations should be given on these.
7. References should be numbered in order of appearance and should be given, on a separate sheet, in the form used in physica status solidi.
8. Units, symbols, etc. should be used in accordance with the recommendations given by the IUPAP.
9. Each paper should contain a proposal for the subject classification according to the classification scheme appearing on the second and third cover of each issue.

The authors will be sent galley proofs of review articles and original papers. Proof reading of short notes and pre-printed titles will be done by the editorial staff.

For each original paper and short note a total of 75 reprints is available free of charge. The number of reprints free of charge per review article is 25. Requests for additional reprints (against payment) should be given together with the submitted manuscript. **The deadline for orders of additional reprints is the 15th and the 30th day of each month** for "physica status solidi (b)" and "physica status solidi (a)", respectively, two weeks before the delivery of the journal.

Kolibri 360 GmbH
Ferdinand-Jühlke-Straße 7
99095 Erfurt, Deutschland
produktsicherheit@kolibri360.de